INTERNAL COMBUSTION ENGINES AND POWERTRAIN SYSTEMS FOR FUTURE TRANSPORT 2019

PROCEEDINGS OF THE INTERNATIONAL CONFERENCE ON INTERNAL COMBUSTION ENGINES AND POWERTRAIN SYSTEMS FOR FUTURE TRANSPORT, (ICEPSFT 2019), BIRMINGHAM, UK, 11–12 DECEMBER, 2019

INTERNAL COMBUSTION ENGINES AND POWERTRAIN SYSTEMS FOR FUTURE TRANSPORT 2019

Editor

Institution of Mechanical Engineers

CRC Press/Balkema is an imprint of the Taylor & Francis Group, an informa business

Typeset by Integra Software Services Pvt. Ltd., Pondicherry, India

Published by: CRC Press/Balkema
Schipholweg 107C, 2316XC Leiden, The Netherlands

First issued in paperback 2023

ISBN: 978-1-03-257100-3 (pbk)
ISBN: 978-0-367-90356-5 (hbk)
ISBN: 978-1-003-02398-2 (ebk)

DOI: https://doi.org/10.1201/9781003023982

Table of Contents

SESSION 4: FUELS AND FUEL INJECTION

SESSION 5: INCREASING EFFICIENCY AND REDUCING EMISSIONS

Preface

With the changing landscape of the transport sector, there are also alternative powertrain systems on offer that can run independently of or in conjunction with the internal combustion (IC) engine. This shift has actually helped the industry gain traction with the IC Engine market projected to grow at 4.67% CAGR during the forecast period 2019-2025. It continues to meet both requirements and challenges through continual technology advancement and innovation from the latest research.

With this in mind, the Internal Combustion Engines and Powertrain Systems for Future Transport conference will not only cover the particular issues for the IC engine market but also reflect the impact of alternative powertrains on the propulsion industry.

The 2019 conference will provide a forum for IC engine, fuels and powertrain experts to look closely at developments in powertrain technology required to meet the demands of the low carbon economy and global competition in all sectors of the transportation, off-highway and stationary power industries.

PRESENTATION HIGHLIGHTS INCLUDE:

- Engines for hybrid powertrains and electrification
- IC engines
- Fuel cells
- E-machines
- Air-path and other technologies achieving performance and fuel economy benefits
- Advances and improvements in combustion and ignition systems
- Emissions regulation and their control by engine and after-treatment
- Developments in real-world driving cycles
- Advanced boosting systems
- Connected powertrains (AI)
- Electrification opportunities
- Energy conversion and recovery systems

- Modified or novel engine cycles
- IC engines for heavy duty and off highway

The Institution of Mechanical Engineers
Powertrain Systems and Fuels Group
One Birdcage Walk
Westminster
London
SW1H 9JJ
T: +44 (0) 20 7973 1251
F: +44 (0) 20 7304 6845
E: eventenquiries@imeche.org
www.imeche.org/events

Organising Committee

Powertrains Systems and Fuels Group

The Institution of Mechanical Engineers

Member Credits

Hua Zhao (Chair)	Brunel University London
Frank Atzler	Technical University of Dresden
Choongsik Bae	Korea Advanced Institute of Science and Technology
Hugh Blaxill	MAHLE Powertrain
Ralph Clague	Jaguar Land Rover
Roger Cracknell	Shell Global Solutions
Sean Harman	Ford Motor Company
David Heaton	Caterpillar
Roy Horrocks	Consultant
Rob Morgan	University of Brighton
Richard Osborne	Ricardo UK
Steve Sapsford	SCE Ltd.
Alan Tolley	JCB
Khizer Tufail	Ford Motor Company
Jamie W.G. Turner	University of Bath
Steve Whelan	HORIBA MIRA

SESSION 1: INTERNAL COMBUSTION ENGINES AND COMBUSTION ELEMENTS

Internal Combustion Engines and Powertrain Systems for Future Transport 2019 – Institute of Mechanical Engineers, ISBN 978-0-367-90356-5

Assessing the low load challenge for jet ignition engine operation

M.P. Bunce, N.D. Peters, S.K. Pothuraju Subramanyam, H.R. Blaxill

MAHLE Powertrain LLC, USA

ABSTRACT

Lean combustion in spark ignition engines is an advanced engine operating mode that has been proven to produce significant increases in efficiency, a requirement for future internal combustion engines in the transportation sector. This operation necessitates the use of advanced ignition or ignition controls concepts in order to achieve acceptable levels of combustion stability.

This study utilizes a Jet Ignition concept that has been under development for several years. MAHLE Jet Ignition® is a pre-chamber-based concept that produces high energy jets of partially combusted species that induce ignition and enable rapid, stable combustion at lambda values in excess of 2. In light duty engines this has resulted in minimum brake specific fuel consumption (BSFC) values of approximately 200 g/kWh with reductions in engine-out nitrogen oxides (NO_x) emissions of 95% compared to conventional gasoline engines.

Historically pre-chamber-based combustion concepts have had limited success achieving acceptable combustion stability under low load operation including idle and catalyst light-off. These conditions require a high degree of spark retard capability, a capability that is typically lacking with jet ignition concepts. The purpose of this study is to evaluate the challenges associated with idle and catalyst light-off jet ignition operation, examine the underlying causes, and explore potential solutions with a goal of achieving similar performance to a conventional spark ignited engine under these conditions. Test results from a dedicated 1.5L 3-cylinder jet ignition engine are provided with comparisons to a conventional spark ignition variant of the same engine. Potential solutions to achieving comparable performance metrics to a conventional spark ignited engine are proposed and evaluated on the testbed. The influence of in-cylinder charge motion is assessed relative to low load performance. The applicability of these results to other jet ignition engine applications is discussed.

1 INTRODUCTION

1.1 Background

The perpetual desire to conserve fuel is being coupled with an increasing modern awareness of the deleterious environmental impact of tailpipe emissions from the transportation sector. In response, increasingly stringent global legislation of greenhouse gas emissions will require a step change in internal combustion engine (ICE) efficiency. Concurrently, the reduction in popularity of diesel engines in the passenger car market is applying pressure on manufacturers' ability to adhere to fleet average fuel economy legislation. It is therefore imperative that technologies that significantly reduce the fuel consumption of gasoline spark ignition (SI) engines are developed and implemented.

A method being increasingly explored to accomplish this goal is dilute gasoline combustion [1-8]. The major limitation in developing dilute combustion systems is the less favorable ignition quality of the mixture. This has necessitated the development of higher energy ignition sources [9,10]. A pre-chamber combustor application is one such technology, having been researched extensively [11-15]. Pre-chamber combustion concepts have demonstrated the potential for stable main chamber combustion at higher levels of dilution than are allowable in typical SI engines [16].

Despite the extensive research and application history of pre-chambers, practical barriers to modern implementation of these systems in passenger car engines remain. An historic key challenge for pre-chambers has been ensuring acceptable low load, idle, and cold start performance [17]. The stringency of tailpipe emissions standards since the last significant commercial implementation of a pre-chamber in a passenger car engine has made acceptable performance at the latter condition particularly critical. Prior research has suggested that pre-chamber geometry must be tailored to encompass acceptable low load performance and to maintain the expected efficiency benefit at part load, and that identifying a common pre-chamber geometry that can accomplish both is challenging [18].

1.2 Jet ignition

MAHLE Jet Ignition® (MJI) is an auxiliary fueled pre-chamber concept that has been under development for several years [17-19]. The concept incorporates elements studied in previous pre-chamber research including: small pre-chamber volume (< 5% of main combustion chamber clearance volume) for minimizing crevice volume and heat loss, small orifice diameter to promote a high degree of flame quenching, and auxiliary fueling in the pre-chamber to allow separate fueling strategies for pre-chamber and main chamber.

A prototype low-flow direct injection (DI) fuel injector provides a separate fueling event in the pre-chamber. This allows precise, effectively de-coupled control over the mixture in both chambers. This low-flow DI injector also enables the use of a common liquid gasoline for both pre-chamber and main chamber injection. Historically, auxiliary fuel injected in the pre-chamber is gaseous due to metering repeatability and impingement issues with liquid gasoline. Both of these issues have been mitigated through pre-chamber injector development (not described in this study). Up to approximately 3% of total system fuel is injected via auxiliary pre-chamber injection. The remainder is delivered to the main combustion chamber conventionally with port fuel injection (PFI) or DI fueling utilizing off-the-shelf fuel injectors.

Pre-chamber combustion creates a rapid pressure increase in the pre-chamber, forcing contents into the main chamber via the orifices in the nozzle. A high degree of flame quenching is accomplished by limiting the diameters of the orifices in the nozzle. This quenching aspect of jet ignition is what differentiates it from a torch ignition system [18,20]. The resulting jets initiate main chamber combustion through chemical, thermal, and turbulent effects. The chemical effect is a product of the radical species present in the jets. These species are highly reactive and readily ignite the air-fuel mixture present in the main chamber. Containing partially or fully burned combustion products, these jets are at an elevated temperature when they enter the main chamber, above the auto-ignition temperature of gasoline, thereby providing a thermal trigger for main chamber combustion. Finally, the jets emerge at a velocity that is proportional to the pressure resulting from pre-chamber combustion. The velocity allows the jets to penetrate into the main chamber, entraining the air-fuel mixture in the main chamber as they proceed. This turbulent effect ensures interaction between the turbulent radical jets and the charge, producing suitable local mixture conditions for ignition and subsequent flame propagation.

The multi-orifice nozzle produces multiple, distributed ignition sites throughout the main chamber, resulting in short burn durations. A previously published study [2] examines the extent to which the ignition sites can be distributed and the importance of this distribution to thermal efficiency.

The jet ignition system is designed to be a low-cost, practical ultra-lean combustion enabling technology. A rendering of pre-chamber placement in a typical cylinder head is shown in Figure 1.

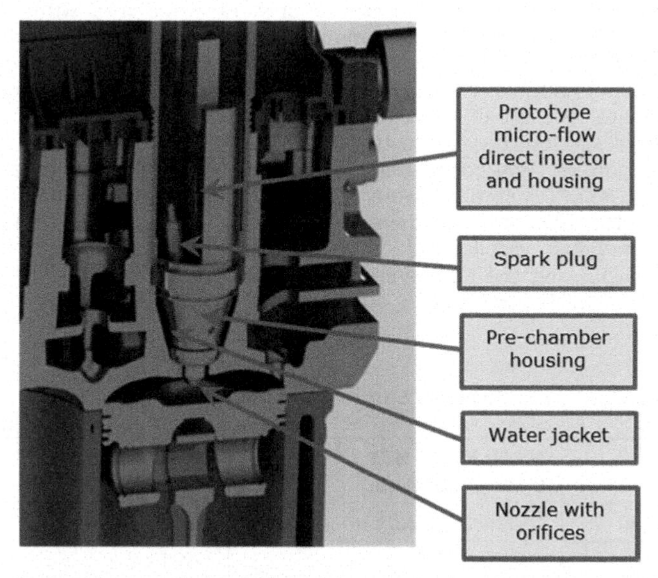

Figure 1. CAD model rendering of a partial cutaway of the pre-chamber assembly in a cylinder head.

1.3 Objective

The objective of this study is to quantify the challenges associated with steady state low load operation, idle, and cold start spark retard (CSSR) operation for catalyst heating with MJI. Mitigations are proposed and explored.

2 APPROACH

Data from a multi-cylinder Jet Ignition engine is presented to assess steady state low load operation at warm temperatures. The idle and CSSR conditions, while also steady state, are evaluated at 20 degree Celsius fluid temperatures to simulate sub-ambient starting temperatures. The targets used for CSSR operation are consistent with those of modern production and pre-production gasoline engines using 3-way catalysts. This data is taken from MAHLE Powertrain's CSSR database. The proposed jet ignition engine operates lean throughout the majority of the engine map, necessitating the use of a lean aftertreatment package, however targets for such an aftertreatment package are not well published in literature. Therefore the decision was made to attempt to adhere to the stability, exhaust enthalpy, and emissions targets of a 3-way catalyst under the assumption that these targets are aggressive compared to those of a lean aftertreatment package.

3 EXPERIMENT

3.1 Jet ignition engine

The MAHLE DI3 Downsizing demonstrator engine is selected as the basis of the dedicated Jet Ignition engine due to the authors' familiarity with the engine, access to the design and underlying analyses, and manufacturer agnostic nature of the platform. The engine is a 3-cylinder engine capable of achieving >30 bar peak brake mean effective pressure (BMEP). While nominally a 1.2 liter displacement, a 1.5 liter variant of the DI3 with an elongated stroke has been developed, and this version is used as the basis for the dedicated engine described here. The engine is repurposed here for a boosted ultra-lean application, with a target peak BMEP of approximately 15 bar. Development of the DI3 engine is well documented [21,22]. The development of the jet ignition variant of this engine was previously published [23]. Table 1 lists the specifications of the base DI3 engine (1.5L) and the MJI3 engine. The DI3 is depicted in Figure 2 and the MJI3 engine is depicted in Figure 3.

Table 1. Engine specifications.

	MAHLE DI3 NTY Specifications	**MAHLE MJI3 Specifications**
Configuration	In-line 3 cylinder	In-line 3 cylinder
Capacity	1497 cc	1497 cc
Bore	83 mm	83 mm
Stroke	92.2 mm	92.2 mm
Compression ratio	9.25:1	10:1-15:1 capable; 14:1 (tested)
Variable valve timing	Inlet and exhaust with 60° CAD authority	Inlet and exhaust with 60° CAD authority
Turbocharger	BMTS with electronic wastegate	Production VGT
EMS	MAHLE flexible ECU (MFE)	MAHLE flexible ECU (MFE)

Figure 2. DI3 engine.

Figure 3. MJI3 engine.

4 RESULTS

Pre-chamber concepts have historically faced challenges under low load operation. This challenge manifests in two distinct ways: poor combustion stability at heavily throttled low loads (less than approximately 2 bar BMEP) and poor spark retard capability at loads consistent with idle and CSSR operation. The well documented efficiency benefits of jet ignition at part load and high load [2] cannot be practically translated to non- and mild hybrid engine applications unless a solution to the low load pre-chamber limitation is identified.

4.1 Pre-chamber gas exchange

Most jet ignition concepts, including MJI, do not include any direct introduction of oxygen in the pre-chamber, instead relying on induced gas exchange between pre-chamber and main chamber to provide sufficient oxygen for combustion. This gas exchange process is driven by pressure differentials amongst intake and exhaust ports, pre-chamber, and main chamber throughout the 4-stroke engine cycle as depicted in Figure 4.

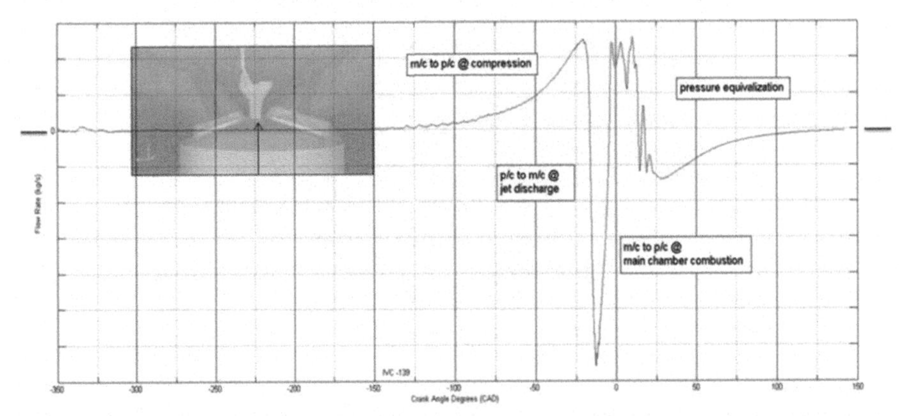

Figure 4. Example mass flow between pre-chamber and main chamber.

Figure 5 displays the oxygen (O_2) and carbon dioxide (CO_2) mass fractions inside the pre-chamber for a representative part load cycle. The intake valve opening event is when O_2 is reintroduced into the system. The O_2 fraction then rises in the pre-chamber despite the downward motion of the piston during this phase. The discontinuities apparent in the O_2 mass fraction trace correspond to events during the intake process, such as valve fully open and start of valve ramp down, indicating that valve position has a substantial influence on O_2 filling of the pre-chamber. With the intake valve closed and piston motion upward during the compression stroke, the remainder of the O_2 filling process occurs, which in turn dilutes the CO_2 mass fraction in the pre-chamber. With sufficient O_2 present, fuel can then be separately added to ensure a pre-chamber lambda within the ignitability limits of the spark plug.

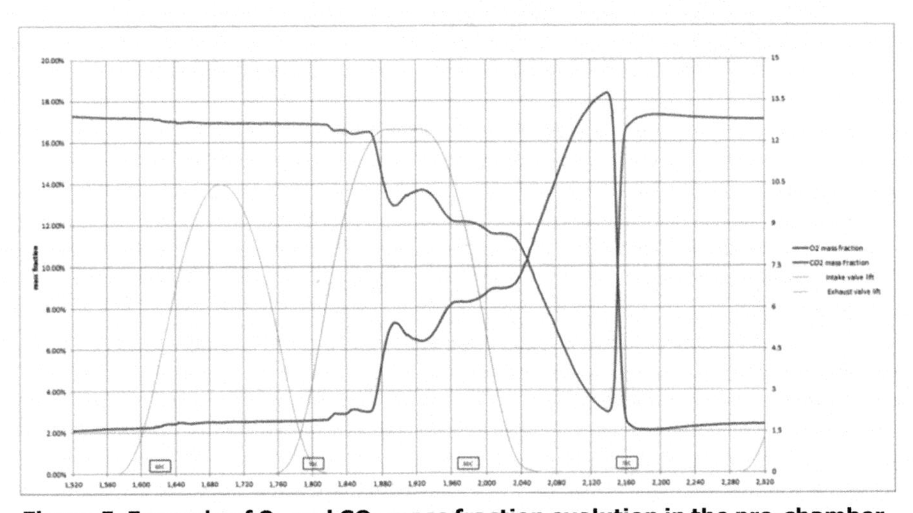

Figure 5. Example of O_2 and CO_2 mass fraction evolution in the pre-chamber.

An experiment was undertaken to measure the residual gas fraction inside the pre-chamber and to determine what impact pre-chamber fuel injection has on residuals. A fast response CO_2 analyzer was used to sample contents directly from the pre-chamber. Two methods of sampling were employed and the results were compared for validation purposes. In the sampling valve method, a fast response solenoid valve was connected to a port that broke through to the pre-chamber volume. The valve was commanded open and closed, allowing a small volume of pre-chamber contents to be sampled by the CO_2 analyzer. Care was taken to ensure that the proper minimum amount of sample gas was provided to the analyzer, and the time during which the valve sampled was swept throughout the cycle until reasonable convergence was achieved.

In the continuous sampling method (Figure 6), the analyzer probe was connected directly to a port at the top of the pre-chamber body. A thin capillary connected this port to the pre-chamber volume. The analyzer constantly sampled pre-chamber contents throughout the cycle. Figure 7 displays the results of the continuous sampling method. The structure of the CO_2 trace mirrors that of the Cambustion guideline CO_2 trace for a conventional SI engine. There is a discernable step in the CO_2 trace that occurs between the measurement of the pre-combustion CO_2 mass fraction and the post-combustion CO_2 mass fraction (immediately after the minimum CO_2 value is reached, accounting for measurement transport delay) that possibly indicates CO_2 mass fraction resulting from the initial pre-chamber combustion event.

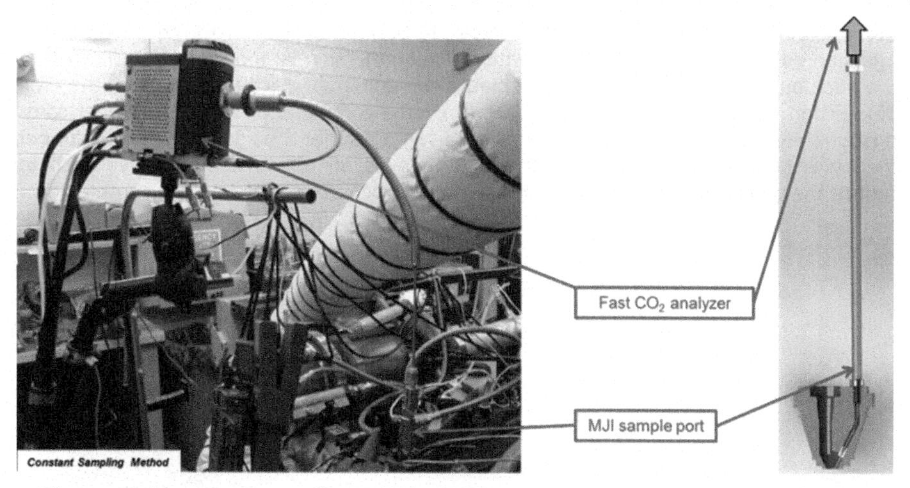

Figure 6. Constant sampling method for measuring pre-chamber residual fraction.

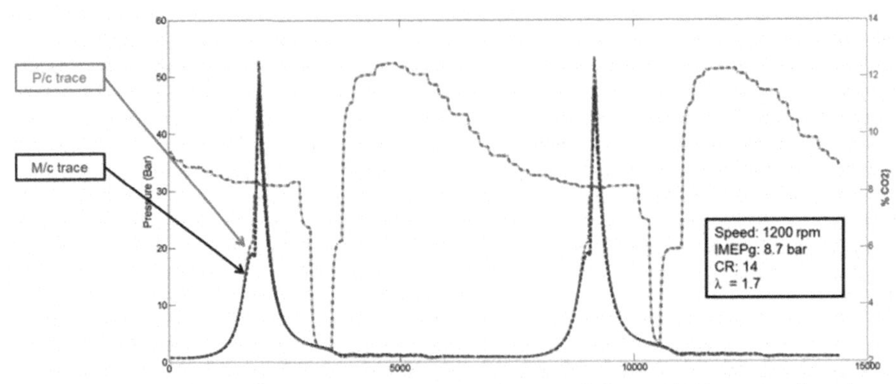

Figure 7. CO_2 evolution in the engine as measured through the pre-chamber sample port.

The engine was operated at lambda=1.4, a stable condition regardless of whether or not fuel is being injected directly into the pre-chamber. As is shown in Figure 8, residual fraction decreases appreciably when fuel is injected directly into the pre-chamber. The residual fraction continues to decrease as the injected fuel quantity is increased. Figure 8 includes a contour graph showing residual fraction trends with pre-chamber fuel injection angle and quantity. While there appears to be little sensitivity to the timing of the injection event, the sensitivity to quantity of fuel injection is apparent. It is likely that the addition of the fuel mass either displaces the dominant residual content at the time of injection, or has a more a complex impact on the completeness of pre-chamber combustion that has yet to be detected. This data suggests that the addition of pre-chamber fuel, even at a constant main chamber lambda, is an effective means by which to reduce the potentially negative impact of residual fraction on pre-chamber combustion.

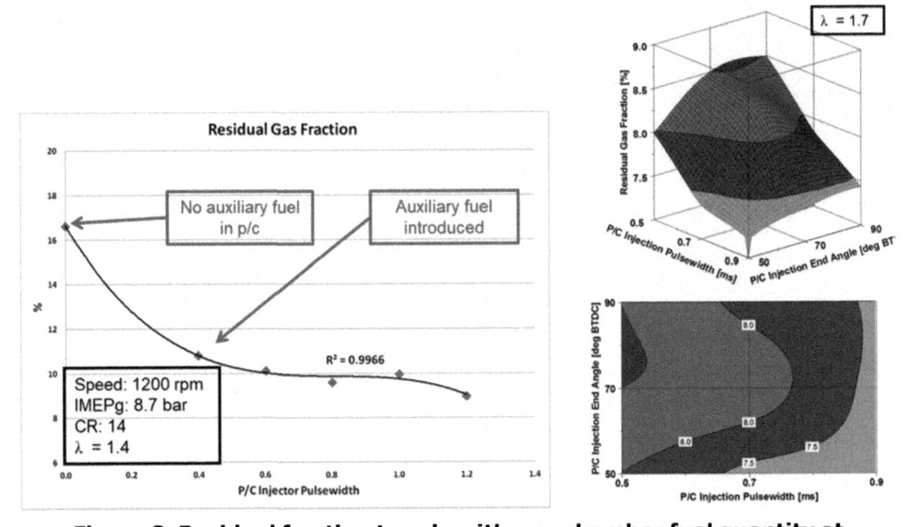

Figure 8. Residual fraction trends with pre-chamber fuel quantity at a representative condition.

4.2 Steady state low load operation

In order to provide correct airflow to maintain acceptable lambda values in the cylinder at low loads, the engine is heavily throttled. As has been established, pre-chambers are reliant on the intake process to facilitate the exchange of residual burned gas from the previous cycle with oxygen-carrying fresh charge. This gas exchange process is pressure-driven and so shifts in this pressure dynamic have a bearing on the proportion of residual gas that is displaced with fresh charge. The reduced intake pressure resulting from heavily throttled operation therefore fails to adequately drive this gas exchange process, producing a pre-chamber with lower proportional oxygen content than is present at less throttled part load conditions. The lower proportional oxygen content present in the pre-chamber is then not adequately mitigated by the pre-chamber filling process that occurs during the compression stroke. The lack of oxygen results in erratic combustion in the pre-chamber, increasing the likelihood of misfires and a corresponding degradation in combustion stability.

Active pre-chambers have the added flexibility to both operate lean in the main chamber and to introduce fuel directly in the pre-chamber. Lean operation provides excess oxygen in the main chamber, increasing the oxygen proportion in the pre-chamber during the compression stroke. To account for this dilution, fuel is then injected directly into the pre-chamber. Data from the measured residual fraction experiment suggests that direct fuel injection in the pre-chamber provides an added advantage that can be exploited at low load conditions: the physical displacement of residuals through the introduction of high pressure fuel.

Figure 9 shows the increase in stable low load extension possible with enleanment. The combustion efficiency challenges associated with burning in a low charge density environment are exacerbated by overly lean operation, as is evidenced by inferior low load extension at lambda=1.7 vs. lambda=1.5, indicating that there is an optimal lambda that balances O_2 filling requirements and combustion efficiency.

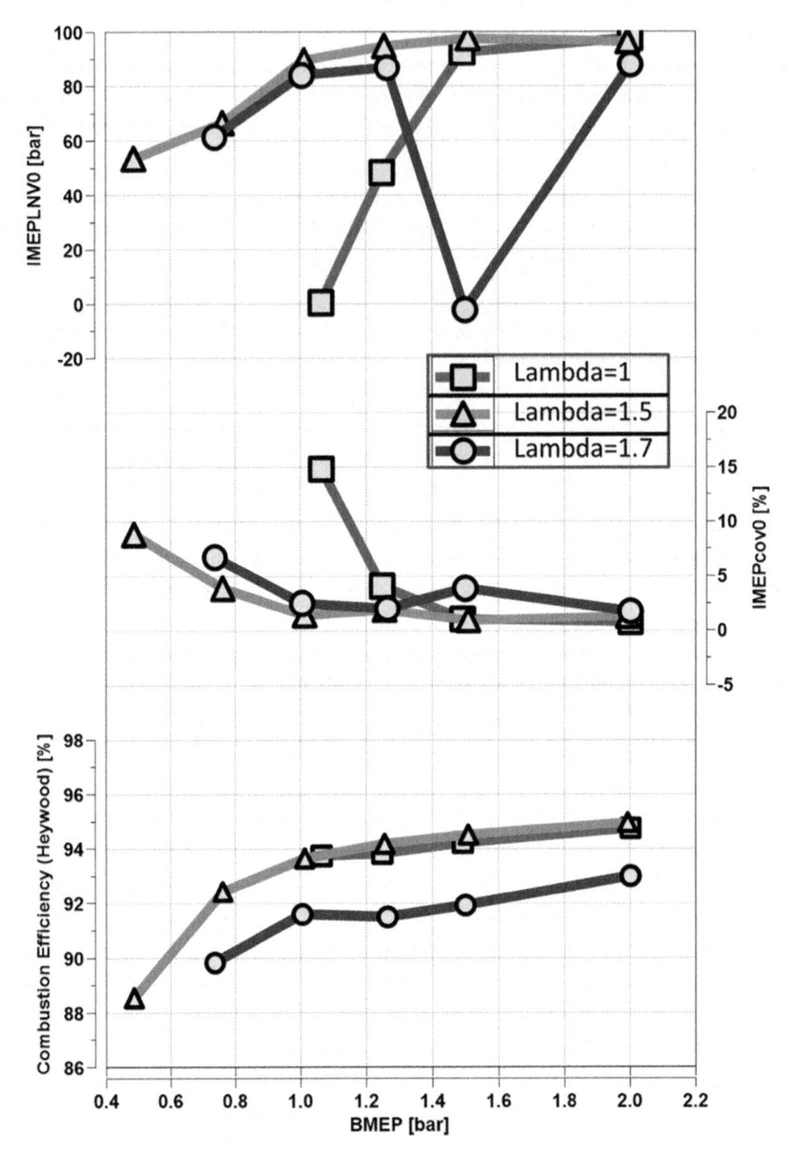

Figure 9. Stability trends at steady state 850rpm low load operation.

4.3 Torque reserve at idle operation

Idle operation is performed at low speed (< 1000 rpm) and low or zero load. A requirement for idle is retarded spark timing. At an idle condition there is an anticipation of sudden torque demand from the operator. The most rapid means by which to increase torque at this condition is to advance spark timing from a retarded location to a location advanced of top dead center (TDC). This is due to the engine controller's ability to adjust spark timing on a cycle-by-cycle basis. This rapid advancement in spark timing corresponds to an equally rapid increase in torque.

The acknowledged pre-chamber spark retard limitation does manifest in MJI. Figure 10 demonstrates the severe deterioration in spark retard capability with decreasing engine load, to the extent that at 2 bar BMEP, the engine is incapable of retarding spark timing beyond TDC. This level of spark retard provides only minimal torque reserve when applied to an idle condition.

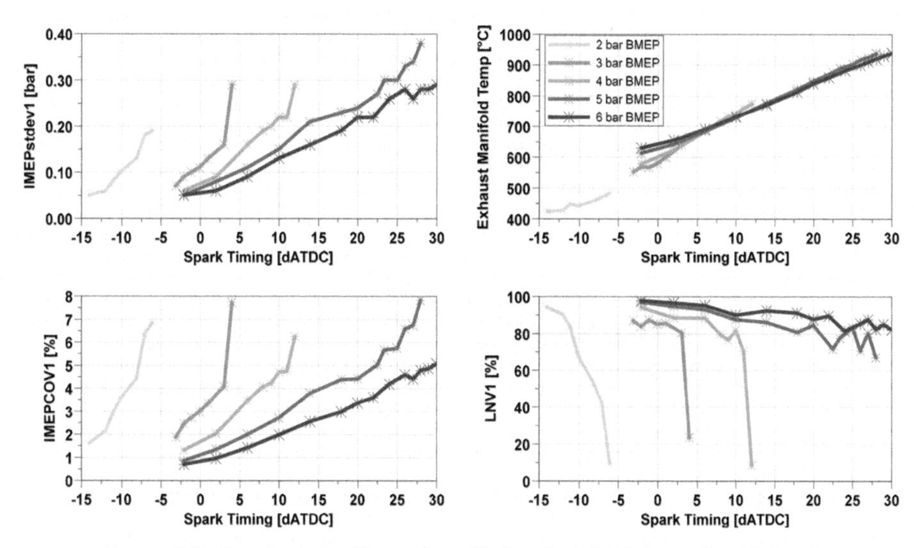

Figure 10. Spark retard trends with load at 1500rpm, lambda=1.

As in the case of low load steady state operation, the active system carries the additional flexibility of operating at a range of lambda values. This offers an advantage for idle operation as well. Fuel injection quantity, similarly to spark timing, can also be adjusted by the engine controller on a cycle-by-cycle basis. Because the active system can operate at a range of stable lambda values at the speed and load necessary for idle, this lambda tolerance can be coupled with the existing minimal spark retard capability to provide ample torque reserve (Figure 11). Sudden torque demand would therefore manifest as a rapid advancement of spark timing and, simultaneously, a rapid increase in fuel quantity injected. The latter results in a transition from a lean lambda to stoichiometric operation.

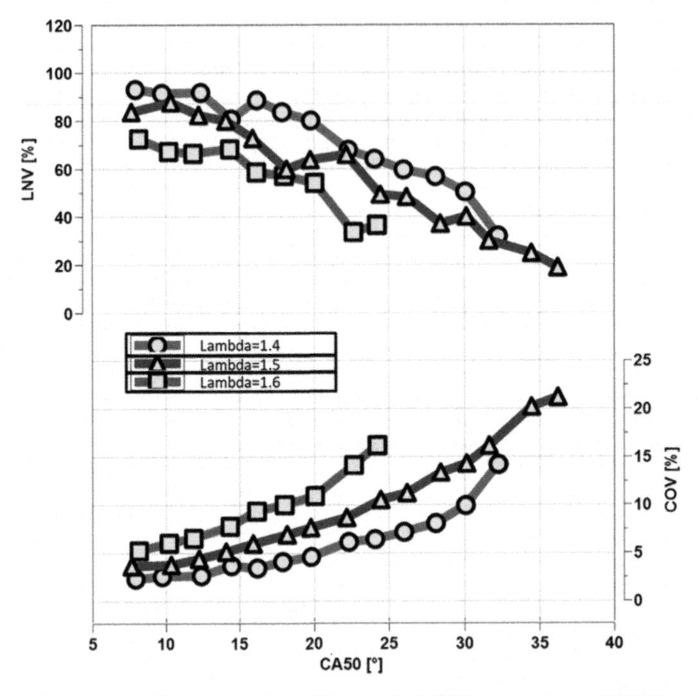

Figure 11. CA50 trends with load at 850rpm, 1 bar BMEP.

4.4 CSSR Operation

The most impactful challenge posed by the low load spark retard limitation concerns the ability to heat the aftertreatment catalyst upon cold start. The tailpipe emissions produced by vehicles are vastly more significant during the startup phase prior to catalyst light-off than they are at any other point in a legislated drive cycle. Catalysts require heat input to work effectively. Prior to achieving a high temperature light-off condition a large proportion of the engine-out emissions pass through un-catalyzed or uncaptured to the tailpipe. Aggressive warm up of the catalyst is therefore critical to ensuring that the vehicle can meet legislated emissions requirements. The common solution to ensure rapid heat input to the catalyst is to retard spark timing to such a degree that combustion occurs exclusively during the expansion stroke. The much later burning process results in both increased exhaust temperature and increased exhaust flow. The latter results from the non-optimal combustion phasing requiring de-throttling to compensate for the poor thermal efficiency. This poor thermal efficiency is purposeful since a large proportion of the combustion process has minimal to no contribution to torque and instead is used largely to generate heat. Spark retard, and its ability to generate high exhaust enthalpy, therefore is an essential element of CSSR operation, which makes pre-chambers' nominal lack thereof a major concern.

Data in Figure 10 is shown at warm oil and coolant temperatures which are not representative of the CSSR condition. Figure 12 demonstrates the further degradation in performance when fluids are conditioned to 20 degrees Celsius.

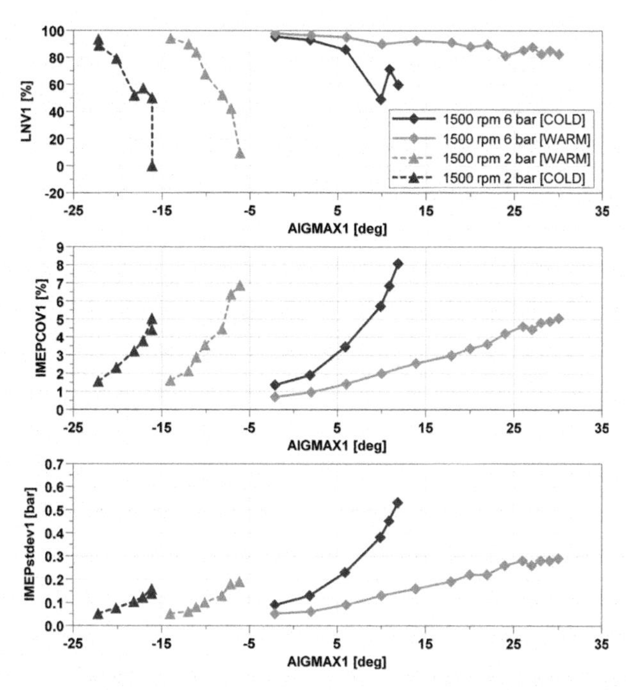

Figure 12. Spark retard trends at warm and cold fluid conditions.

To properly address this issue, an examination of the failure mode is necessary. With jet ignition engines, different segments of the burn curve provide distinct information about combustion progress:

- Early burning, captured in the CA0-10 duration, encompasses the pre-chamber combustion process from spark through time of radical jet introduction into the main chamber, as well as the ignition process in the main chamber [24].
- Mid-burning, reflected in the CA10-50 duration, encompasses the immediate post-jet ignition process and therefore still bears the influence of jet characteristics such as velocity and reactivity.
- Late burning, CA50-90, occurs long after the jet ignition process has concluded and is therefore largely uninfluenced by characteristics of pre-chamber combustion or the resulting jets.

An examination of the different burn duration segments (Figure 13) shows linear trends observed in burn duration versus spark timing. These durations are shorter with MJI compared to the SI engine, which is counterintuitive to the spark retard limitation result. However inspection of the standard deviation of these burn durations indicates an increasing instability with MJI as spark timing is retarded. In this case, the 300-cycle average burn duration provides no indication of the instability. This implies that the unstable cycles occur rarely but their magnitude is severe. It is also telling that the instability is present in all three burn duration segments, including CA0-10. This implies that the pre-chamber combustion event is experiencing misfires or partial burns which cause the main chamber to misfire.

Here the ability of the active system to operate with fuel injected directly into the pre-chamber was again exploited. Auxiliary fuel injection provides direct control over both the fuel quantity and its relative location. In order to prevent over-fueling, the engine is operated lean. Figure 13 compares the spark retard capability of the SI engine, MJI at lambda=1 with no direct pre-chamber fuel, and MJI at lambda=1.4 with fuel injected into the pre-chamber. In this latter configuration, the MJI was able to operate at a retarded spark timing identical to that of the SI engine. The combustion instability present under lambda=1 conditions is no longer present with lean operation.

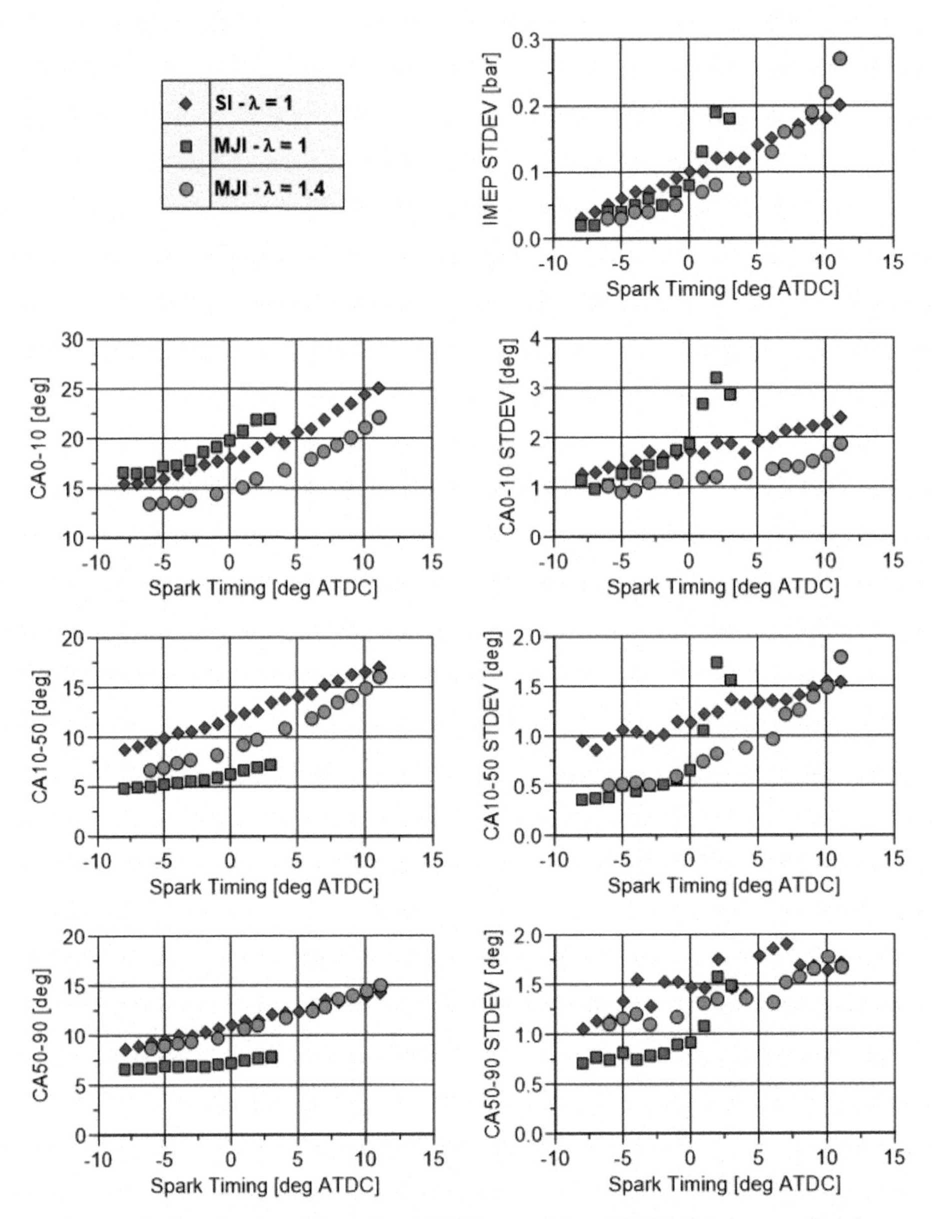

Figure 13. Spark retard trends at 1500rpm, 2 bar NMEP, 20 degree Celsius fluid temperature.

Table 2 lists the categories of sensitivities that were evaluated in order to achieve an exhaust enthalpy target. The final result of the optimization activities is shown in Figure 14. A proprietary combination of calibrated parameters produces a MJI system that can achieve the CSSR target enthalpy. Crucially, a common pre-chamber geometry is used to attain both successful CSSR operation and high load knock mitigation with no excessive compromise of lean limit extension or peak efficiency. A single pre-chamber design coupled with a moderately featured engine is used to span the entire conventional engine operating range and therefore has the potential to be practically applied in a production context.

Table 2. Sensitivities explored as part of CSSR experiments.

Sensitivities Explored
Cam timing
Fuel injection configuration – main chamber
Fuel injection pressure – main chamber
Fuel injection timing – main chamber
Fuel injection pulse strategy – main chamber
Lambda
Pre-chamber fuel injection quantity
Pre-chamber fuel injection timing
Pre-chamber fuel injection pressure
Pre-chamber fuel injection pulse strategy
Pre-chamber geometry
Pre-chamber orientation
Spark strategy
Spark plug gap

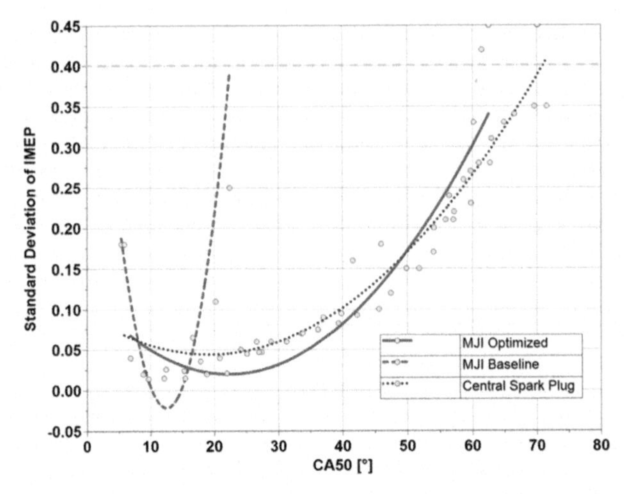

Figure 14. CA50 vs. Combustion Stability at 1500rpm, 2 bar net mean effective pressure (NMEP), 20 degree Celsius fluid temperature.

The ability to adjust lambda is used to minimize engine-out emissions during this period. A calculated 10 second cumulative mass of hydrocarbons (HC) and nitrogen oxides (NO_x) emissions is presented in Figure 15. Enleanment can produce counter HC and NO_x trends but these are nonlinear and therefore a minimum cumulative mass is clearly evident in the near-lean region.

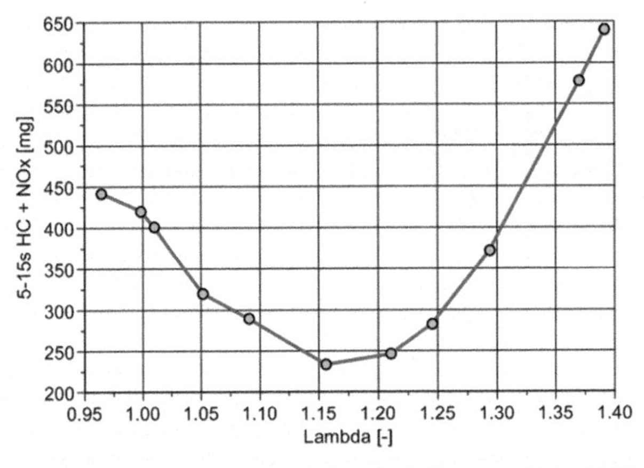

Figure 15. Emissions trends with lambda at 1500rpm, 2 bar NMEP, 20 degree Celsius fluid temperature.

The effect of charge motion on jet ignition performance is not well understood, particularly its impact at retarded spark timing. Four charge motion cases were evaluated: baseline, increased tumble, introduction of swirl, and a combination of swirl and tumble. Charge motion differences from the baseline were induced through the use of plate inserts into each of the intake ports (Figure 16).

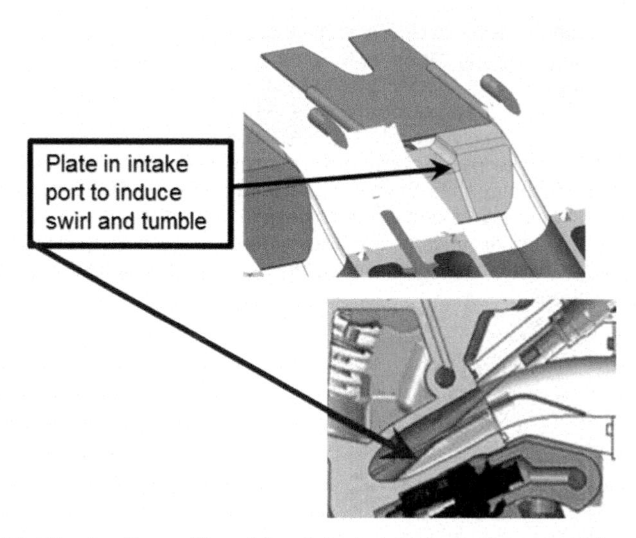

Figure 16. Illustration of tumble plate insert in 1.5L jet ignition engine.

For these experiments, the effect of charge motion was evaluated at a representative low load condition at a stoichiometric lambda. Figure 17 shows distinct differences in combustion stability amongst the charge motion variants across the spark timing range presented. The addition of tumble provides the lowest standard deviation of indicated mean effective pressure (IMEP) during spark retard operation, though spark timing range is not improved. The comparatively poor pre-chamber lowest normalized value (LNV) of the swirl case, with only muted impact on the main chamber LNV implies that the swirl case produces erratic pre-chamber combustion.

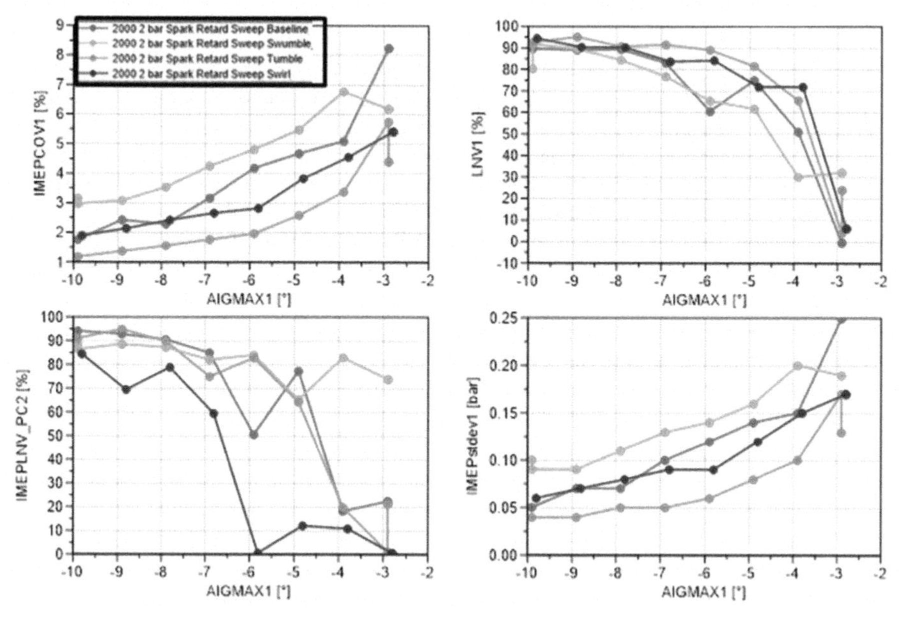

Figure 17. Stability (AIGMax1) vs. spark timing at 2000rpm, 2 bar BMEP, lambda=1.

The baseline and tumble variants clearly show a significant increase in combustion efficiency across the spark timing range versus the two swirl-based variants (Figure 18). This translates to the thermal efficiency results as well, with the swirl-only variant displaying anomalous results in indicated thermal versus brake thermal efficiency. This is possibly due to an abnormally elevated gross IMEP level necessary to maintain the target BMEP level for the swirl-only variant. No charge motion variant, however, exhibits higher combustion efficiency than that of the baseline variant. With no improvement in spark retard capability or emissions at stoichiometric conditions, the addition of charge motion offers no clear advantages in CSSR operation. An upcoming study will focus on the effect of charge motion addition on CSSR performance across the spark timing range at lean conditions.

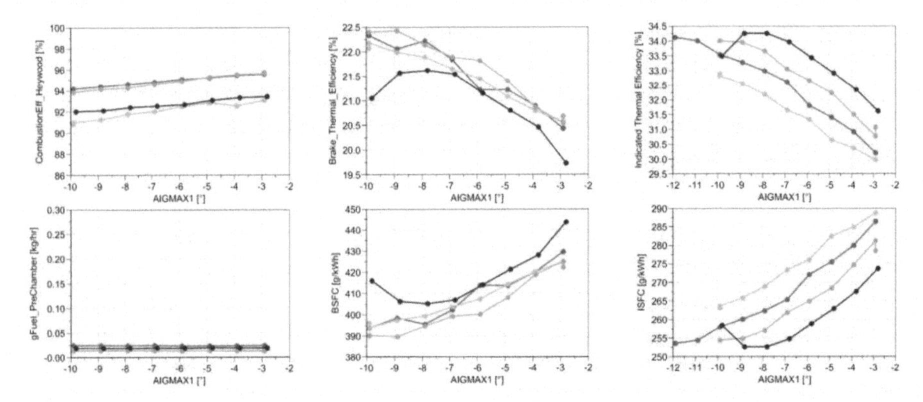

Figure 18. Efficiency vs. spark timing at 2000rpm, 2 bar BMEP, lambda=1.

The steady state results of the active MJI CSSR operation are listed in Table 3 against the targets. These targets are consistent with emissions and heat flux targets of vehicles operating at or near lambda=1 with a 3-way catalyst for emissions control. Commercial lean aftertreatment packages available in diesel passenger car applications generally light-off at lower temperatures than 3-way catalysts due to the lower exhaust temperatures associated with diesel combustion. It is therefore likely that the heat flux target to activate an optimized lean aftertreatment catalyst is lower than the target displayed in Table 3, and the additional heat flux available would simply result in more rapid activation times.

While MPT is continuing to develop a transient CSSR calibration, steady state targets have been achieved at stable conditions.

Table 3. CSSR result status.

Metric	Unit	Target	MJI3-March 2019
HC	g/h/l	5.6	8.3
NO_x	g/h/l	9	4.7
$HC+NO_x$	g/h/l	14.6	13.0
CO	g/h/l	150	57.3
Heat flux	KW/l	>5	5.27
Speed	rpm	<1500	1500
NMEP	bar	2	2
SD-$IMEP_g$	bar	0.4	0.33
LNV	%	>40	44
Coolent Temp	°C	25	28.8

5 CONCLUSIONS

Previous studies [16-19] have demonstrated the potential of jet ignition as an enabling technology for ultra-lean SI combustion to significantly increase thermal efficiency, reduce fuel consumption, and reduce engine-out NO_x emissions. However, pre-chambers have historically had challenges at low load conditions. This includes poor combustion stability under steady state low load conditions and severely limited spark retard capability. The latter is necessary both for ensuring adequate torque reserve at idle and for producing acceptable heat flux to activate the aftertreatment catalysts during CSSR operation.

Active MJI offers the flexibility to operate the main chamber at lean lambda values and to independently add fuel directly to the pre-chamber. This flexibility is leveraged to mitigate the deleterious effects on gas exchange caused by heavily throttled operation, and to supplement limited spark retard capability at idle to provide adequate torque reserve. Under CSSR conditions, the addition of pre-chamber fuel provides more direct control over pre-chamber mixture preparation resulting in reduced instability in the pre-chamber combustion event. Lean operation, in conjunction with other optimization activities, results in MJI achieving CSSR results comparable to those of modern SI engines.

More generally, MAHLE Jet Ignition® will continue to be developed on various engine platforms, including the engine described in this study. Upcoming studies will focus on tailoring the MJI engine for a variety of different end-use applications.

ACKNOWLEDGEMENTS

The authors wish to acknowledge Ian Reynolds and Graham Irlam of MAHLE Powertrain. The authors wish to thank Dr. Mark Peckham of Cambustion for his support.

REFERENCES

[1] Bunce, M. and Blaxill, H., "Sub-200 g/kWh BSFC on a Light Duty Gasoline Engine," SAE Technical Paper 2016-01-0709, 2016.

[2] Bunce, M., Blaxill, H., Kulatilaka, W., and Jiang, N., "The Effects of Turbulent Jet Characteristics on Engine Performance Using a Pre-Chamber Combustor," SAE Technical Paper 2014-01-1195, 2014.

[3] Quader, A. A., "Lean Combustion and the Misfire Limit," SAE Technical Paper 741055, 1974.

[4] Husted, H., Piock, W., Ramsay, G., "Fuel Efficiency Improvements from Lean Stratified Combustion with a Solenoid Injector," SAE Technical Paper 2009-01-1485, 2009.

[5] Germane, G., Wood, C., Hess, C., "Lean Combustion in Spark-Ignited Internal Combustion Engines – A Review," SAE Technical Paper 831694, 1983.

[6] Heywood, J., Internal Combustion Engine Fundamentals, McGraw-Hill, 1988.

[7] Yamamoto, H., "Investigation on Relationship Between Thermal Efficiency and NO_x Formation in Ultra-Lean Combustion," SAE Journal Paper JSAE 9938083, 1999.

[8] Dober, G. G., Watson, H. C., "Quasi-Dimensional and CFD Modelling of Turbulent and Chemical Flame Enhancement in an Ultra Lean Burn S.I. Engine," Modeling of SI Engines SP-1511, 2000.

[9] Ward, M., "High-Energy Spark-Flow Coupling in an IC Engine for Ultra-Lean and High EGR Mixtures," SAE Technical Paper 2001-01-0548, 2001.

[10] Qiao, A., Wu, X., "Research on the New Ignition Control System of Lean- and Fast-Burn SI Engines," SAE Technical Paper 2008-01-1721, 2008.

[11] Ricardo, H. R., Recent Work on the Internal Combustion Engine, SAE Transactions, Vol 17, May 1922.

[12] Gussak, L. A., Karpov, V. P., Tikhonov, Y. Y., "The Application of the Lag-Process in Pre-chamber Engines," SAE Technical Paper 790692, 1979, doi:10.4271/790692.

[13] Robinet, C., Higelin, P., Moreau, B., Pajot, O., Andrzejewski, J., "A New Firing Concept for Internal Combustion Engines: "l'APIR"," SAE Technical Paper 1999-01-0621, 1999.

[14] Murase, E., Ono, S., Hanada, K., Oppenheim, A., "Pulsed Combustion Jet Ignition in Lean Mixtures," SAE Technical Paper 943048, 1994.

[15] Toulson, E., Schock, H., Attard, W., "A Review of Pre-Chamber Initiated Jet Ignition Combustion Systems," SAE Technical Paper 2010-01-2263, 2010, doi:10.4271/2010-01-2263.

[16] Attard, W., Toulson, E., Fraser, E., Parsons, P., "A Turbulent Jet Ignition Pre-Chamber Combustion System for Large Fuel Economy Improvements in a Modern Vehicle Powertrain," SAE Technical Paper 2010-01-1457, 2010, doi:10.4271/2010-01-1457.

[17] Cao, Y., Li, L. "A novel closed loop control based on ionization current in combustion cycle at cold start in a gdi engine," SAE Technical Paper 2012-01-1339, 2012. doi:10.4271/2012-01-1339.

[18] Sens, M., Binder, E., Reinicke, P.B., Riess, M., Stappenbeck, T., Woebke, M., "Pre-Chamber Ignition and Promising Complementary Technologies," 27th Aachen Colloquium Automobile and Engine Technology, 2018.

[19] Attard, W., Kohn, J., Parsons, P., "Ignition Energy Development for a Spark Initiated Combustion System Capable of High Load, High Efficiency and Near Zero NO_x Emissions," SAE Journal Paper JSAE 20109088, 2010.

[20] Attard, W., Blaxill, H., "A Gasoline Fueled Pre-Chamber Jet Ignition Combustion System at Unthrottled Conditions," SAE Technical Paper 2012-01-0386, 2012, doi:10.4271/2012-01-0386.

[21] Attard, W., Blaxill, H., "A Lean Burn Gasoline Fueled Pre-Chamber Jet Ignition Combustion System Achieving High Efficiency and Low NO_x at Part Load," SAE Technical Paper 2012-01-1146, 2012, doi:10.4271/2012-01-1146.
[22] Bassett, M., Hall, J., Cains, T., Underwood, M.et al., "Dynamic Downsizing Gasoline Demonstrator," SAE Int. J. Engines10(3):2017.
[23] Bassett, M., Hall, J., Hibberd, B., Borman, S.et al., "Heavily Downsized Gasoline Demonstrator," SAE Int. J. Engines9(2):729–738, 2016.
[24] Bunce, M. and Blaxill, H., "Methodology for Combustion Analysis of a Spark Ignition Engine Incorporating a Pre-Chamber Combustor," SAE Technical Paper 2014-01-2603, 2014, doi:10.4271/2014-01-2603.

Effects of injection timing of DME on Micro Flame Ignition (MFI) combustion in a gasoline engine

Y. Feng[1], T. Chen[1], H. Xie[1], X. Wang[2], H. Zhao[2]

[1]State Key Laboratory of Engine, Tianjin University, China
[2]Brunel University, London, UK

ABSTRACT

In order to achieve high dilution low temperature combustion and overcome the instability of spark ignition (SI)-controlled auto-ignition (CAI) hybrid combustion, the micro flame ignition (MFI) combustion technology was studied on a port fuel injection (PFI) gasoline engine. A small amount of Dimethyl Ether (DME) was directly injected into the cylinder to form multiple micro flame kernels, which initiated and formed the subsequent gasoline combustion process. The single cylinder engine results showed that the optimization of the MFI strategy increased the thermal efficiency from 33.65% to 42.87% at 2000 rpm and IMEP 4 bar. The NOx emission was reduced to nearly zero and the HC and CO emissions were 60% lower than the traditional SI or CAI/HCCI combustion. The in-cylinder mixture formation and combustion processes of the MFI technology was studied in a single cylinder optical engine by the particle image velocimetry (PIV), laser Rayleigh scattering, Mie scattering and high-speed imaging combined with heat release analysis. In addition to the optical measurements, three dimensional (3D) Computational Fluid Dynamics (CFD) simulations were performed to further explain the experimental observations. The results showed that the typical MFI hybrid combustion process is characterized by three distinct stages of DME auto-ignition followed by a flame propagation and then a final multi-point auto-ignition combustion process. As the DME injection timing was advanced, the MFI hybrid combustion process was shifted from a 2-stage heat release process to a single peak of heat release rate. The early injection of DME at the start of the compression stroke could improve overall reactivity of the fuel/air mixture in the cylinder and subsequent flame propagation process, and the late injection of DME near TDC acted as high-energy ignition kernels for enhanced ignition process of highly diluted mixtures.

1 INTRODUCTION

In recent years, Homogenous charge compression ignition (HCCI) combustion, also called Controlled Auto-Ignition (CAI) combustion, is considered as an important combustion technology in the future gasoline engine due to its high-efficiency, low fuel consumption and low emissions advantages [1,2]. Significant research and development works have shown that the pumping loss was reduced and the engine efficiency was improved through high dilution combustion [3,4]. However, the application of HCCI/CAI combustion is limited by a host of difficulties, such as the ignition and combustion process control and limited operation range. In order to overcome these limits, Spark Ignition (SI) – Controlled Auto-Ignition (CAI) Hybrid Combustion (SCHC), also called Spark Assisted Compression Ignition (SACI) hybrid combustion [5,6] can be a useful way to control the combustion process under high dilution conditions.

Although SCHC can introduce a certain degree of control on the ignition timing and hence adjust the combustion process, the single ignition site and its limited ignition energy render the technology to have limited impact on the combustion process of high diluted mixtures. Reuss et al. [7] found that the spark assisted auto-ignition combustion suffered from high cycle-to-cycle variation because of the cyclic variation in the single flame kernel development. The research result from Xu et al. [8] shows that there were considerable changes of the initial flame shape, area and speed in the SCHC among consecutive cycles in an optical engine. Similar results had been found in other researches [9,10]. Wheeler et al. [11] pointed out cooled EGR (Exhaust Gas Recirculation) could cause slower rates of flame propagation, decreased engine stability and even lead to misfire. Due to the dilution effect of EGR, both the flame propagation speed and shape varied markedly in continuous cycles at the condition of SI-CAI hybrid combustion [12]. Based on the analyses of above studies, it is recognized that more ignition energy in large space would be required for improving the combustion stability and broadening the dilution limit of combustion. Therefore, the micro flame ignition (MFI) technology was proposed and studied [13] and it involves the direct injection of a small quantity of highly ignitable fuel, such as dimethyl ether (DME), into the premixed lean/diluted fuel and air mixture. DME was chosen because it is easy to evaporate at low temperature and low injection pressure when liquid-phase DME is directly injected into the cylinder [14], as well as its high Cetane number. Besides, DME has a high oxygen content and no C-C bond [15], which leads to smokeless combustion. The detail fuel properties of DME is shown in Table 1. The high Cetane number of DME is beneficial to induce the initial spontaneous combustion and accelerate the flame propagation, which helps to stabilize and expand the dilution boundary of hybrid combustion at high dilution conditions.

Table 1. fuel properties of DME.

Chemical formula	**CH3OCH3**
Stoichiometric air/fuel ratio	9.0
Triple point (K)	131.7
Critical temperature (K)	400.1
Normal boiling point (K)	249.0
Ignition energy (mJ)	45
Ignition temperature (K)	508
Cetane number	55-60
Low heating value (MJ/kg)	27.6

Previous studies by Cha et al. [16] and Xu et al. [17] indicated that the DME inject timing could directly influence the distribution of DME in the cylinder and affect the initial auto-ignition timing. Zhang et al [18] showed that the direct injection of DME was effective to control the process of hybrid combustion. The result of Fu's experiments [19] showed that the utilization of MFI strategy could decrease the variation of the heat release process in SCHC and increase the thermal efficiency, as well as little NOx emission.

The experiment results above indicated that the MFI strategy could reduce the cyclic variation and emissions and broaden the dilution boundary of hybrid combustion, which led to further increase of thermal efficiency. In order to maximize the benefits of the MFI combustion control technology, the injection timing and quantities need to be optimized. The ignition and combustion process will depend on the spatial distribution of the DME and in-cylinder conditions and need to be controlled according to the engine load. On the one hand, it is necessary to achieve reliable ignition and stable combustion of highly diluted mixtures. On the other hand, the heat release rate should be controlled to prevent too rapid combustion. In this study, the effect and mechanism of combustion process control by different kinds of micro flame ignition strategy will be further investigated by means of in-cylinder studies of mixture formation, ignition and combustion using CFD and engine experiments.

2 ENGINE AND SIMULATION SPECIFICATIONS

2.1 Thermodynamic single-cylinder engine test bench

The thermodynamic single-cylinder engine was used to study the potential thermal efficiency of MFI combustion technology. The specifications of this single-cylinder engine are given in Table 2. The engine comprises a Ricardo Hydra engine block and a specially designed cylinder head equipped with two sets of identical mechanically fully variable valve actuation (VVA) systems on the intake and exhaust valves. Each VVA system integrates a BMW VANOS variable valve timing device and a BMW Valvetronics continuously variable valve lift device. The engine could be operated with port fuel injection (PFI) or direct injection (DI) or both together. In this study, gasoline was injected into the intake port through a PFI injector at 3 bar and a small amount of DME was injected directly into the cylinder at 40 bar by a 6-hole DI gasoline injector located below the intake valve in the cylinder head. The piston has been redesigned to consider the requirements of wall-guide strategy and fuel distribution control.

Table 2. Engine specifications of thermodynamic single cylinder engine.

Engine Type	4-stroke single cylinder
Bore	86 mm
Stroke	86 mm
Displacement	0.5 l
Compression ratio	14.09
Combustion chamber	Pent roof/4 valves
Fuel	Gasoline(93 RON)/DME
Gasoline injection pressure	3 bar
DME injection pressure	40 bar
Inlet pressure	Naturally aspirated
Coolant temperature	80 °C
Oil temperature	50 °C

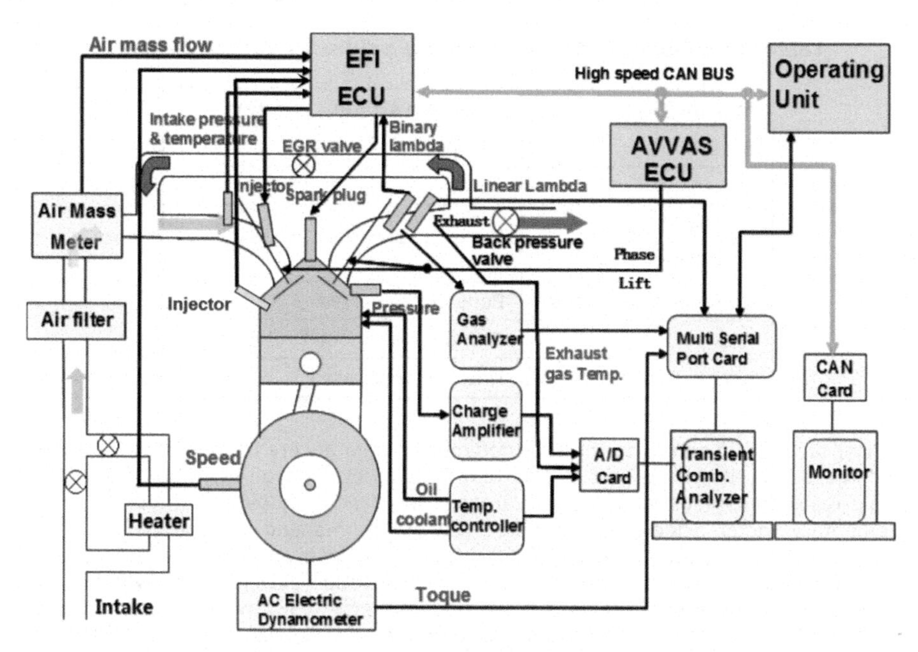

Figure 1. Schematic of the thermodynamic single cylinder test bench.

The schematic of the test bench was shown in Figure 1. A linear oxygen sensor (with an accuracy of ±1.5%) was mounted in the exhaust pipe to ensure precise control of the air/fuel ratio. The air flow rate was measured by an air mass meter and checked with the lambda and fuel injection mass. At each experimental point, the in-cylinder pressure was measured with a Kistler 6125B piezoelectric transducer and a 5011B charge amplifier. The exhaust gas temperature was measured by a K-type thermocouple installed in the exhaust pipe. The RGF (Residual Gas Fraction) was determined from the residual gas mass calculated from the exhaust gas temperature and the in-cylinder gas pressure at the exhaust valve closing time using the state equation of the ideal gas. The emissions were measured by the Horiba MEXA-7100DEGR emission analyzer. The flow rate of direct injected DME was measured by a positive displacement piston type flow meter MAX 213 with uncertainty of ± 0.2%. The amount of port fuel injection of gasoline in a cycle was obtained through the calibration relationship between injection pulse and the quantity of gasoline injected. further information on the single cylinder engine can be found in the reference [20,21].

2.2 Single-cylinder optical engine test bench

In order to investigate the ignition and combustion process of the MFI hybrid combustion, in-cylinder optical studies were carried out in an optical engine as detailed in Table 3. The optical access is provided by a flat-fused silica with a diameter of 71 mm mounted on the piston crown. Due to the lower compression ratio, the basic fuel of optical engine was changed to PRF40 to simulate the auto-ignition of gasoline under high compression ratio. At the same time, the external EGR was applied by the mixture of N_2 and CO_2.

Table 3. Engine specifications of single cylinder optical engine.

Engine Type	**4-stroke single cylinder**
Bore	95mm
Stroke	95mm
Displacement	0.67 l
Compression ratio	9.24
Combustion chamber	Pent roof/4 valves
Fuel	Primary Reference Fuel 40/DME
Inlet pressure	Naturally aspirated

Similar fuel supply setups to the thermodynamic engine were used with PFI gasoline fuel injection in the intake port at 3bar and DME direct injection at 40 bar. The crank-angle-resolved combustion luminosity images were recorded by a Phantom 7.1 complementary metal oxide semiconductor (CMOS) high-speed camera (HSC) or an Intensified Charge-coupled Device (ICCD). Qualitative laser Rayleigh Scattering (LRS) of the DME vapour was employed to visualize the location of DME in the cylinder. The laser beam of the second harmonic of a Nd:YAG laser (532 nm) was formed into a sheet with a combination of cylindrical and spherical lenses. The FlowMaster PIV system by LaVision was employed for the measurement of the in-cylinder flow in cylinder prior to the combustion in the optical engine. The second harmonic output (532 nm wavelength) from two SOLO 120 Nd:YAG lasers was used as the light source for the PIV measurement. More detailed descriptions of the optical engine and measurement techniques can be found in [17].

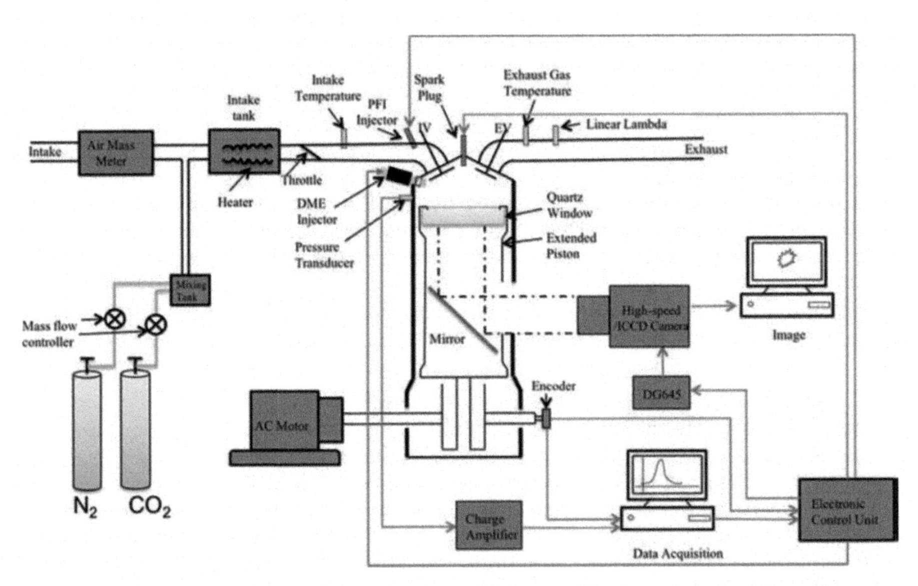

Figure 2. Schematic of the single cylinder optical engine test bench.

For all optical tests, the engine was operated at 600 r/min. The gross indicated mean effective pressure (IMEP) of all studied cases ranged from 2.5~3 bar. The low speed, light-load conditions were limited by the mechanical strength of the optical engine and the concern of damaging the large piston crown window for some conditions explored with high heat release rates.

2.3 CFD simulation platform

To study the characteristic of DME micro flame ignition and the effect of different DME injection timings on the MFI hybrid combustion, numerical simulations were performed using the Converge software platform. The computational grid with a grid spacing of 4 mm and the minimum grids of 0.25mm was generated for the optical engine. Simulations were run from 300° BTDC to 40° ATDC. Gasoline was modeled by a 2-component surrogate comprised of PRF93 (Primary Reference Fuel, 93% iso-octane (iC8H18), and 7% n-heptane (nC7H16)). Reynolds-Averaged Navier Stokes (RANS) approach was applied with RNG k-ε turbulence model in the simulations. The simulated injection used a stochastic collision model (O'Rourke) [22]. The liquid fuel droplet breakup was modeled using the Kelvin-Helmholtz Rayleigh-Taylor (KHRT) droplet breakup model [23]. The droplet evaporation was simulated by using Frossling Drop Evaporation module coupling with Boiling model [24]. Constant temperature boundary conditions were used for solid surfaces. Surface temperatures were set to the coolant temperature, except for the piston, which was set 70 K above the coolant temperature. The initial and intake inflow compositions were the premixed composition specified. The initial temperature of the intake region was the intake temperature, and the initial in-cylinder and exhaust temperatures were determined by1-D simulations and experiments.

As for combustion process, the SAGE model was coupled with the chemical kinetic reaction mechanism for blend fuel [25]. In this study, the PRF surrogate, a mixture of iso-octane and n-heptane, was used to represent gasoline in order to obtain a relatively small size model by considering less components and pathways. The chemical kinetic skeletal mechanism was built by Chen et al by the DRGEP using a Jacobian pairwise relation between coupled species combined with H radical method from a 348-species detailed mechanism. For further information and the verification of mechanism, reference [26-28] were useful. The auto-ignition delay time and laminar flame speeds results was shown in Figure 3 and Figure 4. The CFD simulation result shows that the skeletal mechanism could capture the engine phenomena accurately.

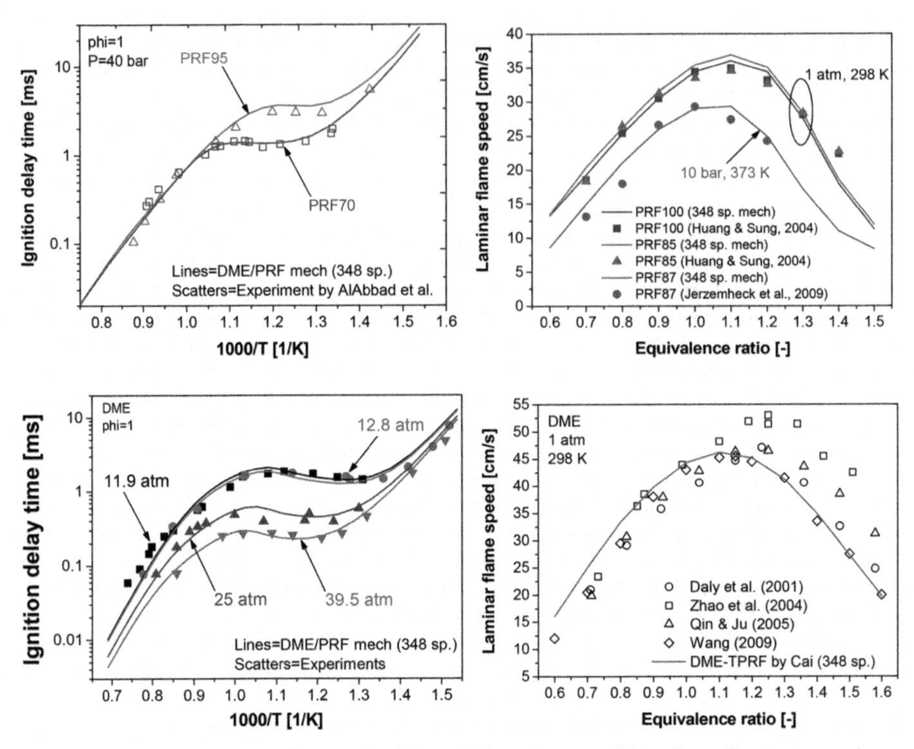

Figure 3. Validation of auto-ignition delay time and laminar flame speeds between the 348-species detailed mechanism and experiments.

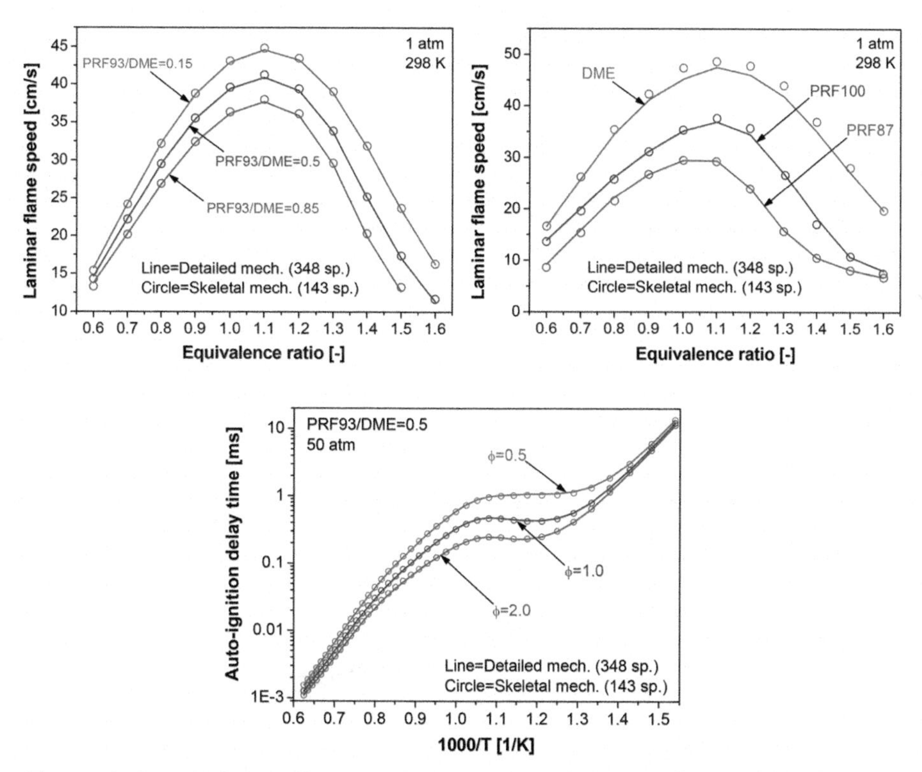

Figure 4. Comparison of laminar flame speeds between 348-species detailed and 143-species skeletal mechanisms.

3 RESULTS AND DISCUSSION

3.1 The effect of MFI combustion on the thermal efficiency and emissions

As shown in Figure 5, the indicated thermal efficiencies were increased by 20-30% between 2-6 bar IMEP by the employment of MFI combustion compared to the conventional spark ignition combustion operation at 1500rpm and 2000rpm. The maximum indicated thermal efficiency reached 40% at 4bar IMEP and 43.98% at 7bar IMEP at 2000rpm. The highest indicated thermal efficiency of 47% was achieved with MFI combustion of lean mixture ofλ =2.0 using split DME injections and effective compression ratio(ECR) 13.0 as shown by the lower graph in Figure 5, where IL stands for the inlet valve lift and EL the exhaust valve lift. For the IMEP (Indicated Main Effective Pressure) 4bar @2000 rpm (Revolution per Minute), the EVO (Exhaust Valve Open) and EVC (Exhaust Valve Close) were 201.9 and 348.1 CAD ATDC (After Top Dead Center). The IVO (Intake Valve Open) and IVC (Intake Valve Close) were 435.0 and 554.1 CAD ATDC.

As reported previously [17], the NOx emission was nearly zero at early and late DME injection timings . The HC and CO emissions were 60% lower than the traditional SI or CAI/HCCI combustion due to the more complete combustion of more diluted air/fuel mixture.

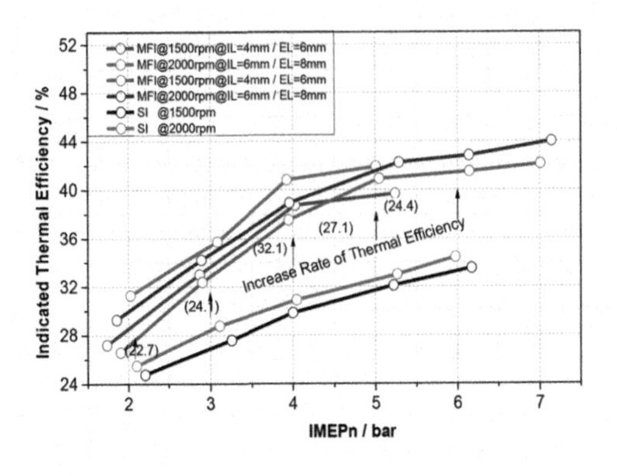

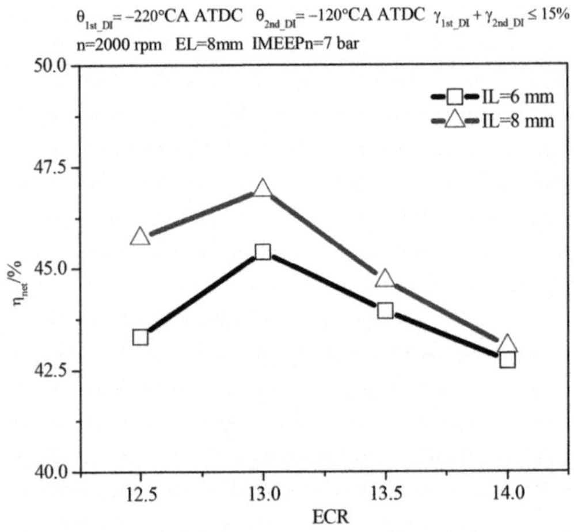

Figure 5. Comparison of indicated thermal efficiency between SI-CAI and MFI strategy.

3.2 Effect of different DME injection timing to MFI hybrid combustion

In order to understand the mechanism of MFI strategy and its effect on hybrid combustion at different injection timings, two optical engine cases with the start of DME injection timing (SOI) at 120 CAD (Crank Angle Degree) BTDC (Before Top Dead Center) and 25 CAD BTDC respectively were studied at the engine operation conditions in Table 4. The results of heat release are shown in Figure 6. The early injection case was characterized by a single-peak heat release rate curve, similar to that of the spark ignition and controlled autoignition hybrid combustion (SCHC) process which comprises spark ignited flame propagation followed by auto-ignition of unburned mixtures. With the delay of DME injection timing, the feature of heat release curve was changed to two peaks. The thermodynamic single cylinder showed the same result with PRF93 and DME fuel. The operation conditions were also shown in Table 5 and the heat release rate of SOI 120 and 40 were presented. In the following sections, optical measurement and CFD simulation results from optical engine will be used to explain such changes in the heat release process brought by the change in the DME injection timing.

Table 4. Optical engine operation conditions.

Revolution	600rpm
IMEP	2.5-3 bar
Intake valve open timing	396.0 CAD ATDC
Intake valve close timing	508.0 CAD ATDC
Intake valve lift	1.00 mm
Exhaust valve open timing	180.0 CAD ATDC
Exhaust valve close timing	328.0 CAD ATDC
Exhaust valve lift	2.00 mm
Gasoline per cycle	14.8 mg
DME per cycle	1.8 mg
Relative Air/Fuel ratio (λ)	1.0
External EGR rate	27 %
Estimated Residual Gas	19 %
Intake temperature	40±1 °C
Coolant temperature	85±2 °C

Table 5. Thermodynamic single cylinder engine operation conditions.

Revolution	1500rpm
IMEP	4bar
Intake valve open timing	425.6 CAD ATDC
Intake valve close timing	534.4 CAD ATDC
Intake valve lift	5.01 mm
Exhaust valve open timing	171.5 CAD ATDC
Exhaust valve close timing	278.5 CAD ATDC
Exhaust valve lift	4.83 mm
Gasoline per cycle	9.23 mg
DME per cycle	2.55 mg
Relative Air/Fuel ratio (λ)	1.98
Estimated Residual Gas	34%
Intake temperature	40±1°C
Coolant temperature	85±2°C
Throttle	Full Open

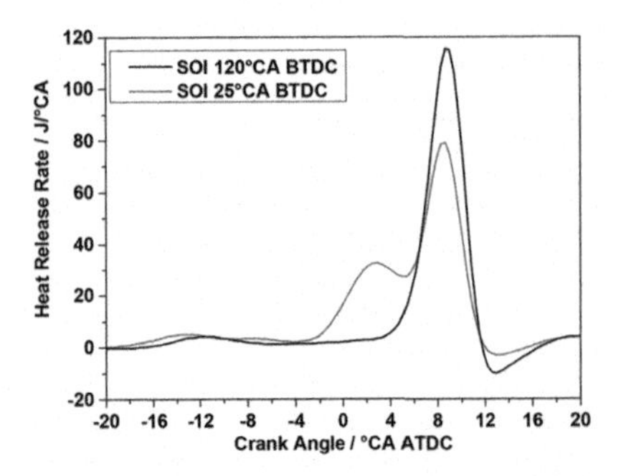

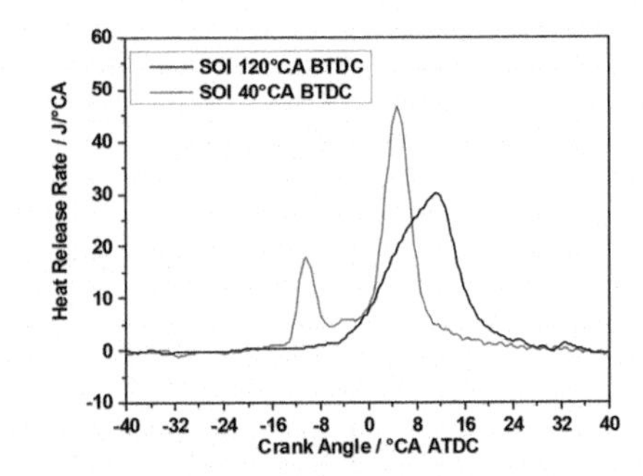

Figure 6. Comparison of heat release rate and pressure in cylinder under different DME injection conditions from optical engine (left) and thermodynamic single cylinder engine (right).

3.2.1 *Effect on the DME distribution by DME injection timing and in-cylinder flow*

The distribution of DME would determine the location of initial ignition sites and could be directly affected by its injection timing and in-cylinder flow. Due to the limited optical access through the piston crown, the PIV measurement in Figure 7 was restricted to a horizontal plane where a large anti-clockwise swirling flow from the intake side was detected.

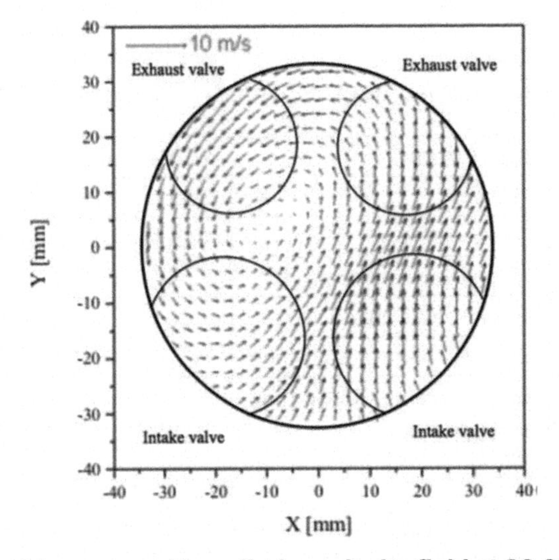

Figure 7. Ensemble averaged in-cylinder velocity field at 20 CAD BTDC in the studied DME MFI case (100 cycles)

More comprehensive flow results were obtained by CFD simulations. As shown in Figure 8, at the IVC timing (200 CAD BTDC), the flow field was dominated by an anti-clockwise tumble flow structure whose tumble ratio was 1.1. At the same time, a 4mm vortex core generated in the bottom of the combustion chamber by the structure of the chamber. With the piston moving up, the tumble was compressed to the left-hand side and an anti-clockwise swirling flow formed neat the intake side as the PIV image.

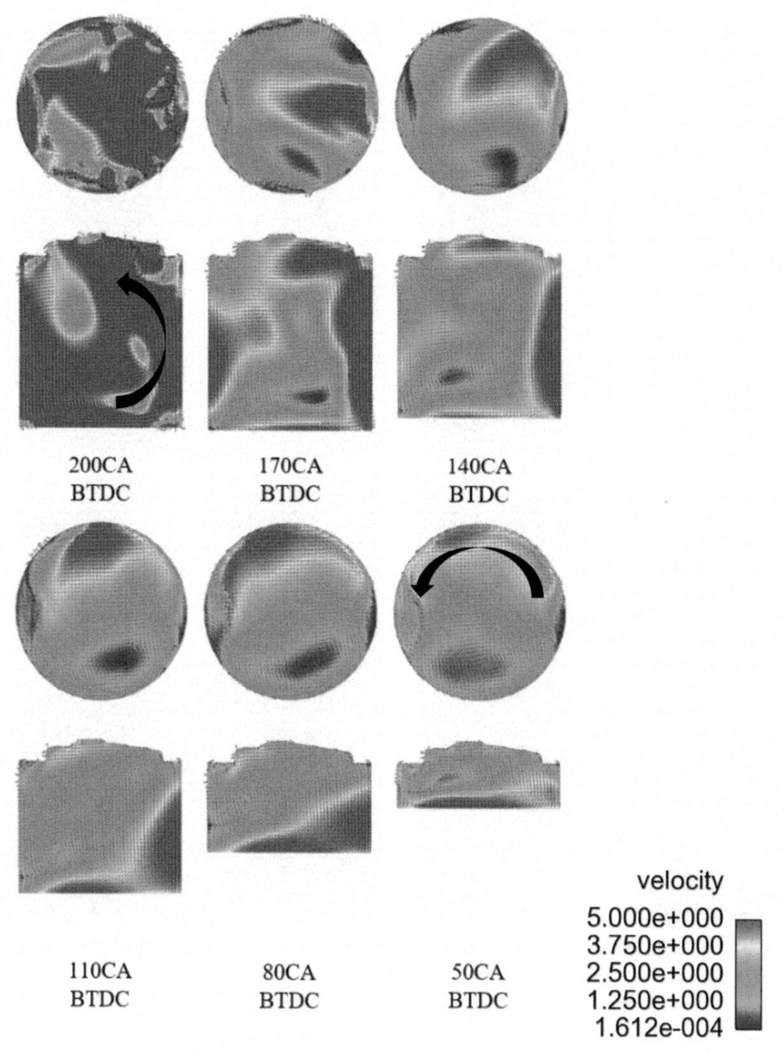

Figure 8. The simulation of flow field development

Figure 9 shows the in-cylinder DME distribution after its injection at 25 CAD BTDC. Strong Mie scattering signals from the DME spray were observed at 1.44 CAD and 2.16 CAD after the start of injection (ASOI). Then the DME spray evaporates rapidly, leading to much weaker signal at 2.88 CAD ASOI. The DME completed its evaporation by 4.32 CAD ASOI (20.68 CAD BTDC), where no obvious bright spot from Mie scattering could be observed. Instead, it was dominated by the more uniformly distributed Rayleigh scattering of the DME vapour with more smoothly varying gradients between regions of high and low intensity. The DME spray started from the intake side (lower part of the image) and its vapour concentrated at the middle-left part beneath the intake valves at about 20 CAD BTDC.

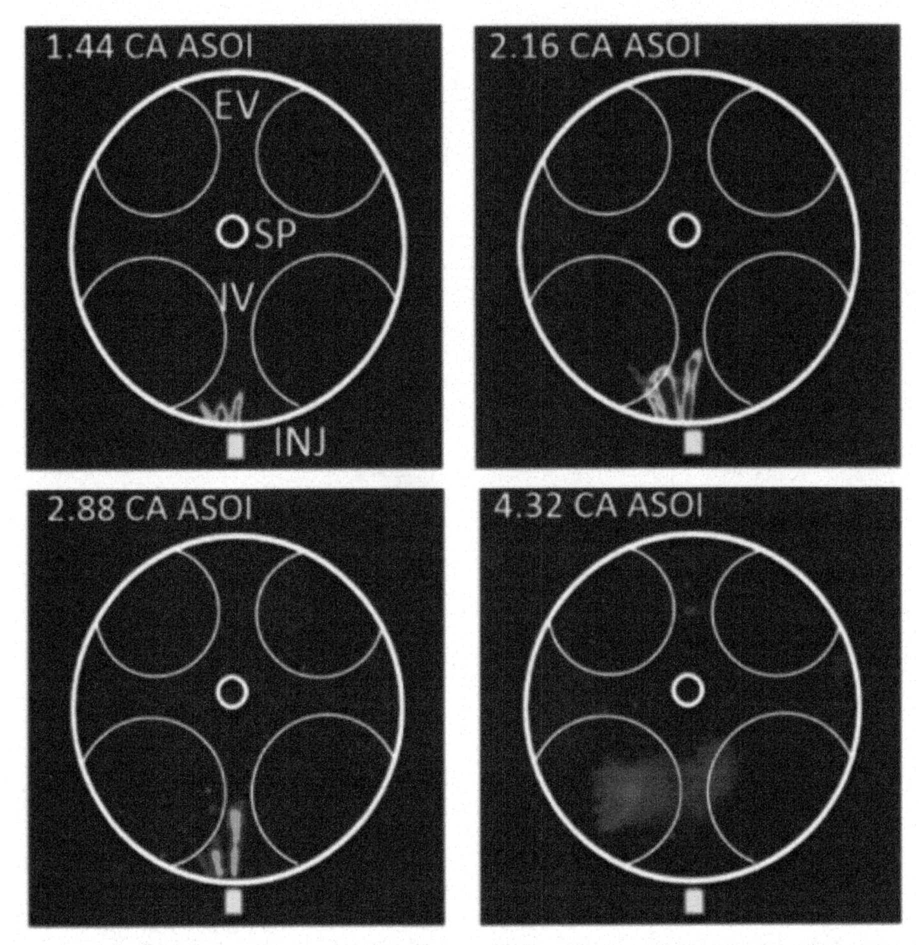

Figure 9. DME spray development with DME SOI 25 CAD BTDC (averaged over 50 cycles)

As the piston rose, it prevented laser Mie/Rayleigh scattering measurement from being carried out near TDC. Instead, the distribution of DME and in-cylinder flow near TDC was obtained from CFD simulation, as shown in Figure 10. Due to the anti-clockwise rotating vortex, the edge of spray was convoluted by the flow in cylinder. At 5 CAD BTDC, the DME injected in to cylinder at 25 CAD BTDC was moving to almost the middle of the cylinder with a concentrate distribution. At the same time, the vortex core moved to the right side of the picture due to the combined effect of velocity of spray and the anti-clockwise rotating vortex.

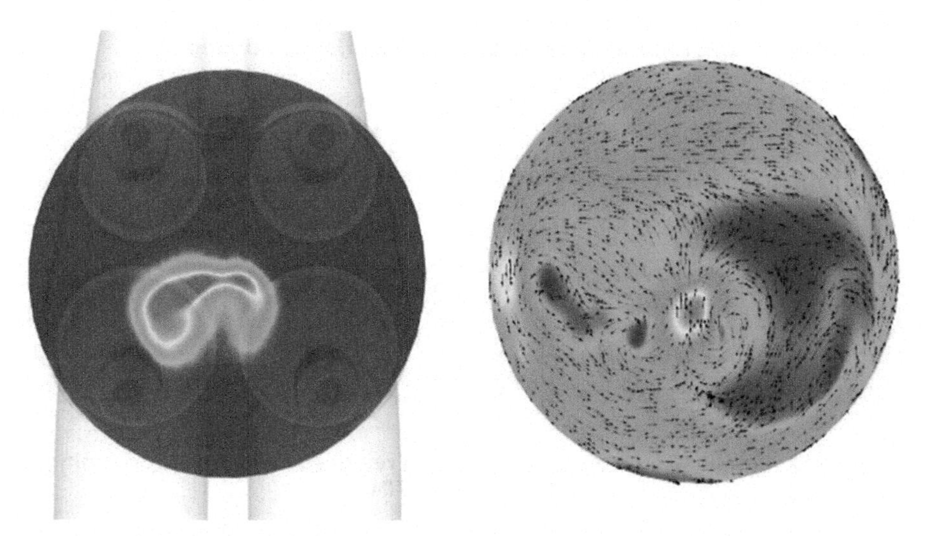

Figure 10. DME distribution and flow field at 5 CAD BTDC with DME SOI 25 CAD BTDC from simulation result

In the case of early DME injection, Figure 11 shows that the DME in cylinder showed a more homogeneous distribution in the left-lower side, due to the tumble motion after the injection. The velocity of DME spray broke the original vortex so that the vortex core was smaller than the vortex in Figure 10 at the same crank angle. As a result, the DME showed an almost homogeneous distribution due to the long mixture formation time and its interaction with the in-cylinder flow.

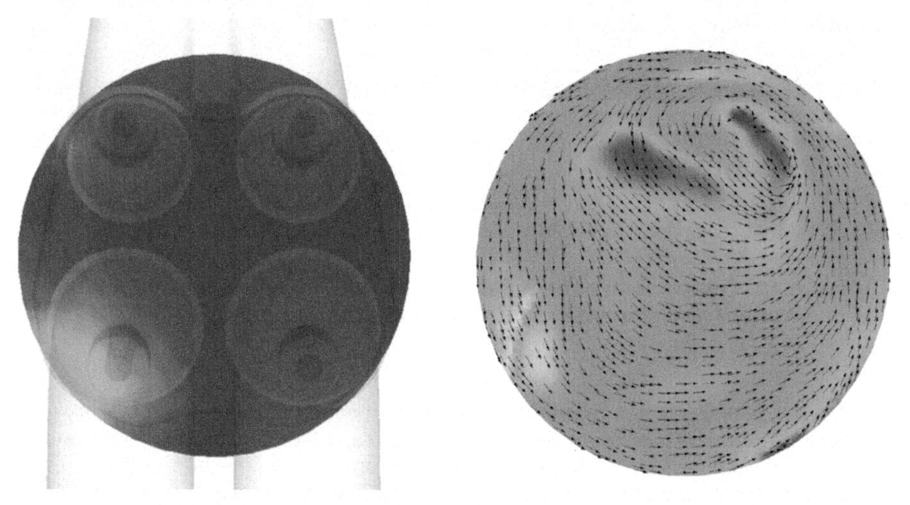

Figure 11. DME distribution and flow field at 5CA BTDC with DME SOI 120CA BTDC from simulation result

3.2.2 *Effect on the combustion process by DME injection timing*

For the early injection case, the images from the high-speed camera in Figure 12. showed that the auto-ignition took place in the left-bottom side at 6.48 CAD ATDC, where there was the high concentration DME in Figure 11. After the initial auto-ignition, there were more ignition points in the field of view. The flame propagation filled up the optical window at 9.36 CAD ATDC, corresponded with the measured heat release characteristic.

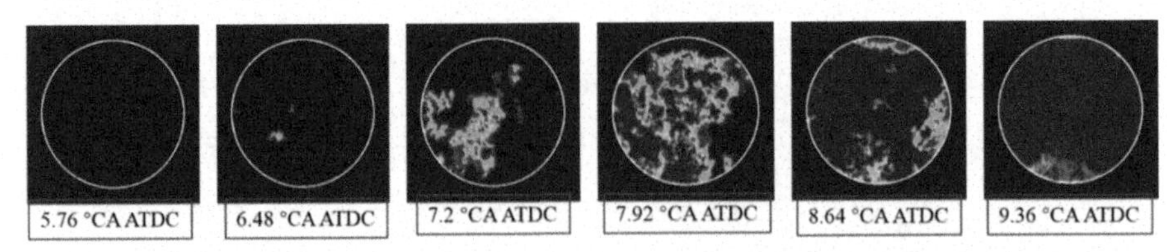

Figure 12. Combustion high-speed imaging results of DME injection timing SOI 120 CAD BTDC

The late injection case showed a rather different combustion process. The result of high-speed imaging of SOI 25 CAD BTDC case is shown in Figure 13. The initial auto-ignition took place at 1.4 CAD BTDC in the middle-left area, where the DME distribution was concentrated in Figure 10. The auto-ignition process was weak from 0.7 CAD ATDC. The flame front propagated to the right side until 5.8 CAD ATDC. A bright auto-ignition point then appeared in the upper-right of the optical window. After 5.8 CAD ATDC, the mixture around the flame front started to auto-ignition rapidly. The combustion of DME late injection case showed a 3 -stage combustion process included initial auto-ignition, flame propagation and auto-ignition process.

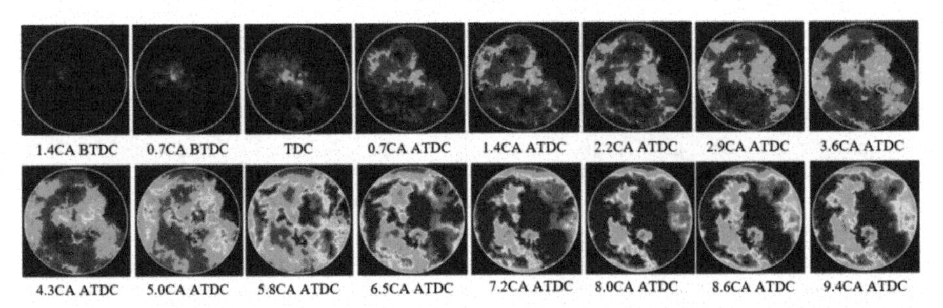

Figure 13. Comparison of heat release rate and in-cylinder pressure of experiment and simulation with DME injection timing 25 CAD BTDC.

The above results illustrated that in the early DME injection case, the high activity fuel DME raised the activity of premixed fuel/air mixture, which led to multi-point auto-ignition combustion with a few micro flames. In comparison, the late injection case showed a different kind of combustion process of initial auto-ignition, flame propagation and auto-ignition process.

The strong correlation of DME distribution and the combustion process with the heat release pattern were further investigated by the 3D CFD simulation. The simulation boundary was the same as the experiment operation specifications in Table 5, and the result in Figure 14 indicated that the simulation result was able to replicate the salient feature of the experiment result.

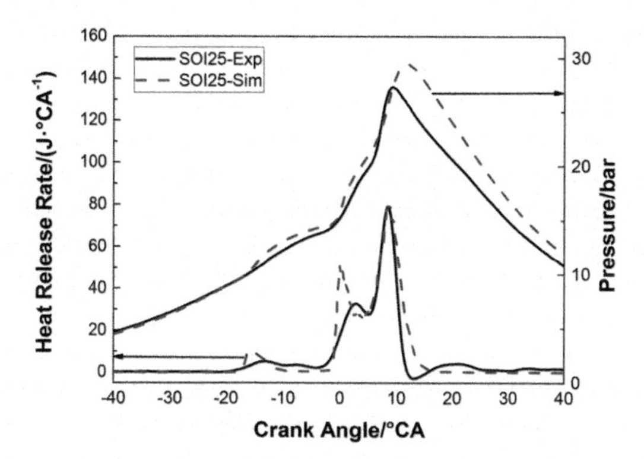

Figure 14. Comparison of heat release rate and in-cylinder pressure of experiment and simulation with DME injection timing 25 CAD BTDC

In Figure 14, the heat release process can be divided into 3 process. The first stage was a rapid heat release process. The heat release rate rose to 30J/°CAD during 5 CAD around TDC. In the second stage, the heat release rate dropped a little and stabilized. At last, a rapid heat release process occurred and the value of heat release rate reached its peak value.

The correlation between the concentration of DME and autoignition sites was further investigated by plotting DME distribution and the flame front contour as shown by the 1000K iso-surface in Figure 15. The initial auto-ignition process started at 2 CAD BTDC in the middle of DME region. There were two small auto-ignition points. One CAD later, the two auto-ignition sites expanded and merged together. It took another CAD for the flame propagated to the rest of DME region. The flame propagation process continued for about 5 CADs until new auto-ignition points appeared in the upper-right side and lower-right side at 6 CAD ATDC. The DME area was consumed by the flame at TDC, the combustion at 5 CAD ATDC and 6 CAD ATDC was mainly characterized by the flame propagation with the start of autoignition of premixed air/fuel mixture which led to the major heat release peak. Therefore, the combustion process can be summarized as DME auto-ignition, flame propagation and auto-ignition combustion of premixed air/fuel mixture. In the process, the multi auto-ignition sites of DME and their subsequent merger enabled the stable ignition and complete combustion of the highly diluted mixture.

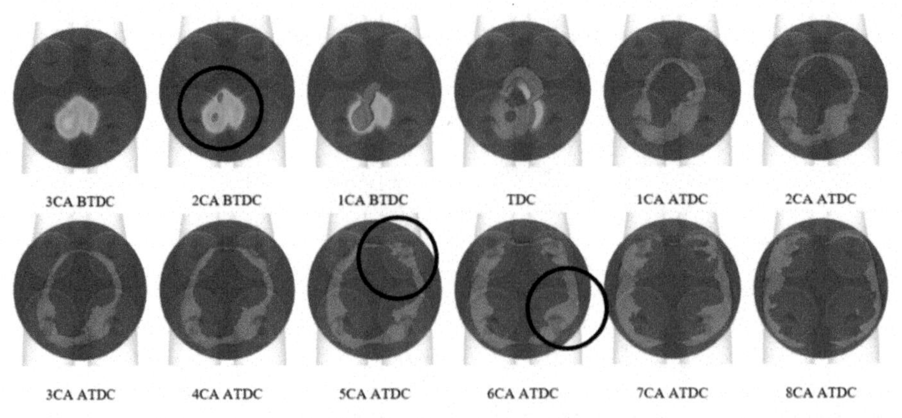

Figure 15. Simulated DME distribution and 1000k isosurface of simulation result with DME SOI 25CA BTDC

4 SUMMARY

The experiment carried on a thermodynamic single cylinder engine proved that MFI strategy was an effective way to increase the thermal efficiency and reduce emissions. The indicated thermal efficiency increased to 42.87% at 2000 rpm and IMEP 7 bar with nearly-zero NOX emission and 60% decrease of HC and CO emissions.

The heat release analysis showed that the change in the DME injection timing could lead to different combustion processes. The early injection of DME during the start of the compression stroke resulted in 2-stage heat release process with one peak heat release rate. When the DME injection was delayed to the end of the compression, there were three stages of combustion characterized by two major peaks in the heat release rate curve.

The in-cylinder optical studies and CFD simulation results indicated that the DME distribution was largely determined by the injection timing and to a lesser extend by the in-cylinder flow field. The early DME injection produced a homogeneous distributed DME in the premixed air/fuel mixture which increased the reactivity of the premixed mixture and led to the fast multi-point auto-ignition combustion and higher single peak heat release rate. Compared with early injection case, the late injection produced a stratified DME region where the first stage auto-ignition combustion started at multiple sites which merged to form a larger flame front to consume the rest of the DME mixture. The expanding flame front then compressed the rest of the premixed air/fuel mixture to autoignition, producing the second peak of the heat release rate curve. Therefore, it can be concluded that the combustion process of the MFI operation was directly dependent on the DME distribution which was mainly determined by its injection timing and the interaction with the in-cylinder flow field.

REFERENCES

[1] Zhao Fuquan (Frank), Asmus Thomas W., Assanis Dennis N., et al., Homogenous Charge Compression Ignition (HCCI) Engine: Key Research and Development Issues [M], Society of Automotive Engineers, 2002.

[2] M.-B. Liu, B.-Q. He, H. Zhao, Effect of air dilution and effective compression ratio on the combustion characteristics of a HCCI (homogeneous charge compression ignition) engine fuelled with n-butanol, Energy 85 (2015) 296–303.
[3] J.B. Heywood, Internal Combustion Engine Fundamentals, McGraw-Hill, Inc., 1988.
[4] S. Richard, Introduction to Internal Combustion Engines, 4th ed., SAE International, 2012.
[5] Zigler B, Keros P, Helleberg K, et al. An experimental investigation of the sensitivity of the ignition and combustion properties of a single-cylinder research engine to spark-assisted HCCI. International Journal and Engine Research. 2011, 12: 353–375.
[6] Persson H, Hultqvist A, Johansson B, et al. Investigation of the Early Flame Development in Spark Assisted HCCI Combustion Using High Speed Chemiluminescence Imaging. SAE Technical Paper, 2007-01-0212.
[7] D.L. Reuss, T-W. Kuo, G. Silvas, V. Natarajan, V. Sick, Experimental metrics for identifying origins of combustion variability during spark-assisted compression ignition, Int. J. Engine Res. 9 (2008) 409–434.
[8] Effect of Valve Timing and Residual Gas Dilution on Flame Development Characteristics in a Spark Ignition Engine. SAE Int. J. Engines 7(1):488–499, 2014.
[9] Polovina, D., McKenna, D., Wheeler, J., Sterniak, J. et al., "Steady-State Combustion Development of a Downsized Multi-Cylinder Engine with Range Extended HCCI/SACI Capability," SAE Int. J. Engines 6(1):504–519, 2013, doi: 10.4271/2013-01-1655.
[10] Mendrea, B., Chang, Y., Akkus, Y., Sterniak, J. et al., "Investigations of the Effect of Ambient Condition on SACI Combustion Range," SAE Technical Paper 2015-01-0828, 2015, doi: 10.4271/2015-01-0828.
[11] Daisho, Y., Yaeo, T., Koseki, T., Saito, T. et al., "Combustion and Exhaust Emissions in a Direct-injection Diesel Engine Dual-Fueled with Natural Gas," SAE Technical Paper 950465, 1995, doi: 10.4271/950465.
[12] Xie Hui, Xu Kang, Wan Minggang, Chen Tao and Zhao Hua, Investigations into the influence of internal and external exhaust gas recirculation on the combustion stability in an optical gasoline spark ignition engine, Proc IMechE Part D: J Automobile Engineering 2015, Vol. 229(11) 1514–1528
[13] Xie Hui, Xu Kang, Chen Tao, Zhao Hua; Investigations into the Influence of Dimethyl Ether (DME) Micro Flame Ignition on the Combustion and Cyclic Variation Characteristics of Flame Propagation-Auto-ignition Hybrid Combustion in an Optical Engine, Combustion Science and Technology, In Press, doi: 10.1080/00102202.2016.1223060
[14] S.H. Park, C.S. Lee, Applicability of dimethyl ether (DME) in a compression ignition engine as an alternative fuel, Energy Convers. Manag. 86 (2014) 848–863.
[15] Angelo Basile, Francesco Dalena, Methanol Science and Engineering, Elsevier, 2018
[16] Cha J, Kwon S, Kwon S, et al. Combustion and emis-sion characteristics of a gasoline–dimethyl ether dual-fuel engine[J]. Proceedings of the Institution of Mechan-ical Engineers Part D Journal of Automobile Engineering, 2012, 226 (12):1667–1677.
[17] Xu Kang. Study of DME Micro Flame Ignition Hybrid Com-bustion under High Dilution Conditions in a Gaso-line En-gine[D]. Tianjin: School of Materials Science and Engineering, Tianjin University, 2017.
[18] Zhang H. Experimental Investigation of Gasoline – Di-methyl Ether Dual Fuel CAI Combustion with Internal EGR [D]. Unit-ed Kingdom: Brunel University, PhD The-sis, 2011.
[19] Xueqing Fu, Bangquan He, Hongtao Li, Tao Chen, Sipeng Xu, Hua Zhao. Effect of direct injection dimethyl ether on the micro-flame ignited (MFI) hybrid

combustion and emission characteristics of a 4-stroke gasoline engine. Fuel Processing Technology, 2017.
[20] Yan Zhang, Hui Xie, Nenghui Zhou et al., "Study of SI-HCCI-SI Transition on a Port Fuel Injection Engine Equipped with 4VVAS". SAE Technical Paper 2007-01-0199, 2007
[21] Tao Chen, Hui Xie, Le Li, et al., "Expanding the Low Load Limit of HCCI Combustion Process Using EIVO Strategy in a 4VVAS Gasoline Engine". SAE Technical Paper 2012-01-1121, 2012
[22] Richards K J, Senecal P K, et al. Converge(Version 2.3)Manual [R]. Middleton, WI: Convergent Science, Inc, 2016.
[23] O'Rourke P J, Amsden A A. A spray/wall interaction submodel for the KIVA-3 wall film model [C]. SAE Pa-per. Detroit, MI, USA, 2000, 2000–01–0271.
[24] Schmidt David P, Rutland C J. A new droplet collision algorithm[J]. Journal of Computational Physics, 2000, 164(1): 62–80.
[25] Endrullis J, Hendriks D, Bodin M. An introduction to combus-tion: [M]. WCB/McGraw-Hill, 2000.
[26] Chen Y, Chen J Y. Application of Jacobian defined direct interaction coefficient in DRGEP-based chemical mecha-nism reduction methods using different graph search al-gorithms [J]. Combustion & Flame, 2016, 174: 77–84.
[27] Chen Y, Chen J Y. Towards improved automatic chemical kinetic model reduction regarding ignition delays and flame speeds [J]. Combustion & Flame, 2018, 190: 293–301.
[28] Chen Y, Chen T, Feng Y, et al. H radical sensitivity assisted automatic chemical kinetic model reduction for laminar flame chemistry retaining: A case study of gasoline/DME mixture under engine conditions [J]. Energy and Fuels, 2019, 33: 3551–3556

SESSION 2: HYBRID APPLICATIONS

The development and testing of a free-piston engine generator for hybrid electric vehicle applications

A.J. Smallbone, S. Roy, K.V. Shivaprasad, B. Jia, A.P. Roskilly

Department of Engineering, Durham University, Durham, UK

ABSTRACT

In this work, we present some of the first experimental results along with simulation results obtained in developing and testing a novel dual-piston free-piston engine generator (FPEG), designed for electric- vehicle (EV) range-extender or hybrid powertrain applications. The benefits of a high-efficiency, compact and lightweight design of the proposed range-extender are presented. The technical details and experimental set-up of a two-cylinder prototype and its instrumentation are also outlined. Results are presented for simulation and recent test programmes carried out across both 2-stroke and 4-stroke operational modes. The methods associated with engine control are detailed alongside key post-processed engine characteristics.

1 INTRODUCTION

Today's global energy production aims at cleaner production to preserve and protect environmental resources. In recent decades, the world has started to address the environmental changes caused by emissions of hydrocarbon fuelled vehicles and is instigating solutions to reduce these pollutants. Also, the faster depletion of fossil fuels has led to research on alternative fuel engines in replacing existing conventional fuels with a clean, economical and efficient energy source. To this end, the current perception is that conventional engines will be replaced by more electrified vehicle powertrains. The main challenge for electric vehicles (EVs) designers is limited driving range. One option is to have a small on-board installation of an electric generator such that the driving range can be increased. This has led to disruptive advancements in the development of range extenders. One such technology is the use of a free piston engine generator (FPEG).

The FPEG has the potential to be an effective integrated engine and electricity generator. The piston moves along the engine cylinder which is driven by combustion gases and kinetic energy of the piston movement is converted into electrical energy by a linear electric machine. When using a FPEG, the use of crank shaft and connecting rod is eliminated resulting in several advantages such as higher efficiency, compact design and piston motion is directly converted to electrical power; due to the simplicity of the system. Along with these, the compression ratio can be easily varied, and the stroke length is independent as there is no crankshaft. In a FPEG, the electrical energy is achieved from chemical energy by means of a combustion process.

In 1928, an engineer and inventor R.P Pescara from Argentina presented the free-piston engine (FPE) [1] and since several designs for free-piston engines have been projected. In all these designs, the piston is free to move between its endpoints which makes the free-piston engine to run with variable stroke length and high control requirements [2]. Zhang and Sun [3] stated that when seven renewable fuels like ethanol, biodiesel, hydrogen etc. are considered for trajectory-based combustion control enabled by FPEG, it was found that FPEG has the highest flexibility of fuel compared to others. The work of Roman Virsik [4] compared FPEG with other range extender technologies available and then has concluded that FPEGs with the use of ICEs (internal combustion engines) are feasible and has higher efficiencies with lower emissions and NVH values. This can be implemented in vehicles. Heron and Rinderknecht [5] in their work compared several range extender technologies for electric vehicles. They compared the FPEG, polymer electrolyte fuel cells, TRE (traditional reciprocating engine) and ICEs. They concluded that the FPEGs have higher efficiency compared to TREs and ICEs. In the 1940s, the German navy used the free-piston air compressors to supply compressed air for launching torpedoes. The free-piston engines were used to feed hot gas to a power turbine. This was employed in stationary and marine powerplants, most successful being a model developed by SIGMA in France [6].

Researchers at Toyota central laboratory, Japan, developed the FPEG prototype comprised of a linear generator, a gas spring chamber and two-stroke combustion chamber with hollow circular step shaped piston. In this study, the characteristics of FPEG motion were studied both numerically and experimentally for enabling stable continuous operation. In the simulation, the researchers evaluated the FPEG with SI and premixed charged compression ignition modes of combustion. The output power of 10kW was obtained with both combustion mode whereas premixed charged compression ignition mode of combustion delivered higher thermal efficiency of 42% compared to SI combustion mode. Experimental results confirmed that the FPEG prototype operated stably for quite a long period of time, despite of the abnormal combustion during the test. The researchers have explored the unique piston motion, which causes its impacts on combustion and power generation in the FPEG [7], [8]. The detailed review on free-piston engine on its history and development was presented by Hanipah, Mikalsen and Roskilly [9], [10].

Free-piston engines are broadly classified into three types based on piston arrangement as single piston, opposed piston and dual piston [11]. The working principle is identical for each type, but variances between them are the compression stroke realization and combustion chambers design. FPEGs consists of different modules. They are central combustion module, central gas spring module, central combustion with integrated gas spring module, central combustion with branched linear generators module and dual module system, either of which can be used according to the user. These modules are chosen accordingly based on their NVH behaviour. However, with the implementation of the FPEG the NVH is observed to be low. The engine noise and vibrations occur because of the mechanical forces, and combustion process can be controlled by synchronizing these modules. Due to the presence of higher degrees of freedom for the FPEG system, Homogenous Compression Ignition (HCCI) combustion can also be implemented [4].

This article presents the experimental outputs on FPEG prototype system based on a dual-piston free-piston engine design. This design and its operation is built upon a series of robust numerical modelling studies carried out over the years by the researchers of the centre [12], [13], [14], [15].

2 DESCRIPTION OF THE ENGINE CONFIGURATION

2.1 Dual-piston FPEG concept engine

The schematic configuration and prototype of FPEG established at the Sir Joseph Swan Centre for Energy Research is shown in schematic in Figure 1 and photos shown in Figure 2 respectively. This design of the engine generator is in accordance with the patent by Mikalsen and Roskilly [16]. The adopted design parameters are tabulated in Table 1. The FPEG concept mainly comprises two opposing internally combusted FPE and a linear electric machine. Each of the free-piston engine is consisted combustion chamber, spark plug, piston and set of poppet valves. The engine employs an actuated control system for intake and exhaust. The electric machine, generally called the stator, is located between the engines and it can be operated as a generator, or a motor. The pistons of FPEG are connected using moving part of the system called mover. The linear electric machine initiated the starting process by operating as a motor and then switched as generator mode once the system attains steady state. Combustion takes place alternatively in each cylinder of engines, which drives the piston and mover assembly to oscillate back and forth motion. The electric generator then converts the movement of the piston into electrical energy.

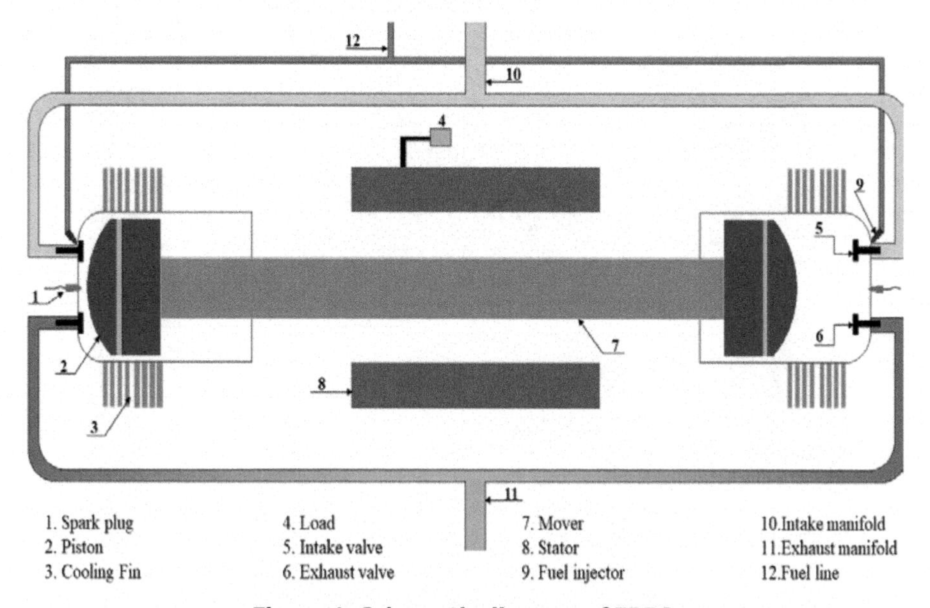

Figure 1. Schematic diagram of FPEG.

Table 1. Specifications of prototype FPEG.

Parameters	Value	Unit
Maximum stroke	40.0	mm
Cylinder bore	50.0	mm
Intake/exhaust valve lift	4.0	mm
Intake valve diameter	20.0	mm
Moving mass	7.0	kg
Exhaust valve diameter	18.0	mm
Intake manifold pressure	1.3	bar
Exhaust manifold pressure	1	bar
Ignition position from cylinder head	5.0	mm
Load constant of the generator	810	$N/(ms^{-1})$

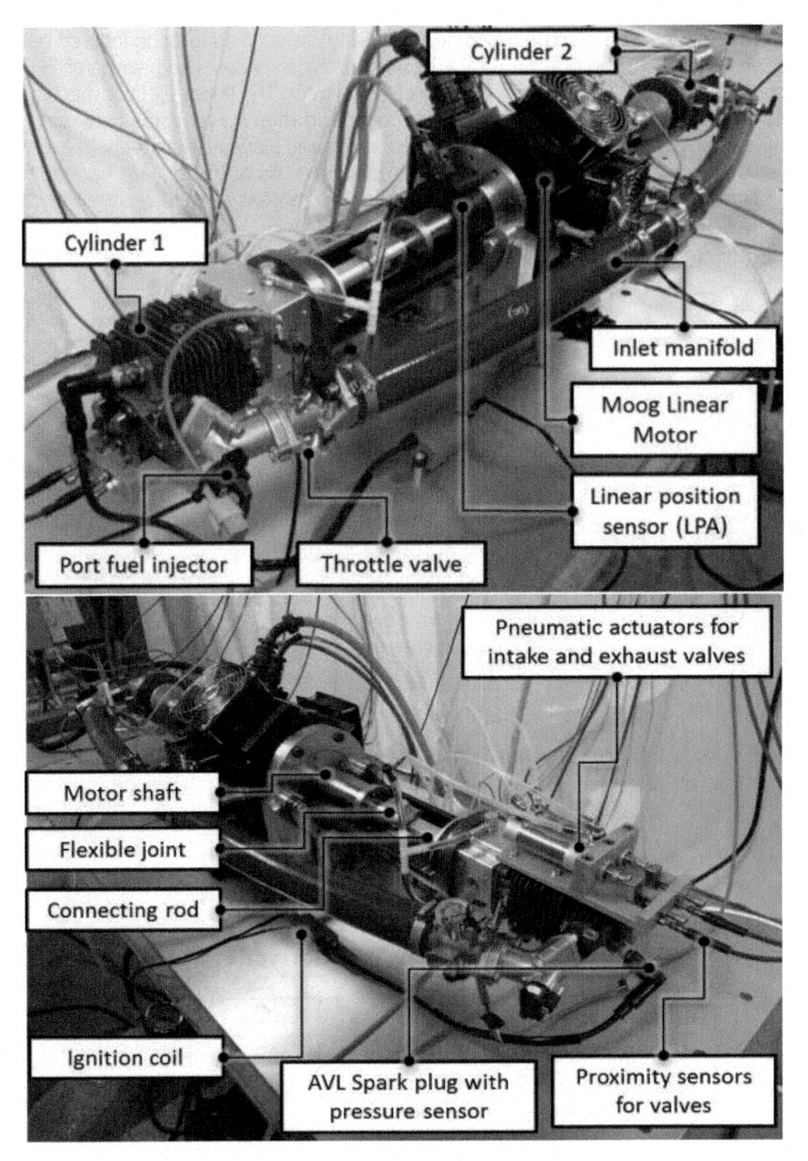

Figure 2. FPEG prototype.

2.2 Four-stroke and two-stroke control mode

To operate the FPE in four-stroke mode, each cylinder requires two oscillation cycles or four oscillatory motion of its piston to complete the sequence of operations that finishes a single power stroke. The processes of four-stroke cycle are as follows [17].

1) Suction/Intake stroke: During this stroke, the piston will be at its TDC and the valve control system opens the intake valve. This stroke is continuous as piston reaches to its BDC and ended by closing of intake valve. If there is not enough force behind the piston to drive this event then the electric machine acts as a motor.

2) Compression stroke: During this stroke, both the inlet/intake and outlet/exhaust valves are in closed positions and the piston compresses air–fuel mixture as it moves from its corresponding BDC to TDC. If there is not enough force behind the piston to drive this event then the electric machine acts as a motor. At the end of the compression stroke, the spark-plug produces the spark and initiates the combustion.
3) Power stroke: The combustion occurs at the end of compression stroke produces enormous amount heat energy with high pressure which pushes the piston towards its BDC. The kinetic energy of the piston movement is converted in to electricity by linear electric generator.
4) Exhaust stroke: This stroke initiated as the outlet valve opens and piston reciprocates from its corresponding BDC to TDC. The burnt gases expel out from the cylinder chamber and the cycle is repeated as outlet valve closes.

Similar with the working of FPEG in two-stroke mode, alternatively each cylinder gets the power stroke and throughout the generating process, the linear electric machine is functioned as generator. The processes of two-stroke cycle are explained as follows.

1) Compression stroke: Throughout this stroke, the valve control system of FPEG closes both the inlet valve and outlet valve. As the piston moves from BDC to its corresponding TDC, the fuel-air mixture is compressed in the cylinder. This will continuous until the system achieves its compression ratio. At the end of this stroke, spark plug produces the spark for the combustion.
2) Power stroke: The initiation of power stroke is determined by spark timing and expansion of combustion gases. Expanded gas pushes the piston from TDC to its corresponding BDC. The control system opens the outlet valve followed by inlet valve and the exhaust gas expel out from the cylinder and fresh air fuel mixture is enters the cylinder.

The Table 2 provides the sequence of processes in each cylinder and corresponding modes of linear machine operation for the four-stroke and two-stroke cycles. During four-stroke mode of engine operation, the linear machine switched to motor as well as generator according to strokes and it act as a generator throughout the two-stroke mode of engine operation.

Table 2. Sequence of processes of FPEG in two-stroke and four-stroke mode.

	Four-stroke			Two-stroke		
	Right cylinder	**Left cylinder**	**Linear machine**	**Right cylinder**	**Left cylinder**	**Linear machine**
Stroke⟶	Air exhaust	Air intake	Motor	Gas exchange + compression	Power + gas exchange	Generator
Stroke⟵	Air intake	Compression	Motor	Power + gas exchange	Gas exchange + compression	Generator
Stroke⟶	Compression	Power	Generator	Gas exchange + compression	Power + gas exchange	Generator
Stroke⟵	Power	Air exhaust	Generator	Power + gas exchange	Gas exchange + compression	Generator

2.3 Valve actuating mechanism

The designed FPEG is facilitated with a Festo pneumatic system to activate the overhead intake and exhaust valves. Nominally, the valve lift is 4 mm, though this is changeable, which permits for further control, improvement and optimization of the valve operation. To operate the FPEG in four-stroke mode, the valve control system opens inlet valve towards the end of exhaust stroke and closes at the end of suction stroke. This gives the maximum compression stroke of about 40mm. The outlet valve opens at the end of the power stroke and it remains open throughout the exhaust process. In two-stroke mode, the compression process starts after the closing of inlet valve which result in reduction of compression stroke to 34mm. The scavenging process of two-stroke mode of operation combines intake and exhaust gas exchange process. The poppet valves are used for both intake and exhaust processes rather than scavenging ports. During the process exhaust valve opens before opening of intake valve and it closes after the closing of intake valve. A compressor is employed in FPEG for the two-stroke mode of operation to enhance the manifold pressure and each outward stroke relates to a power stroke. As the valves are actuated based on the piston position, the scavenging durations for both operating modes will be significantly affected by the engine speed and piston profile. In order to avoid mechanical contact between the piston and cylinder head, the working mode of the linear electric machine is switched at a specific point during Stroke 1 and Stroke 2.

2.4 Numerical model description

The numerical model was built on many of the assumptions and principles of thermodynamic models employed regularly across the conventional ICE research and development community. A full description can be found elsewhere [11,13,14,15]. An engine dynamic model was formed to determine the piston motion. The three sub models of in-cylinder gas thermodynamic processes, linear electric machine force and mechanical friction force were developed and fed into the engine dynamic model. The in-cylinder gas thermodynamic process comprises heat transfer to the wall, piston's expansion or compression process, gas leakage, gas exchange process, scavenging process for two-stroke engine mode and heat release during combustion. The linear electric machine force considered is either resisting force or driving force depending on its operating mode whereas the friction sub model defines the friction force acting on the piston rings.

The model was developed in Matlab/Simulink. The developed model is used for both starting and steady operation of FPEG. In starting mode, the combustion model was disabled, and linear electric machine was enabled to run as a motor. In steady operation, the linear electric machine operates as a generator and combustion model was enabled. The computed piston velocity, in-cylinder pressure, and piston displacement at the end of the starting process were taken as the preliminary values for the steady operation of the FPEG.

The design parameters for the prototype and initial boundary conditions are fixed as the two and four-stroke mode of operations are employed on the same prototype. The considerable changes are the working mode of the electric machine and the valve timing strategy. The prototype specifications and the input parameters for both mode of engine operation is tabulated in Table 2.

3 RESULT AND DISCUSSIONS

3.1 Simulation results

The objective of the simulations was to identify different ways in which the engine could be controlled to achieve stable and steady operation. In doing so, it would become clearer as to what hardware would be required and how to control the system. The same analysis aimed to explore the sources of power losses in its operation and the sensitivity of these parameters on performance.

3.1.1 *Exploring engine operational performance*

The simulation of the FPEG engine was carried out at stoichiometric air–fuel ratio (λ=1.0) with the target piston dead centre is +17.0mm and -17.0mm from the middle position of the stroke and clearance from the cylinder head is 4.0mm. The engine performance of both four-stroke and two-stroke is shown in Table 3. The two operational modes were compared with the motor force shown in Table 1.

With a fixed motor force, it was noted the system operated at different engine speeds depending on the operating mode. Furthermore, the indicated power is much greater for two-stroke engine cycle compared to four-stroke engine mode of operation. The engine speed for the two-stroke cycle mode was higher, and the power stroke took place in every cycle. In the meantime, in four-stroke mode of operation, the peak cylinder pressure and the movement of the piston would produce higher friction between piston ring and cylinder liner interface. in overcoming. This results in almost half the indicated power is consumed in overcoming the friction along with the pumping losses of the motoring process.

Table 3. Predicted FPEG performance parameters in four and two-stroke mode.

Performance parameters	Four-stroke	Two-stroke
Mean speed (rpm)	900	2000
Operating Frequency (Hz)	15	31
Maximum piston velocity (m/s)	3.8	3.1
Maximum cylinder pressure (bar)	76	47
Thermodynamic efficiency (%)	44.9	34.5
Consumption of fuel (kg/kW h)	0.20	0.22
Indicated power (W)	1230	3900
Electric power output (W)	640	3760
Maximum compression ratio	16.2	7.36
Power to weight ratio (W/kg)	0.22	0.070

3.1.2 *Engine throttle position*

The Figure 3 demonstrates the effect of engine throttle opening positions on indicated power and electric power production. For two-stroke cycle, a reduction in indicated power from 5 kW to 1.8 kW is occurred when the throttle opening mode was changed from fully-open to half-open. In the interim, the electric power generation is somewhat lesser than the indicated power. It also observed that, with the same throttle opening positions, both indicated power and electric power of four-stroke cycle are considerably lesser than that of the two-stroke cycle. The indicated power decreases from 1.8 kW to 0.8 kW when the throttle opening area reduces from 100% to 50%. It is recommended to operate the engine at higher loads rather than below 50%, as the energy conversion efficiency is getting minimal for the four-stroke mode.

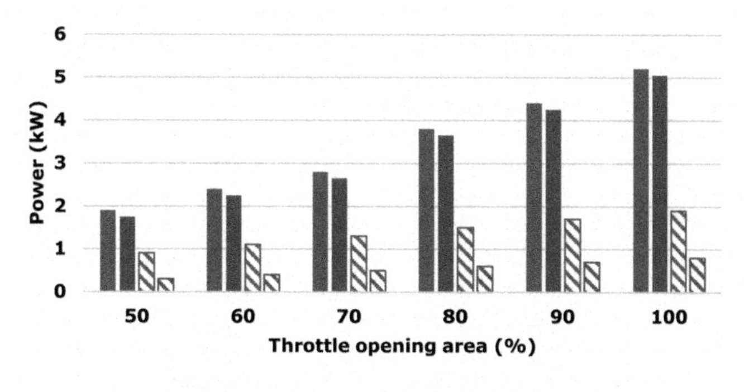

Figure 3. Variation of power with throttle opening area.

3.1.3 *Ignition timing*

Figure 4 displays the variations of indicated power with ignition timing for two-stroke and four-stroke engine mode when it runs at wide open throttle. As it can be seen from the plot, for the two-stroke engine mode, the electric power slightly increases with the increase in ignition timing. Advancing ignition timing can reduce the period of post combustion and improve the in-cylinder pressure. However, with the ignition timing advance, no substantial variation was detected neither in indicated power nor in electric power when FPEG is operated in four stroke engine mode.

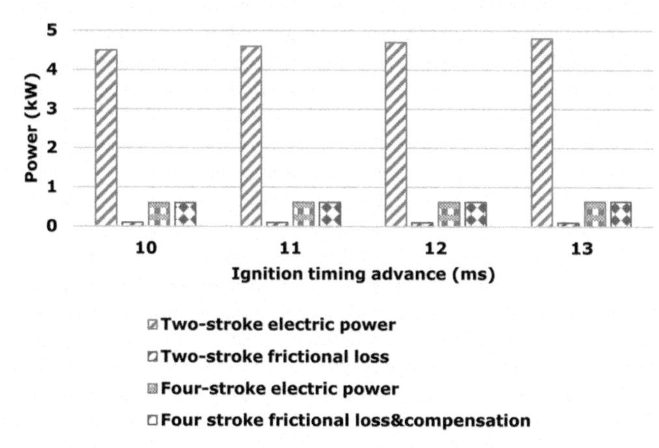

Figure 4. Variation of power with ignition timing.

3.1.4 *Power circulation*

Indicated power distributions in two and four-strokes cycle of FPEG when it runs at wide open throttle are depicted in Figure 5. As there is removal of crankshaft and connecting rod in the FPEG, it helps in the reduction of frictional losses as compared to conventional engines, which in-turn reduces the fuel consumption. Compensation is the electric power used to offset the overall power consumptions during the motoring process. Compensation was zero for two-stroke cycle and was 42% for four-stroke cycle, *i.e.* during the motoring processes for supplying additional electric power, 42% of the indicated power was consumed.

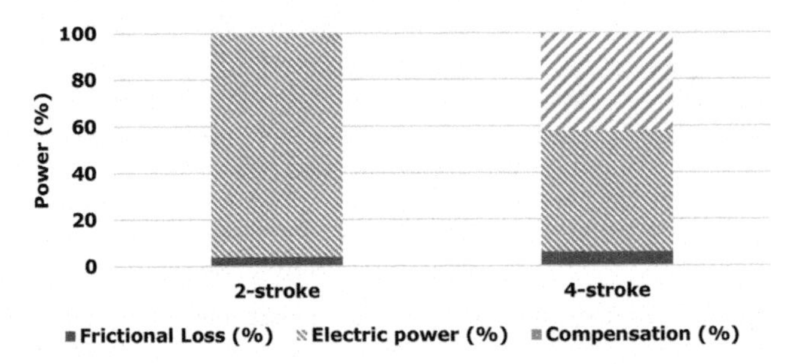

Figure 5. Power distribution in two-stroke and four-stroke cycle.

3.1.5 *Effect of motor force on power of four-stroke cycle*

During four-stroke mode of FPEG operation, the linear electric machine primarily acts as a motor and later as a generator. An important parameter for four-stroke cycle which influences on FPEG performance is the motor force. The Figure 6 illustrates the different motor force effect on FPEG's electric power, frictional loss and compensation at wide open throttle. As seen from the plot, the electric power increases with the increase of motor force. The higher motor force reduces the period between pumping and combustion which raises the engine speed and thus further development of indicated power. The plot also depicts that, as motor force increases, engine consumes more power to recompense the supply of electric power during motoring process.

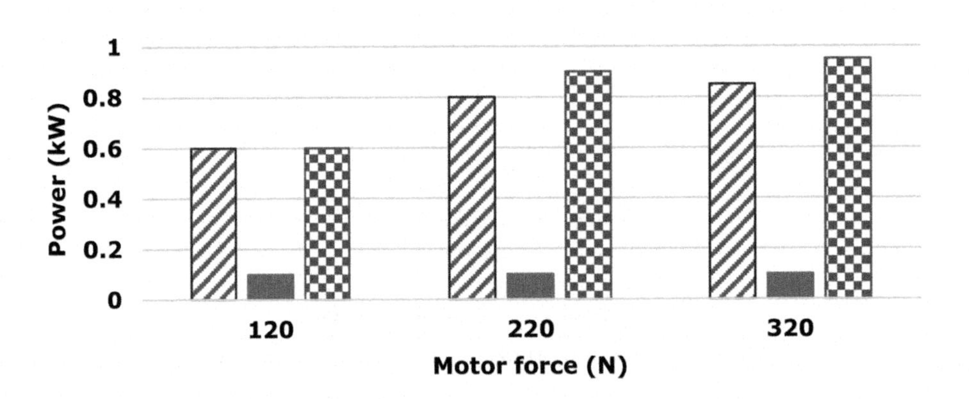

Figure 6. Power distribution with different motor force.

3.2 Experimental results

The numerical modelling indicated that manipulation of the motor force itself could be used to control the engine with a high degree of fidelity. As a result, the decision was made to utilise a more complex hardware and control system which could control the motor force in real time to achieve a sinusoidal motion profile at a pre-defined frequency.

3.2.1 *Control signals*

The experimental results of two-stroke mode of FPEG operated at 5Hz have been presented in this section. During suction phase of FPEG, the compressed air is supplied to the engine through an intake manifold to facilitate gas exchange and to boost the intake manifold pressure. Nominally, the engine was operated at stoichiometric air–fuel ratios (λ=1.0).

The timing of spark and valve opening/closing are initiated based on piston displacement and velocity. Figure 7 illustrates the control signals for the spark and valve timing for a cylinder of FPEG. The spark plug produces the spark to initiate the combustion before the piston reaches its TDC. The exhaust valve opens before opening of intake valve and intake valve closes after the closing of exhaust valve as seen in the diagram.

3.2.2 *Pressure and volume*

Variation of pressure and velocity with the displacement is depicted in Figure 8. In this figure, the obtained piston velocity curve is close to a sinusoid and pressure curve depicts the work done by the gas expansion. At the middle stroke of the piston, the highest velocity ensues, and value of velocities are zero at the dead centres where the value of velocity changes its sign.

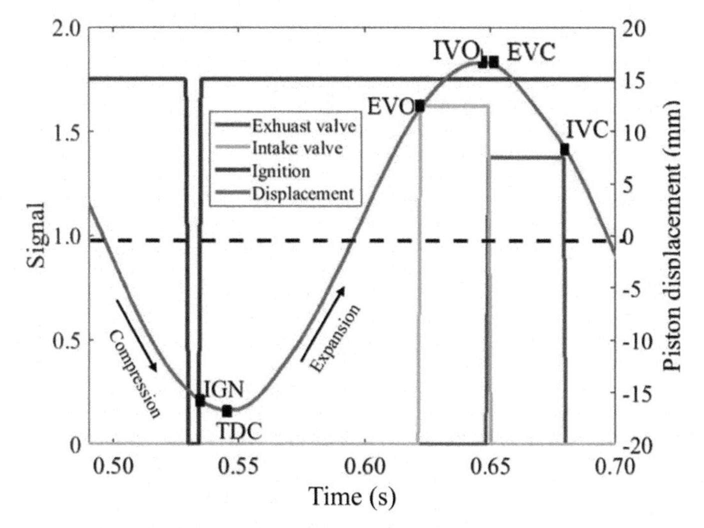

Figure 7. Control signals in a cylinder.

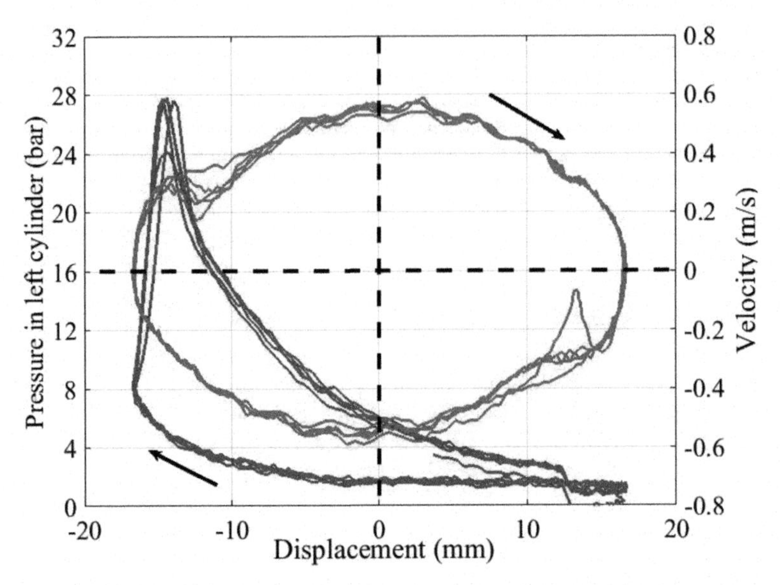

Figure 8. Cylinder pressure with displacement.

3.2.3 *Piston dynamics*

Figure 9 illustrates piston velocity and its displacement attained during the test. Figure represents the five separate compression and expansion events occurred in the engine. As expected, the value of piston velocity and displacement were increased. The profile typically achieves a ± 17mm displacement and the impact of combustion is more apparent in the plot.

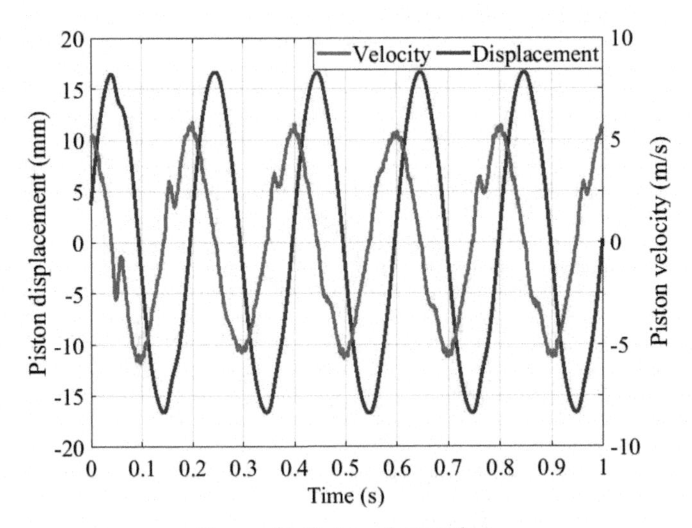

Figure 9. Piston dynamics.

3.2.4 *Cylinder pressure*

Variation of in-cylinder pressure for each cycle and its corresponding ignition signal with respect to time is shown in Figure 10. From the plot, it can be concluded that the in-cylinder pressures are rapidly increasing and decreasing when combustion occurs. The peak pressure of 27 bar attained by the engine cylinder during the combustion as seen from the plot.

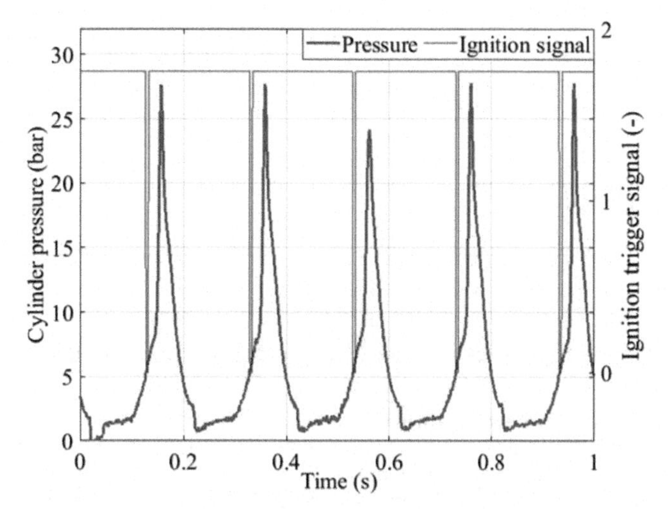

Figure 10. In-cylinder pressure.

3.3 Effect of intake and exhaust valve timings on IMEP

During the above testing, the engine was operating continually without issue. It was then considered that a parameter sweep would be carried out on valve timing. The objective being to explore the sensitivity of the system stability.

3.3.1 *Intake valve closure timing*

The duration of opening was fixed to 40 ms (milli-seconds) and the intake valve closing (IVC) timing set from -6 ms to +6ms aBDC (after Bottom Dead Centre). The results of variation of an intake valve closure (IVC) timing on indicated mean effective pressure (iMEP) are displayed in Figure 11.

As IVC timing is delayed, the in-cylinder pressure is reduced as less combustible mixture is trapped within the cylinder chamber in turn lowers the pressures at spark ignition. The reduced pressure of less amount of combustible mixtures produces lower indicated mean effective pressure as depicted in the plot.

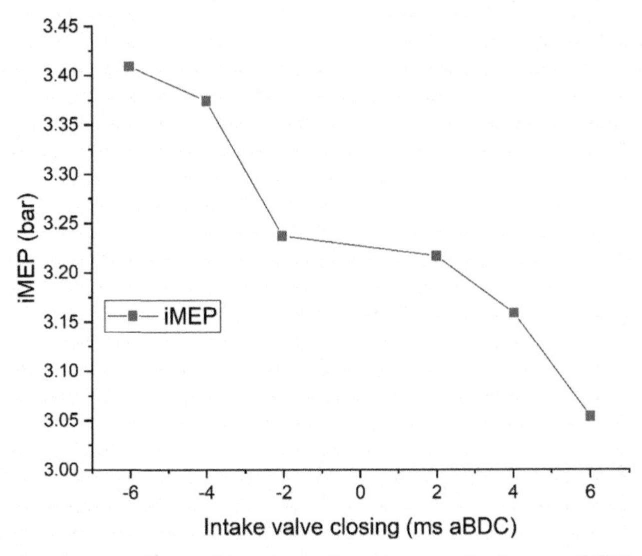

Figure 11. Effect of intake valve closure timing on iMEP.

3.3.2 *Exhaust valve opening timing*

The results of variation of an exhaust valve opening (EVO) timing on indicated mean effective pressure are presented in Figure 12. The open duration was fixed to 39 ms and EVO timing was varied from 33 ms to 39 ms aTDC (after top dead center).

Delaying the EVO generally resulted in higher in-cylinder pressures with later opening times. However, the Figure 12 shows only minor differences in iMEP with the variation of EVO timing. It was observed that beyond 33 ms to 39 ms EVO timing limits, the engine failed to operate with the combustion due to ineffective scavenging process.

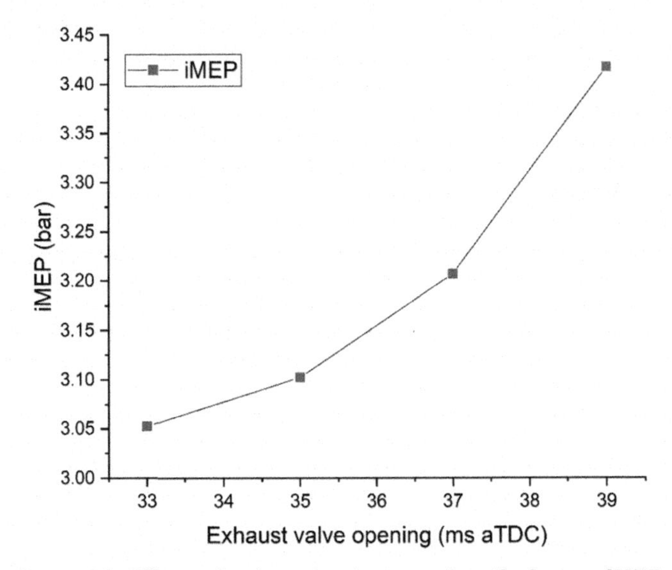

Figure 12. Effect of exhaust valve opening timing on iMEP.

4 SUMMARY

In the present study, the performances of two-stroke and four-stroke cycles of a FPEG are compared. The work can be summarised as:

- A novel twin-opposed free-piston engine prototype has been design, built and tested.
- A numerical model of the system has been developed which was used to understand and determine its core control algorithms
- Using the model, motor force was identified as a key control parameter able to ensure stable operation.
- The prototype was then upgraded to enable control of motor force in real time and thus to enable for the piston motion to follow a sinusoidal profile.
- The first experimental results showing operation of the prototype are presented and discussed.

REFERENCES

[1] R.P. Pescara, Motor compressor apparatus. US Patent 1,657,641, 1928.

[2] W.T. Toutant, The Worthington–Junkers free-piston air compressor, Journal of the American Society of Naval Engineers 64 (3) (1952) 583–594.

[3] C. Zhang and Z. Sun, "Trajectory-based combustion control for renewable fuels in free piston engines," Applied Energy, vol. 187, pp. 72-83, 2017.

[4] A. H. Roman Virsik, "Free piston linear generator in comparison to other range-extender technologies," World Electric Vehicle Journal, 2013.

[5] A. Heron and F. Rinderknecht, "Comparison of Range Extender technologies for Battery Electric Vehicles," in Eighth International Conference and Exhibition on Ecological Vehicles and Renewable Energies (EVER), Monte Carlo, Monaco, 2013.

[6] R. Huber, Present state and future outlook of the free-piston engine, Transactions of the ASME 80 (8) (1958) 1779–1790.

[7] Kosaka, H., Akita, T., Moriya, K., Goto, S., Hotta, Y., Umeno, T., and Nakakita, K. (2014). Development of Free Piston Engine Linear Generator System Part 1 - Investigation of Fundamental Characteristics. https://doi.org/10.4271/2014-01-1203

[8] Goto, S., Moriya, K., Kosaka, H., Akita, T., Hotta, Y., Umeno, T., and Nakakita, K. (2014). Development of Free Piston Engine Linear Generator System Part 2 - Investigation of Control System for Generator. https://doi.org/10.4271/2014-01-1193

[9] Mikalsen R, Roskilly AP. The design and simulation of a two-stroke free-piston compression ignition engine for electrical power generation. Appl Therm Eng 2008;28(5);589–600. https://doi.org/10.1016/j.applthermaleng.2007.04.009

[10 Hanipah, M. R., Mikalsen, R., and Roskilly, A. P. Recent commercial free-piston engine developments for automotive applications. Appl Therm Eng 2015;75; 493-503. https://doi.org/10.1016/j.applthermaleng.2014.09.039

[11] B. Jia, R. Mikalsen, A. Smallbone, A. P. Roskilly, A study and comparison of frictional losses in free-piston engine and crankshaft engines, Appl Therm Eng,140,217-224, doi:https://doi.org/10.1016/j.applthermaleng.2018.05.018.

[12] Hanipah, Mohd Razali. Development of a spark ignition free-piston engine generator. Diss. Newcastle University, 2015.

[13] Jia B, Zuo Z, Tian G, Feng H, Roskilly AP. Development and validation of a free-piston engine generator numerical model. Energy Conversion and Management. 2015 Feb 1;91:333-41.

[14] Jia B, Mikalsen R, Smallbone A, Zuo Z, Feng H, Roskilly AP. Piston motion control of a free-piston engine generator: A new approach using cascade control. Applied energy. 2016 Oct 1;179:1166-75.
[15] Jia B, Smallbone A, Zuo Z, Feng H, Roskilly AP. Design and simulation of a two-or four-stroke free-piston engine generator for range extender applications. Energy conversion and management. 2016 Mar 1;111:289-98.
[16] Mikalsen Rikard, Roskilly AP. Free-piston internal combustion engine. U.S. patent application 13/698,569.
[17] Heywood JB. Internal combustion engine fundamentals. McGraw-Hill Press. 1988.

Internal Combustion Engines and Powertrain Systems for Future Transport 2019 – Institute of Mechanical Engineers, ISBN 978-0-367-90356-5

Optimising gasoline engines for future hybrid electric vehicles

Roscoe Sellers, Richard Osborne, David Sweet

Ricardo UK Limited

ABSTRACT

The selection and optimisation of Hybrid Electric Vehicle (HEV) powertrains and architectures creates new challenges and opportunities. The position and size of the electric machines in conjunction with the transmission type can alter the engine operating profile to enable enhancements to the cost and efficiency of the engine. To achieve such benefits requires a systematic evaluation of the powertrain system at the beginning of a development cycle, to understand how the engine may be altered to match the hybrid system. Ricardo has developed an approach to review the performance and operating profile of a wide range of candidate HEV configurations without the need for a detailed control system to be developed as part of the assessment. This allows consideration of what is possible for future powertrains towards achieving the maximum benefits from electrification and optimisation of the engine.

1 INTRODUCTION

The number of hybrid topologies for consideration in possible future vehicle applications is vast, which makes defining the development path of engine technology a challenge. To help inform this decision, Ricardo has developed an Architecture Independent Modelling (AIM) approach to assess and rank a large number of HEV configurations, and down select most promising candidates. This approach brings a number of benefits, including both the speed and quality of the initial decision making process for hybrid vehicle powertrains. This is critical considering the continuous pressure on right-first-time hardware specification, and achieving a reduction in physical testing during the concept phase.

This increased agility to deploy the latest technology in an effective manner can help maintain a competitive product in an evolving marketplace, as well as working towards the demanding requirements for fuel consumption reduction. The results from AIM studies so far indicate the great potential in deploying dedicated hybrid engines (DHE) coupled with a dedicated hybrid transmission (DHT) to maximise the synergies within a hybrid vehicle.

2 POWERTRAIN DIVERSITY LED BY ELECTRIFICATION

Electrification is creating new opportunities for powertrain system integration, and the dramatic increase in investment and focus on Battery Electric Vehicles (BEVs) also brings technology development for the batteries, electric machines and the power electronics at a much faster rate than the typical lifecycle of engine technology.

The challenge, therefore, is to consider new ways of assessing these interactions and opportunities such that:

1. Appropriate architectures are selected against targets
2. New opportunities for integration and optimisation are found

3. Concept selection is done quickly without excessive investment
4. Output can support the next step in analysis or system calibration to reduce development time and cost.

Figure 1 illustrates a subset of the potential configuration options that may need to be considered when defining future vehicle powertrain architectures, and considers a scenario where a vehicle and engine platform may span a number of segments and markets.

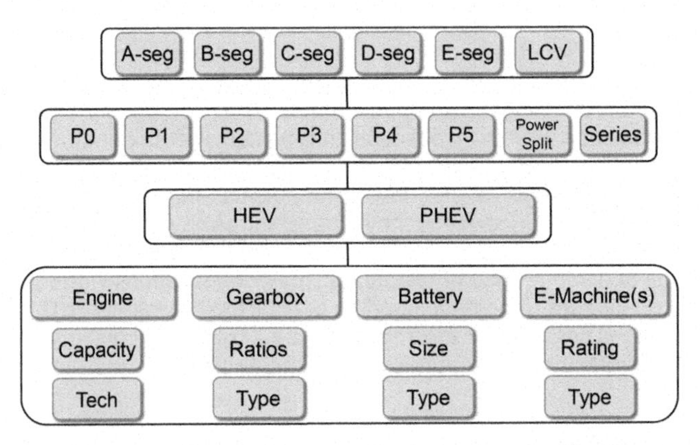

Figure 1. Example powertrain system options.

3 ARCHITECTURE INDEPENDENT MODELLING APPROACH

Within the Virtual Product Development (VPD) toolbox at Ricardo [1][2], a new approach to tackling these challenges has been developed. This Architecture Independent Modelling (AIM) tool allows a wide range of the HEV options to be reviewed without a complex control strategy being developed for each case. This not only enables many options to be considered, but it can be performed quickly and with individual tuning of the inputs and variables tailored to a specific scenario. As shown in

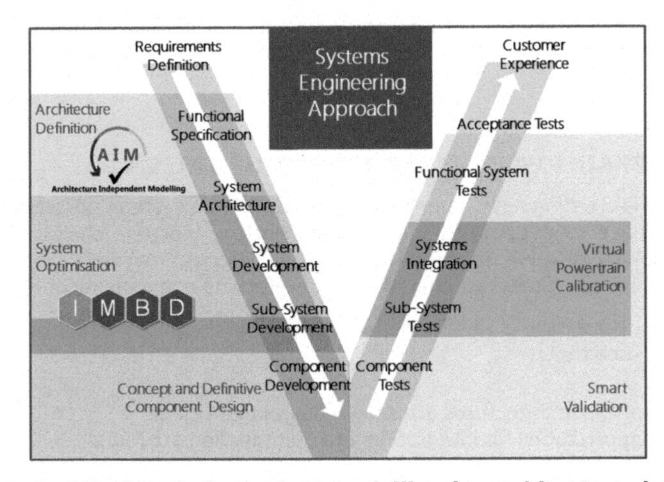

Figure 2. Architecture independent modelling for architecture definition.

Figure 2 with its position on a V-cycle, AIM is primarily intended for the initial phase of a development programme within architecture definition.

The purpose of AIM is the evaluation of HEV systems to define how the powertrain can work together, enabling a comparison of the architectures and components, and identify the opportunities for optimisation.

The basis of AIM is an energy controller that can determine the most effective use of the hybrid system over a drive-cycle while maintaining the target battery State-Of-Charge (SoC). This controller uses DoE techniques to rapidly compare different operating profiles for the engine, gearbox and E-Machine(s) to find the optimum for each candidate for comparison. To achieve this, a range of inputs are required to characterise the system and define boundaries for evaluation including vehicle specification, engine rating and BSFC maps, motor characteristics and the target battery size range. Using this information AIM can then assess when to use the torque assist, generation or Electric Vehicle (EV) modes of the hybrid system and also which gear to use and the resulting impact on the engine speed and load.

The cycle optimisation process can consider the most appropriate gear ratio selection, battery size and E-machine sizes as part of the assessment, along with specifying a minimum EV range (if applicable) and SoC neutrality. The output from the model provides a breakdown of the powertrain components' performance over the specified driving route, such as fuel consumption, average engine BSFC, E-machine utilisation, along with the optimised specifications for the system as illustrated in Figure 3.

➢ DoE Architecture Optimisation

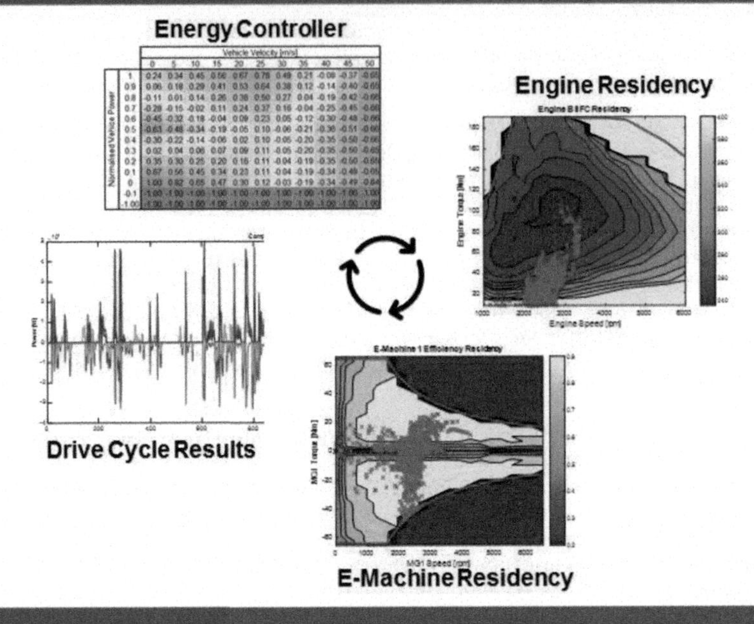

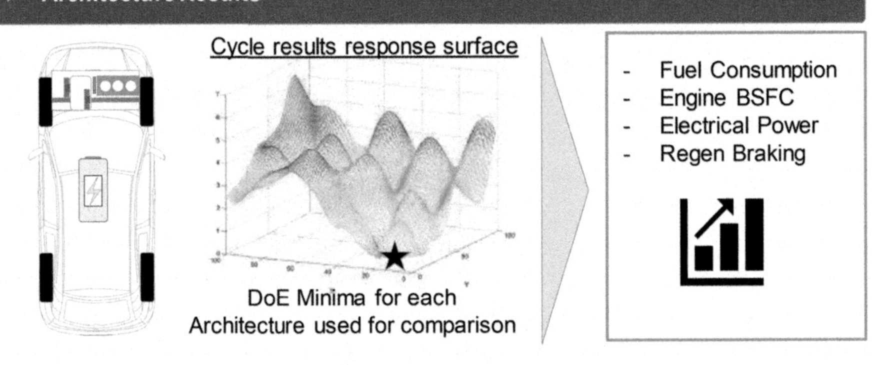

Figure 3. Illustrative AIM process flow.

One of the outputs from an engine perspective is the shift in speed-load residency for a given HEV architecture. This provides significant insight into how best to further optimise the engine characteristics to suit a particular hybrid system.

4 HYBRIDISATION IMPACT ON ENGINES

With the wide range of possible hybrid architectures, vehicle applications and system configurations, it is difficult to generalise the potential change to an engine operating profile, so a specific context is required for an example. One of the most interesting areas is the cost sensitive mass market C-segment that dominates sales volumes which is increasingly biased towards SUVs.

The key for this segment is to achieve high efficiency along with cost competitiveness for the powertrain. As the degree of electrification increases, some hybrid cost and complexity is ideally offset by engine and transmission simplification. For the engine we can consider the architecture itself, the operating range and the region for efficiency optimisation.

4.1 C-Segment cast study

A mid-specification C-segment HEV for 2025 is considered for the case study with a target specification as shown in Table 1. A non-hybrid platform is used for reference with the same vehicle attributes and engine to highlight the impact from hybridisation.

Table 1. Vehicle specification.

	Unit	Value
Vehicle type	-	C-segment HEV
Vehicle mass	kg	1320
ICE power	kW	80
E-machine power	kW	40
Battery capacity	kWh	1.5
System voltage	V	400

To demonstrate how the opportunities for engine optimisation can be identified using AIM, a small range of vehicle architectures have been selected for the case study as shown in Figure 4. This includes a single E-Machine P3 layout, and two E-Machine power-split type arrangement that enables parallel and series modes to be utilised. For each of these cases the number of gears was also varied to understand if a reduction could be considered.

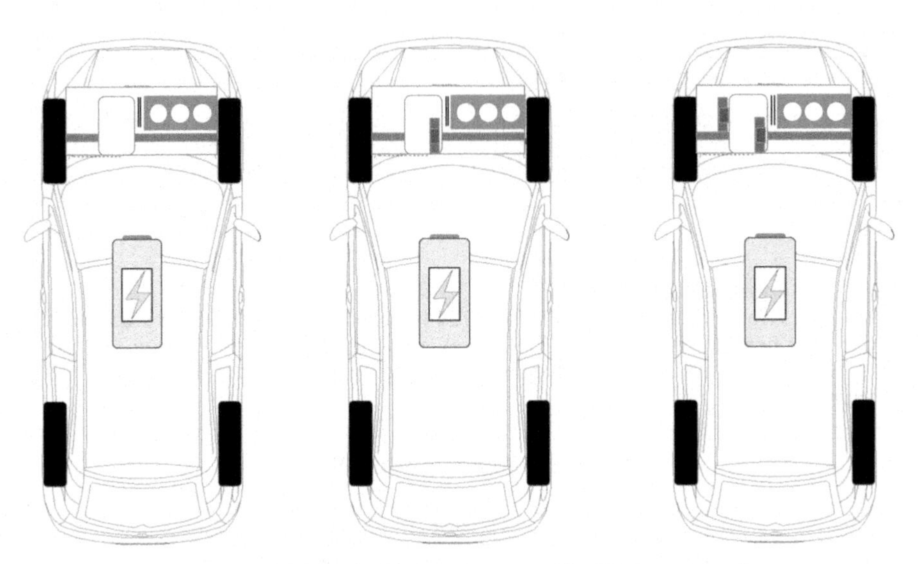

Figure 4. Vehicle architectures used in the case study.

The key area of interest for the engine is how different technology packages and tailoring of efficiency and operating areas can be combined with the hybrid system to deliver the optimum package. Therefore a number of options have been considered, a wide-range, stoichiometric TGDI VGT engine is compared with a narrower range FGT engine, and finally a high-efficiency Dedicated Hybrid Engine (DHE) with a summary of details in Table 2.

Table 2. Engine specifications.

	Wide VGT	Narrow FGT	DHE
Peak power speed [rev/min]	5000	5000	4750
Peak torque min. speed [rev/min]	1500	3000	3000
Capacity [litres]	1.5	1.5	1.5
Cylinders	4	4	3
Bore-to-stroke (B/S) ratio	0.87	0.87	0.75
Compression ratio	12.5	12.5	15
Balancer	No	No	No
Turbocharger	VGT	FGT	High Eff FGT
Fuel pressure [bar]	350	350	350
Variable valvetrain	Dual VVT	Dual VVT	Dual VVT
Miller Cycle	EIVC	EIVC	EIVC
LP cooled EGR	No	No	Option

The characteristics of these engine types are represented through the torque curves and BSFC maps as shown in Figure 5. The VGT variant has the widest operating area and low

BSFC zone, whereas the FGT engine assumes a reduced operating range and low BSFC area. The DHE combines the lower torque curve with a lower absolute level of BSFC achieved but over a reduced area. This is intended to be exploited by the HEV by means of altering the engine operating profile towards this area. On each of the baseline maps the operating area for a non-hybrid vehicle over a WLTP is plotted for reference.

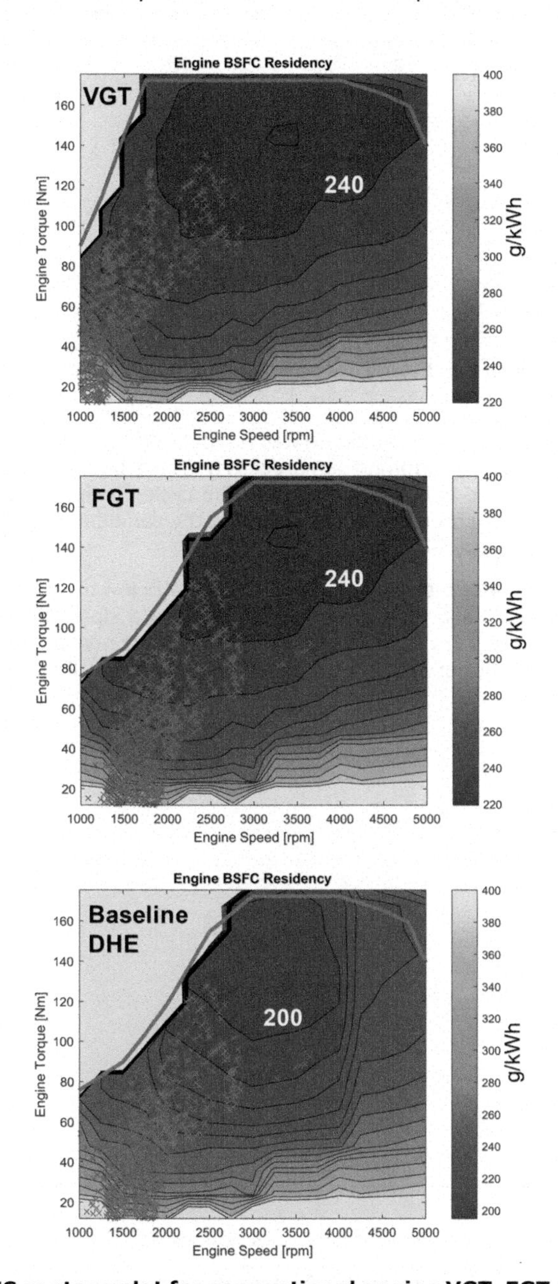

Figure 5. BSFC contour plot for conventional engine VGT, FGT and DHE with WLTP residency overlay in non-hybrid vehicle.

For this case study the battery and E-machine sizes and characteristics are held constant, but for other AIM studies they can be included as optimisation variables. The subset of variables considered for the study is summarised in Figure 6.

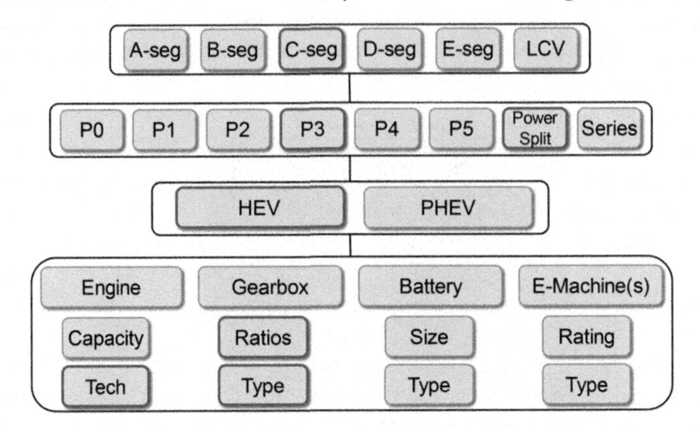

Figure 6. Elements considered in case study.

4.2 AIM analysis

With the inputs prepared, each candidate was run over WLTC and RDE cycles using the AIM DoE methodology. This optimisation process finds the best approach for using the engine, transmission and hybrid systems over the cycles to judge operating profiles and residencies. This can then be used to compare the candidates and understand the implications for engine optimisation.

A selection of the AIM cycle results are plotted in Figure 7 below to compare the candidate performance over WLTC and RDE-Urban phase, showing a clear differentiation for the hybrid cases as expected. It can also be seen that a distinct number of DHE candidates achieve a significantly lower WLTP fuel consumption, which will be explored in more detail.

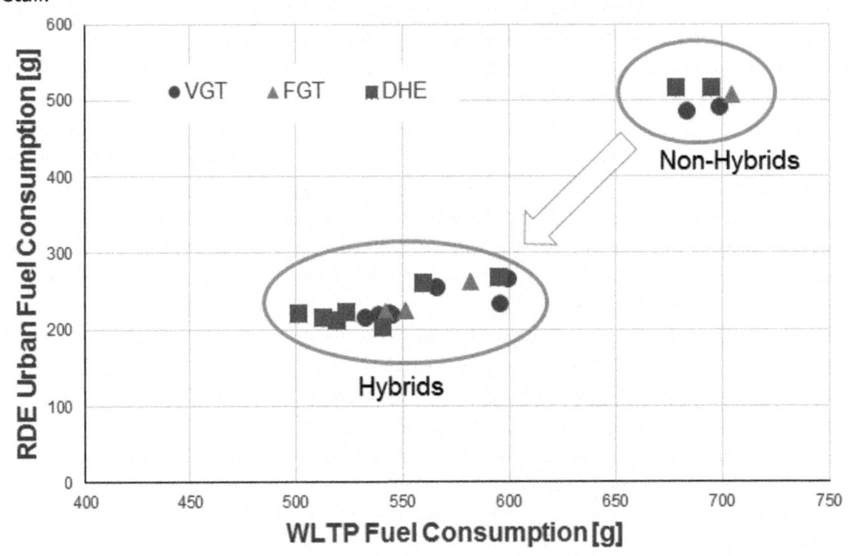

Figure 7. Selection of results from AIM analysis comparing performance of the different engine types over WLTP and RDE cycles.

Using the data generated by AIM we can determine what factors are critical to optimising performance, and below a subset of the results are used to establish the interaction between the hybrid system and engine operation.

As shown in Figure 8, the DHE used in a non-hybrid application does not offer an advantage as the high efficiency zone is not utilised by the standard 6 speed gearbox. Looking at the baseline wide map VGT compared with the higher efficiency DHE, we can also see that increasing hybrid system integration improves the overall CO_2 reduction, and also the incremental benefit of the DHE.

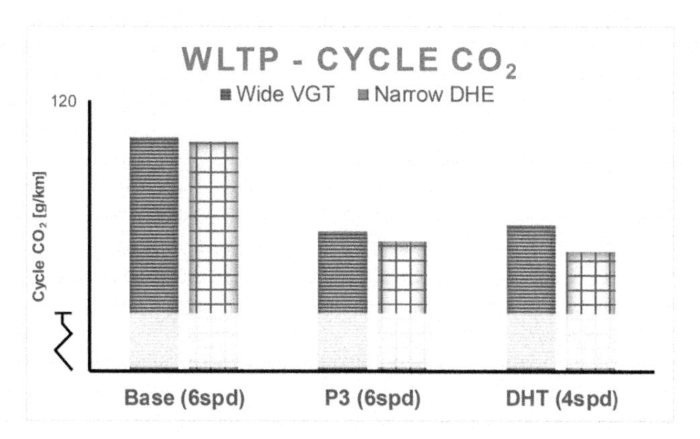

Figure 8. WLTP cycle CO_2 comparison.

CO_2 reduction for the HEVs is achieved through a combination of the regenerative braking energy and utilisation by the E-machine, along with the shift in operating range of the engine. This shift depends on the flexibility of the hybrid architecture, and the 2 E-machine DHT layout offers this to a greater degree. In Figure 9 the improved average BSFC values for the engine over the cycle are shown.

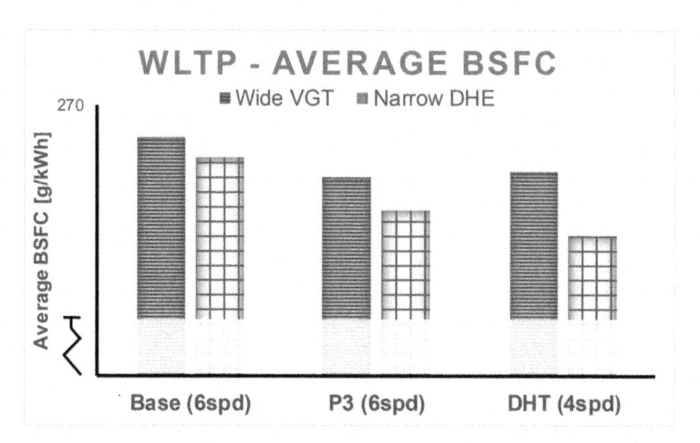

Figure 9. Comparison of average BSFC for different operating profiles.

The activity of the hybrid system can be summarised by the percentage use of the E-Machine to propel the vehicle. With the baseline at 0, we can see in Figure 10 an increasing use as the hybrid system capability increases. It is also notable that the DHE case records higher E-Machine activity, and this is due to the improved trade-off for using the electrical system and associated losses to shift the engine operating region to the peak efficiency zone.

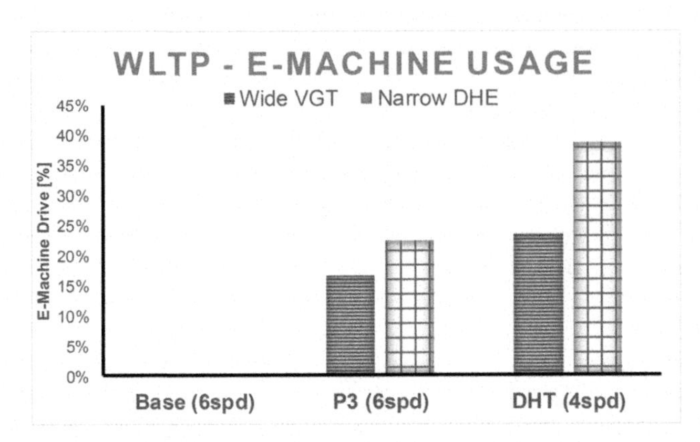

Figure 10. Comparison of E-Machine activity over the cycle.

Looking more closely at the engine operating residency over the WLTP for the DHE case in Figure 11, it is clear how the AIM optimised output has shifted towards the low BSFC zone. This is a critical consideration for combining a highly integrated hybrid powertrain.

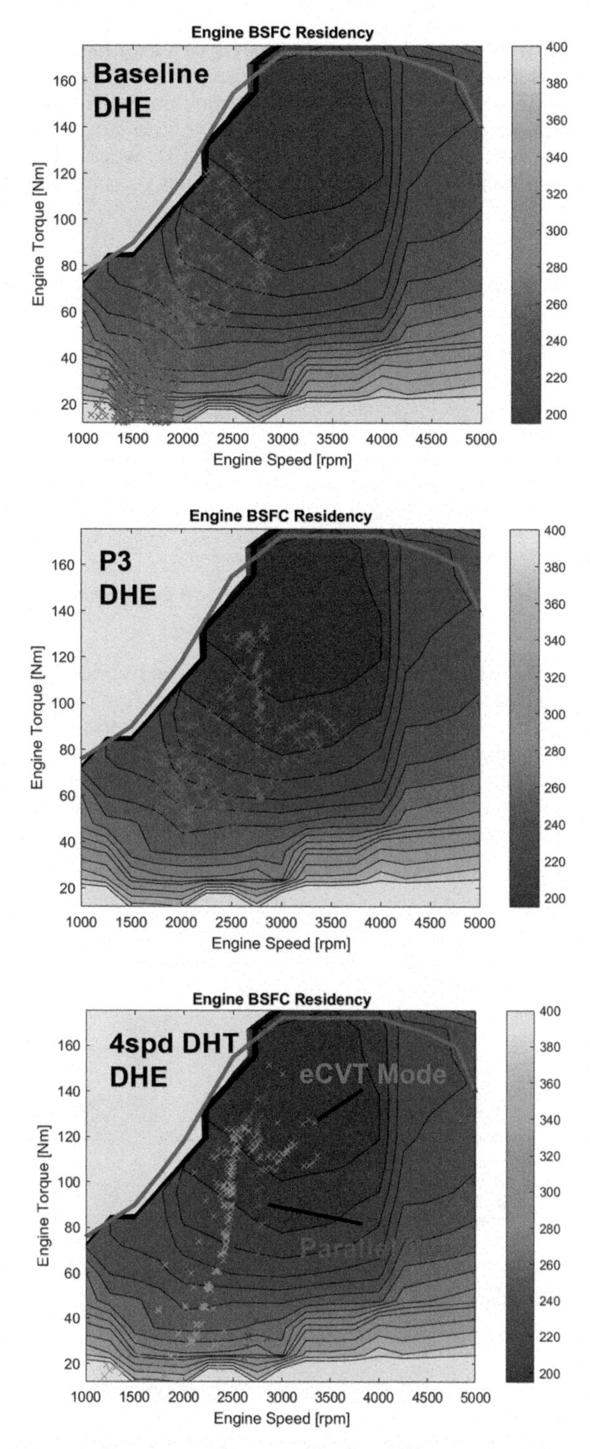

Figure 11. WLTP engine operating profile for different hybrid architectures.

4.3 The benefit of aim in helping define future engines

The case study has shown how a DHE and DHT can be combined to maximise the benefit of a hybrid powertrain technologies. Without looking at full system optimisation, the cost impact of electrification for a C-segment vehicle can become less attractive. To quantify this benefit on a cost basis an example CO_2 walk in Figure 12 shows how this synergistic approach derived from the AIM analysis can improve this trade-off.

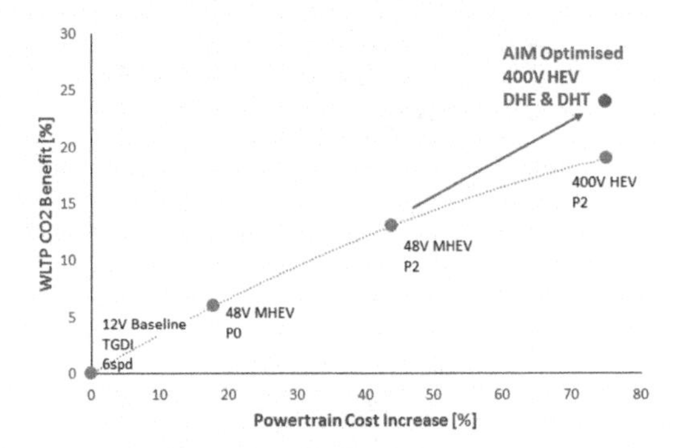

Figure 12. Engine and electrification cost benefit walk.

Using AIM guidance, Ricardo have designed a dedicated hybrid engine (DHE) family to meet an electrified passenger-car platform range. This family is based around a 1.5 litre capacity and three-cylinder architecture. The prime variants are turbocharged in order to provide modularity with different power output levels, and for higher BTE potential EGR and lean burn variants are considered. A naturally-aspirated (NA) variant has also been developed for lower cost, lower power applications where greater battery capacity balances lower ICE efficiency.

The combustion system is based on increased compression ratio, Miller cycle EIVC operation with a direct-injection fuel system, and further features are shown in Figure 13.

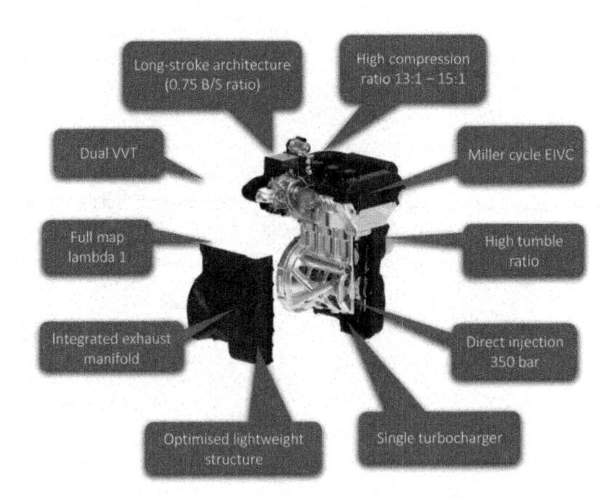

Figure 13. Dedicated hybrid engine (DHE) overview.

It is clear that the opportunity to re-optimise an engine will depend on the hybrid system, but Ricardo see a number of key areas for particular focus. Series-hybrid and range-extender applications provide a further opportunity to tune a dedicated engine concept towards the extremes of efficiency or low cost.

High-Efficiency Combustion System - Ricardo has pursued long stroke and high compression ratio combustion systems with early intake-valve closing for a number of years [3-6], in order to achieve a step change in thermodynamic efficiency.

Reduced Engine Speed Range - Base engine architecture and overall system optimisation can benefit from the engine speed reduction that is enabled by hybrid augmentation.

Low-Speed Torque Reduction - Further benefits of base engine architecture, with potential for 3-cylinder engine balancer deletion and air system optimisation.

Lower-Cost Optimised Turbocharger - The reduced low-speed torque and speed range allows consideration of a re-optimised air path for the revised flow and pressures ratios, and in particular a turbocharger with higher efficiency over a smaller map range.

FEAD Deletion - Higher efficiency on-demand electric ancillaries will be an integral part of enhanced powertrain systems, providing flexible control for engine off/re-start conditions. This can enable a range of further benefits for simplification of the base engine with enhanced encapsulation and thermal performance.

5 CONCLUSIONS AND FUTURE OUTLOOK

The rapid growth in powertrain electrification and increasingly complex architecture options provide both challenges and opportunities. It remains critical that system selection and optimisation considers the most cost-effective path to achieve the maximum benefits. This requires a concept selection phase that can quickly consider a wide range of options before the detailed design optimisation takes place. Ricardo have developed the Architecture Independent Modelling approach to support these steps, and the example presented shows how a range of engine specific enhancements can be considered with the following outcomes highlighted:

- Different HEV architectures impact the engine operating profile in different ways
- Understanding the overall system interaction is important for the initial phase of concept definition and engine specification
- Defining an engine that complements the operating profile of a HEV enables improved fuel consumption benefits for lower cost
- Architecture Independent modelling enables these trade-offs to be evaluated quickly and effectively to support decision making and architecture down-selection
- A mass market C-segment HEV should incorporate an engine derivative specifically for the architecture to maximise the system synergies and benefits
- Ricardo have devised a modular approach to enable a base engine to span a wide range of applications and hybrid types

There are many attributes to consider for Hybrid Electric Vehicles, but the opportunity to enhance the engine through electrification cannot be missed on the path towards future CO_2 reduction.

ABBREVIATIONS

AIM	Architecture Independent Modelling
BEV	Battery Electric Vehicle
BSFC	Brake Specific Fuel Consumption
DHE	Dedicated Hybrid Engine
DHT	Dedicated Hybrid Transmission
DoE	Design of Experiments
FEAD	Front End Accessory Drive
FGT	Fixed-Geometry Turbine
HEV	Hybrid Electric Vehicle
LCV	Light Commerical Vehicle
PHEV	Plug-in Hybrid Electric Vehicle
PS	Power-Split
SoC	State-of-Charge
WLTP	World harmonized Light-duty Test Procedure
VGT	Variable-Geometry Turbine

ACKNOWLEDGEMENTS

The authors would like to thank the directors of Ricardo plc for permission to publish this paper. We would also like to acknowledge the significant contributions to the work made by Tom Stokes, Rob Parkinson and Richard Gordon.

REFERENCES

[1] Seabrook, J., Dalby, J., Shoji, K., Inoue. A., *Virtual Calibration to Improve the Design of a Low Emissions Gasoline Engine*. International Conference on Calibration Methods and Automotive Data Analytics, Berlin, 2019.

[2] Dalby, J., Fiquet, F., Ward, A. et al. *Hybrid Powertrain Technology Assessment through an Integrated Simulation Approach*. 14th International Conference on Engines and Vehicles, 2019.

[3] Pendlebury, K. J., Osborne, R. J., Stokes, J., Dalby, J. *The Gasoline Engine at 2020*. 23rd Aachen Colloquium Automobile and Engine Technology 2014.

[4] Osborne, R. J., Pendlebury, K. J., Stokes, J., Dalby, J. and Rouaud, C.: *The Magma Engine Concept*. JSAE Annual Congress 2015.

[5] Pendlebury, K. J., Osborne, R. J., Downes, T. A. and O'Brien, S., *Development of the Magma Combustion System*. JSAE Annual Congress 2016.

[6] Sellers, R. D., Osborne, R. J., Cai, W., Wang, Y.: *Designing and Testing the Next Generation of High-Efficiency Gasoline Engine Achieving 45% Brake Thermal Efficiency*. 28th Aachen Colloquium Automobile and Engine Technology 2019.

Internal Combustion Engines and Powertrain Systems for Future Transport 2019 – Institute of Mechanical Engineers, ISBN 978-0-367-90356-5

MAHLE modular hybrid powertrain

M. Bassett[1], A. Cooper[1], I. Reynolds[1], J. Hall[1], S. Reader[1], M. Berger[2]

[1]MAHLE Powertrain Ltd, Costin House, St James Mill Rd, Northampton, UK
[2]MAHLE Powertrain GmbH, Fellbach, Germany

ABSTRACT

Vehicle manufacturers are facing increasing pressure by legislation and economics to reduce vehicle emissions and deliver improved fuel economy. By 2025 significant reductions in tailpipe carbon dioxide (CO_2) emissions will need to be achieved to meet these requirements, increasing the interest in hybrid and electric vehicle technologies.

Building on previous research projects, focused on the design of a dedicated range extender (REx), heavily downsized internal combustion engines and pre-chamber combustion system developments, MAHLE Powertrain propose a possible hybrid technology pathway for powertrains optimised for the future global automotive market targeted to meet emissions and CO_2 targets for 2030 and beyond.

1 INTRODUCTION

Vehicle manufacturers are facing increasing pressure by legislation and economics to reduce vehicle emissions and deliver improved fuel economy. By 2025 significant reductions in carbon dioxide (CO_2) emissions will need to be achieved to meet these requirements whilst at the same time satisfying the more stringent forthcoming Euro7 emissions regulations. Since September 2017 Real Driving Emissions (RDE) testing has been adopted as part of the regulatory approval regime (1) to ensure a correlation between emissions in real world driving conditions and those achieved under controlled laboratory test conditions. With the introduction of RDE testing the engine operating region that is under scrutiny during compliance testing has significantly increased. Furthermore, manufacturers may be required to disclose their engine base emissions strategy (BES) and any auxiliary emissions strategy (AES), which may preclude the use of fuel enrichment for component protection in future engines (2). These combined demands increase the focus on the emissions performance of engines over their entire operating map.

Because battery electric vehicles (BEVs) do not generate local pollutants during usage, and they can potentially rely on energy provided by a selection of renewable sources, they are the focus of much current interest. However, due to the present limitations of battery technology (in terms of size, weight and cost) the overall range of such a vehicle is limited in comparison to an equivalent gasoline or diesel fuelled vehicle. An additional consideration is the embedded CO_2 content of large battery packs and their recyclability.

Plug-in hybrid electric vehicles (PHEVs) partly overcome the limitations of current battery technology by retaining an internal combustion engine (ICE). The engine can directly provide motive power when the battery is depleted. Furthermore, once the vehicle has significant electric drive capability, it is possible to remove any dynamic loading from the engine, and allow it to operate in a much more steady-load manner, where the target is to maintain battery state of charge (SOC) once the battery has become depleted. Range extended electric vehicles (REEVs), have a REx unit that converts a fuel, such as gasoline, into electrical energy whilst the vehicle is driving, without a mechanical link to the vehicle wheels (series hybrid). This enables the traction battery storage capacity to be reduced, though still maintaining an acceptable vehicle driving range. MAHLE has developed a dedicated REx engine to identify the requirements and challenges faced in the development of components for such future engines (3).

Gasoline engine downsizing is also an effective technology for CO_2 reduction and is the process whereby the engine operating load point is shifted to a higher, more efficient region, through the reduction of engine swept volume, whilst maintaining the full load performance of the original engine through pressure charging. Further improvements in fuel economy are possible through increased levels of downsizing (4). However, as specific output increases so too do the technical challenges, which include abnormal combustion (pre-ignition and detonation), low speed torque, transient response and durability.

1.1 MAHLE downsizing engine

In order to conduct research into the requirements for advanced downsizing engines and their components, MAHLE developed a demonstrator engine. The resulting heavily downsized engine was a direct injection, 1.2 litre, 3-cylinder inline, turbocharged, gasoline engine (referred to colloquially as the MAHLE Di3). The original version of the Di3 engine achieved a peak brake mean effective pressure (BMEP) of 30 bar, and a peak power output of 120 kW (100 kW/litre) (5-7).

In order to explore the limits of the benefits achievable through engine downsizing, further developments of this engine, including the addition of an electric supercharger, enabled the specific brake mean effective pressure of the engine to be increased to 35 bar BMEP and the specific power increased to 193 kW (161 kW/litre) (8). This engine was then fitted into a C-segment demonstrator, replacing the standard 2.0 litre turbocharged gasoline direct injection (GDI) engine. The resulting vehicle demonstrated a 15 % reduction in fuel consumption over the NEDC, purely due to downsizing (9). Analysis has been undertaken to determine the boosting and exhaust gas recirculation (EGR) system layout required to achieve similar levels of specific torque and power output whilst maintaining lambda=1 operation across the entire engine operating map (10). Further benefits from the 48 V mild-hybrid system are possible, and with the right level of recuperation this could be as much as a further 15 % (11).

MAHLE Powertrain has also been engaged in the development of an industrialised version of the original downsizing engine, shown in Figure 1. Two 3-cylinder engine variants, based on the original version of the Di3 engine, have been developed with capacities of 1.2 and 1.5 litres, sharing maximum commonality (12). The advanced combustion system is combined with a low-friction base engine design and optimized thermal management which combined enables the engines to be capable of delivering competitive real-world economy, the 1.5 litre version achieves a peak brake thermal efficiency (BTE) figure of 36.7 %.

Figure 1. Industrialised 3-cylinder downsizing engine (12).

1.2 MAHLE Jet Ignition (MJI®)

MAHLE Powertrain has been developing the MAHLE Jet Ignition® (MJI) technology for more than 10 years including the development of pre-chambers with both active and passive configurations (13-15). The active system contains both a small spark plug and a low-flow direct injection (DI) fuel injector within the pre-chamber, whereas the passive system, developed initially for motorsport applications, dispenses with the pre-chamber mounted injector and is fuelled by drawing its charge from the main chamber during the compression stroke.

The Active MJI system, depicted in Figure 2, enables reduced fuel consumption and CO_2 through lean gasoline operation. Traditionally this has necessitated more complex and expensive after-treatment system to cope with increased NOx emissions. The Active MJI pre-chamber has the potential to overcome this by pushing beyond established lean limits of combustion to the region beyond Lambda 1.5 where the air dilution is sufficiently extreme to maintain lower gas temperatures throughout the operating cycle, this being less favourable to NOx formation. The Passive MJI system has been developed for applications operating under stoichiometric conditions, relying on EGR to dilute the charge for enhanced efficiency.

Both the active and passive MJI systems use a very small auxiliary pre-chamber which, once ignited, discharges fast-moving, super-heated jets through a multi-orifice nozzle into the main combustion chamber. This chamber design achieves lower crevice volume and heat loss by virtue of its ultra-compact size combined with small nozzle diameters that promote a high degree of flame quenching, allowing the jets to penetrate deep into the main chamber before igniting the charge.

Most recently MJI has been applied to the Di3 engine. Early results from fired tests show stable combustion even under ultra-lean conditions, over a wide range and the capability to operate at Lambda in excess of 1.8, at part-load naturally aspirated conditions, when using active MJI. The passive MJI system has enabled operation at elevated compression ratios, with increased EGR tolerance resulting in excellent BTE figures achieved even when reverting to port fuel injection (PFI), which will be discussed in greater detail later.

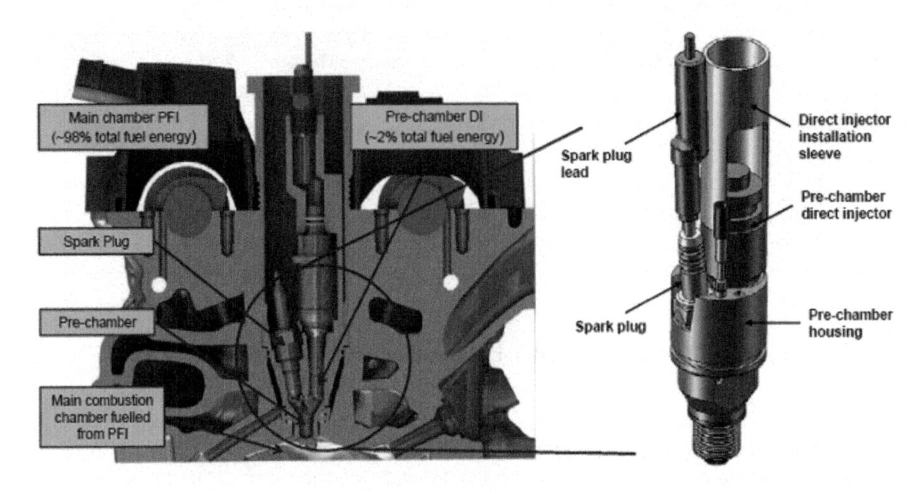

Figure 2. Section through a cylinder head showing a typical active MJI® system installation (15).

2 FUTURE HYBRID POWERTRAIN REQUIREMENTS

The present study examines the powertrain requirements to meet the needs of passenger cars in the 2030 timeframe, and beyond. Drawing on, and combining, the experience gained through the development of the REx and Downsizing demonstrator engines, with the addition of the stoichiometric MJI system, a future high-voltage hybrid electric vehicle powertrain architecture is proposed which has been targeted to meet anticipated future legislative requirements.

Figure 3 shows the manufacturers' fleet average tail-pipe CO_2 targets for the EU. This is based on an average European vehicle with a kerb mass of 1400 kg. The current target of 130 g/km can be readily met using current conventional ICE technologies. The first point, labelled (1) and shown in Figure 3, for 2019, represents the CO_2 figure achievable for a compact crossover SUV (1400 kg kerb mass), using the 1.5 litre productionised version of the Di3 engine. Beyond 2021 the individual vehicle CO_2 limits are extremely challenging without a significant degree of hybridisation. The second point in Figure 3, labelled (2) and aligned with 2020, represents the compact crossover SUV using the same ICE technology, but with the addition of a ~15 kW capable 48 V mild-hybrid system. This enables the 2021 limit to be achieved.

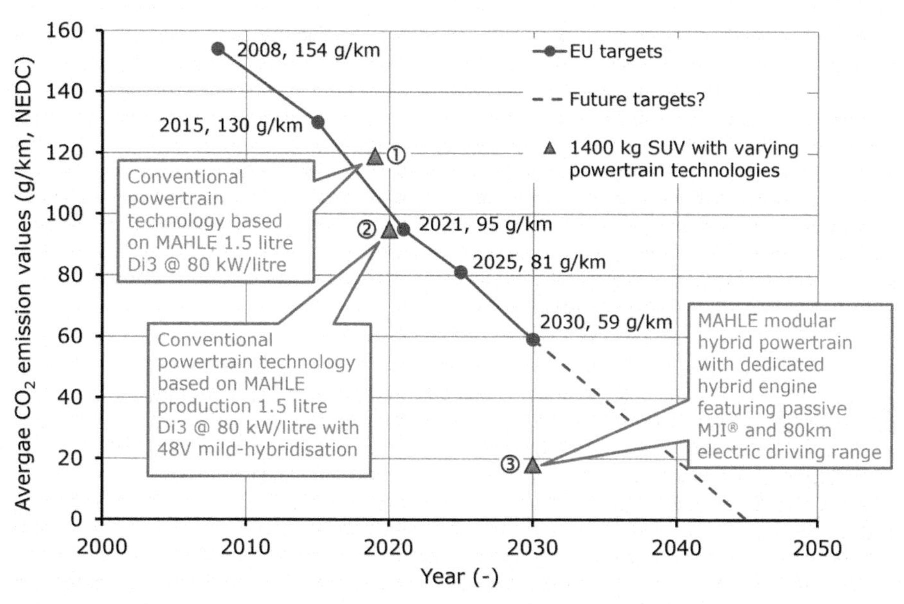

Figure 3. Fleet average CO_2 targets, along with examples of the anticipated CO_2 levels that a 1400 kg compact crossover SUV might achieve with various powertrain technologies.

However, the 2030 limit is only achievable with a significant degree of plug-in hybrid capability (as per point 3 in Figure 3), to enable the vehicle to be eligible to apply CO_2 weighting factors based on pure electric driving capability, as specified in UN/ECE Regulation 101 (16). This regulation enables a vehicle with pure electric driving capability to apply a scaling factor to the measured drive-cycle charge sustaining CO_2, based on the pure electric driving range over the cycle. The weighting factor for plug-in hybrid vehicles tested using the recently introduced world light-duty test procedure (WTLP) shows a similar trend. These weighting factors are based on statistical data for typical daily driving distances and are founded on the assumption that the user will recharge their vehicle every night, thus the weighting factor adjusts the measured tail-pipe CO_2 to allow for the fuel use displaced by pure electric driving.

In reality, manufactures will achieve their fleet CO_2 obligations via the sale of a variety of vehicles featuring a range of technologies including conventional ICE, mild-hybrids, plug-in hybrids and full battery electric vehicles. The key factor then is to sell a sufficient percentage of the low emitting models to achieve the required fleet averaged CO_2 level to meet the targets. Increasing the size of the battery of a PHEV enables the electric driving range to be increased, yielding a more favourable CO_2 tail-pipe weighting factor, and hence a lower vehicle CO_2 rating.

Currently vehicles are assessed on a tailpipe CO_2 emissions basis, and not on a total life cycle CO_2 equivalent basis. In future the total life cycle impact of the vehicle may come under scrutiny, and it is important (from an environmental, as well as legislative, perspective) that we consider the total life cycle impact of a vehicle – otherwise it is very difficult to compare ICE, PHEV and BEVs on an even basis. Of particular note with a BEV or PHEV is the equivalent CO_2 impact of the battery generated during its production. It can be shown, based on current battery technology and UK grid carbon intensity, that there is a minima in life cycle equivalent CO_2 for PHEVs when they have an electric range of around 100 km (17). This arises due to the embedded CO_2 in a larger battery outweighing the marginal benefit of the additional electric driving range, which is only used for exceptional journeys. Furthermore the life cycle equivalent CO_2 of a PHEV with this size of battery roughly equates to that of a pure BEV with a 200 km driving range (BEVs with a greater range having a worse life cycle impact, again due to the embedded CO_2 from the battery pack production).

2.1 Plug-in hybridisation

Beyond 2030, plug-in hybridisation, with a significant electric driving range capability to achieve a favourable tail-pipe weighting factor, will be a key technology enabler to achieving fleet CO_2 targets. Once the vehicle has significant electric drive capability, there are other benefits that can be leveraged to improve the efficiency and emissions performance of the ICE. Firstly it is possible to remove any dynamic loading from the engine, enabling it to operate in a much more steady-load manner, approaching the operation seen for range extender units, where the target is to maintain battery SOC once the battery has become depleted.

Once the electric drive system is capable enough to remove the entire dynamic load requirement from the engine, the engine needs only to be sized to simply maintain battery state of charge during use. Secondly, the removal of the requirement for the engine to respond instantaneously to driver demand, as this is handled via the electric drive unit, means that the engine can be optimised for a narrower operating range, removing the need for much of the complexity encountered on prime-mover ICEs. A final benefit of this type of architecture is that it can be ensured that the ICE is not subjected to any significant loading until the exhaust after-treatment has attained the temperature required for good conversion efficiency. This attribute is especially beneficial when considering the recent addition of real-driving emissions (RDE) testing to vehicle certification, whereby the vehicle could encounter full-load driver demand from a cold soaked starting condition.

3 MAHLE MODULAR HYBRID POWERTRAIN

The MAHLE Modular Hybrid Powertrain (MMHP) has been developed to meet the legislative requirements for passenger cars in the 2030 timeframe, and beyond. The powertrain layout has been guided using experience gained during the development of the MAHLE REx and Downsizing demonstrator engines. The powertrain also benefits from the most recent advances made with the passive MJI® system with stoichiometric operation combined with EGR.

3.1 Driveline architecture

The dual mode hybrid arrangement has been selected for the MMHP. It combines the best features of the series and parallel arrangements. A schematic of the architecture selected is shown in Figure 4. The MMHP has been based around a high-voltage PHEV architecture, with a dedicated hybrid internal combustion engine (DHICE) integrated with a dual-mode hybrid electric drive, which features two electric machines. The traction motor provides full vehicle dynamic performance and vehicle maximum speed without assistance from the DHICE. This enables improved emissions and reduced after treatment complexity by avoiding any requirement for the DHICE to ever perform a cold start at a high power demand, which has been identified as a challenge with PHEV systems that rely on ICE power for full vehicle dynamic performance capability (18). Additionally, the direct drive arrangement enables seamless torque delivery, enabling the use of a simple automated manual transmission. The design of the unit has targeted reduced cost, complexity, package size and weight compared to current hybrid powertrains. The system is also specified to deliver very low drive-cycle CO_2 figures whilst also offering very low real world life-cycle CO_2.

When battery SOC is high, the vehicle can operate as a pure BEV. Once the battery is depleted, the system can operate as a series hybrid at low vehicle speeds, having the NVH and operating flexibility that this arrangement offers. At higher vehicle speeds the engine can be connected directly the wheels, via a gearbox, with a number of ratios enabling some flexibility in engine operating speed. The traction motor is directly connected to the wheels, thus there is seamless torque delivery, even during a gear-shift event. The draw-backs of this arrangement are that it combines the complexities of both the series and parallel architectures by requiring two electric machines in addition to a multi-speed transmission (albeit with a low ratio count).

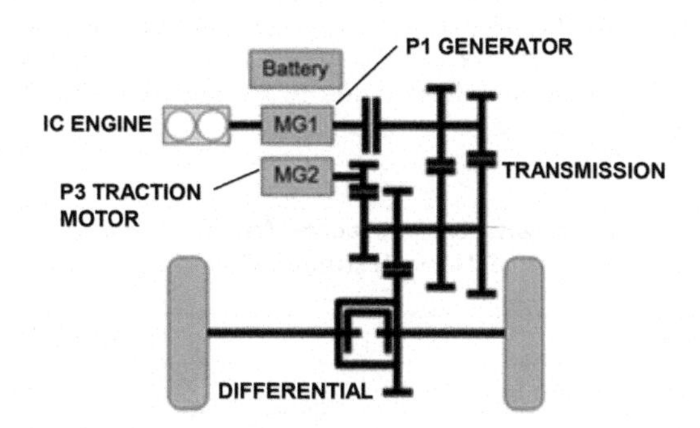

Figure 4. Dual-mode hybrid drive-line architecture of the MMHP.

3.2 Vehicle performance targets and system power requirements

For this study, the target vehicle application is a compact crossover SUV. In order to determine the power requirements of the various components within the hybrid driveline (engine, generator and traction motor) a number of vehicle performance targets have been set. The dynamic performance of the vehicle determines the power requirements for the traction motor (MG2 in Figure 4). The generator (MG1 in Figure 4) sizing is determined by the minimum operating speed of the engine and the transmission ratios selected. The generator must be powerful enough to enable battery SOC to be maintained under all conditions until the vehicle is travelling fast enough to enable the system to switch into parallel hybrid mode. Finally, the engine maximum power and the transmission ratios required can be determined by the maximum speed that the vehicle should be able to maintain battery SOC whilst cruising, which also needs to cover hill-climbing and towing scenarios. The vehicle performance targets, used to determine the powertrain requirements, are summarised in Table 1. Trailer towing and hill-climb analysis was conducted at gross vehicle weight (GVW).

Table 1. Performance targets set for the compact crossover SUV with the MMHP.

Parameter	Units	Target
Acceleration time (0-100 km/h)	(s)	<9.0
Maximum vehicle speed	(km/h)	>180
Charge sustaining speed (level road)	(km/h)	>130
Charge sustaining with 1000 kg trailer (6% grade)	(km/h)	80
Pure electric driving range (WLTP)	(km)	>80
Charge sustaining fuel consumption (WLTP)	(litres/100 km)	5.5
Weighted WLTP CO_2	(g/km)	<18

The performance targets summarised in Table 1 can be used to determine the powertrain component performances required. The high-voltage system components operate on a nominal system voltage of 400 V. The traction motor is water cooled and features a novel stator winding cooling system, which enables the motor to achieve a very high continuous power rating from an extremely compact package size. The traction motor selected for this application has a peak power rating of 157 kW and a continuous rating of 127 kW. The series-hybrid generator unit is capable of operating at a continuous power rating of 20 kW. Finally, the ICE power requirement to meet the charge sustaining operation was determined to be 60 kW. Table 2 summarises the power capability of the components comprising the hybrid powertrain.

Table 2. System parameters selected for the compact crossover SUV with the MMHP.

Parameter	Units	Value
System nominal voltage	(V)	400
Traction motor power (peak/continuous)	(kW)	157/127
Series hybrid generator power (peak/continuous)	(kW)	40/20
DHICE peak power	(kW)	60

3.3 Dedicated hybrid internal combustion engine

The specifications for the DHICE, which is integral to the MMHP, are summarised in Table 3 and have been devised to achieve greatest efficiency, with a low cost architecture and compact package size. The power output target based on vehicle requirements is 60 kW and it was also decided that the maximum engine operating speed should be limited to 4000 rev/min. This was based on experience from the REx engine, where the operation of the engine could be effectively masked by vehicle aerodynamic and tyre noise, if the engine was operated below this speed (16). A two-cylinder layout was chosen to minimise the engine package volume (and reduce part count and weight). A parallel twin, even-firing configuration with contra-rotating balancer shaft layout has been adopted, to minimise vibration and reduce the torque recoil about the crank centre line arising due to cyclic speed fluctuations.

The combustion system is based around the passive MJI system, with PFI fuelling, and the engine features 2 valves per cylinder with fixed valve event timing to reduce cost. To maximise the operating efficiency, the engine has a high degree of Miller-cycle operation and a high geometric compression ratio (CR). The engine operates with a single, fixed geometry turbocharger, using an electrically controlled waste-gate actuator. To enable further efficiency gains an external, cooled, exhaust gas recirculation (EGR) system is also used. The engine has a displacement of 1.0 litres, and the power and maximum speed requirements necessitate that the engine achieves a peak BMEP of 18 bar.

Table 3. MMHP DHICE Specification.

Parameter	Units	Value
Swept volume	(litres)	1.0
Number of cylinders	(-)	2
Bore/Stroke	(mm)	83/92.4
Peak power	(kW)	60
Operating speed range	(rev/min)	1500 - 4000
Maximum BMEP	(bar)	18
Maximum BTE	(%)	>40

3.4 Transmission specification

The driveline architecture of the MMHP was shown schematically in Figure 4. The high-voltage traction motor directly drives the wheels via the differential (in a P3 location, labelled MG2 in Figure 4). There is also a high-voltage, liquid-cooled, generator mounted directly on the engine crankshaft, in the location of a conventional engine flywheel (in a P1 location, labelled MG1 in Figure 4). The inverters for the traction motor and generator are integrated into the housing of the respective devices. The transmission input shaft is also directly engaged with the generator. To save cost, weight and package space, there is no clutch within the system. This is not required as the DHICE is not used for vehicle pull away. The engine is decoupled from the driveline, for pure-electric or series-hybrid operation, by selecting neutral in the simple automated manual transmission unit.

The transmission design uses cylindrical helical gears to provide a cost optimised solution and can be tailored to have 1, 2 or 4 ratios depending upon application requirements. The architecture has been designed to enable all variants, irrespective of number of ratios, to use common ratios and main transmission casing. The ratio options selected for the transmission are summarised in Table 4.

Table 4. MMHP transmission specification.

Parameter	Value
1 speed transmission ratio	3.91 : 1
2 speed transmission ratios	6.31 / 3.91 : 1
4 speed transmission ratios	6.31 / 4.93 / 3.91 / 3.19 : 1
Traction drive ratio	16.45 : 1

The transmission ratios have been selected to enable the DHICE to be operated in direct drive mode over a wide range of vehicle speeds. Figure 5 shows road load curves for the 1400 kg compact crossover SUV operating on both a level road and a 6 % grade. The influence of a 1000 kg trailer and a caravan (1000 kg trailer with a large frontal area) on the road load of the vehicle is also shown in Figure 5. The engine tractive wheel force, for the DHICE in each of the 4 transmission ratios listed in Table 4, are also shown in Figure 5.

For some applications a single gear ratio (3rd gear, 3.91:1) could enable the majority of the vehicle performance targets to be met, such as charge sustain cruising in direct drive mode at up to 130 km/h on a level road and 80 km/h on a 6% grade. However, to enable charge sustaining operation whilst towing on a gradient, an additional lower ratio (1st gear, 6.31:1) would also be needed to enable the DHICE to be engaged at higher wheel torque levels (to effectively achieve a higher power level at a lower vehicle speed). If higher charge sustaining vehicle speeds are required an additional higher ratio can be added to the transmission (4th, 3.19:1), which will enable 160 km/h to be achieved with the DHICE still directly coupled to the wheels. The ratios are paired within the transmission, thus the addition of the higher ratio enables the 2nd gear ratio to be added to the transmission, which enhances the level of engine power assistance that can be achieved at a vehicle speed between 80 to 105 km/h.

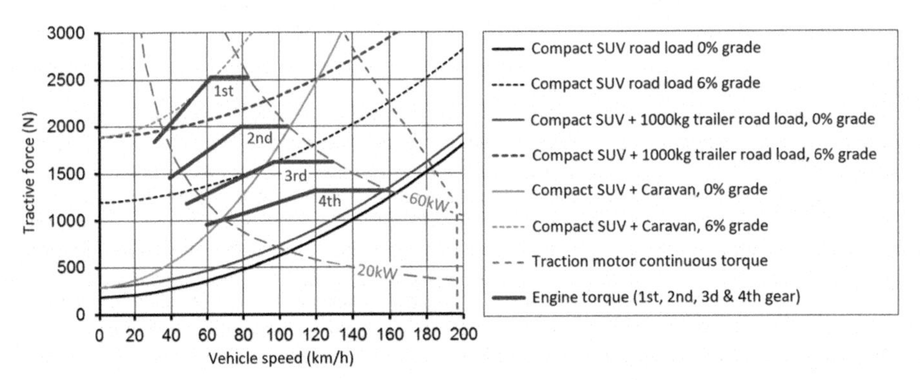

Figure 5. Road load and cascade diagram for the 1400 kg compact crossover SUV with a 4-speed version of the MMHP.

The transmission is configured as two sets of two ratios with a neutral position between each of the ratios. In the 4-speed version, when the DHICE is in direct drive mode, one set of gears will be in neutral, whilst the other is in the selected drive ratio. This layout means that even for the simple sequential layout used, the transmission is only one shift away from neutral in any condition. The 4-speed transmission requires two electrically actuated worm-drive rotary shift barrels, and is capable of non-sequential gear selection. The 2-speed transmission only requires a single electrically controlled actuator. The transmission unit is shown in Figure 6, where a comparison of both the 4-speed and 2-speed units is shown and the common main transmission casing can be clearly seen.

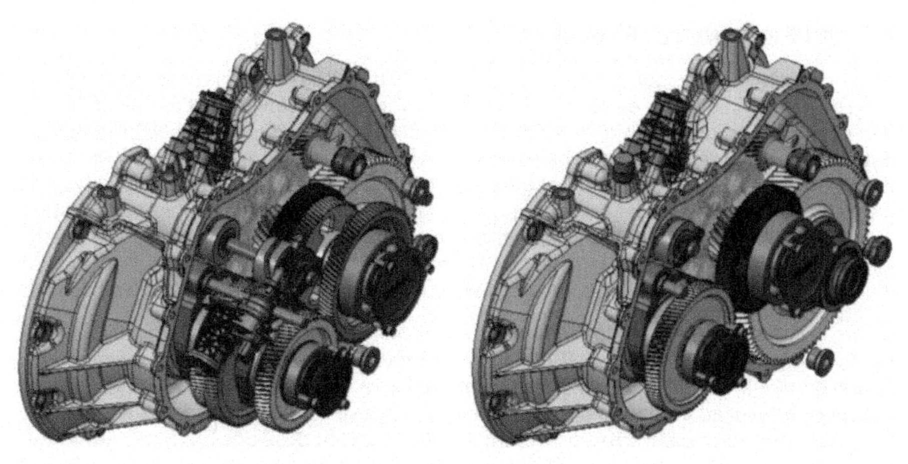

Figure 6. 4-speed and 2-speed versions of the transmission for the MMHP.

3.5 Fully integrated unit

The complete integrated hybrid unit has been designed to be as compact as possible. Figure 7 shows views of the MMHP unit with the 2-speed transmission, and includes the transmission, engine, generator, traction motor and the inverters for the motor and generator. Due to the compact two-cylinder engine layout, the total width of the unit (ignoring the exhaust after treatment system, which would be designed to suit a particular vehicle application) is only 721.5 mm (762.2 mm for the 4-speed variant), which is a crucial dimension when considering a transverse vehicle installation. Similarly the total mass of the unit has also been kept as low as possible and is 180 kg for the 2-speed variant and 191 kg for the 4-speed variant (this includes engine, transmission, generator, traction motor and both inverters).

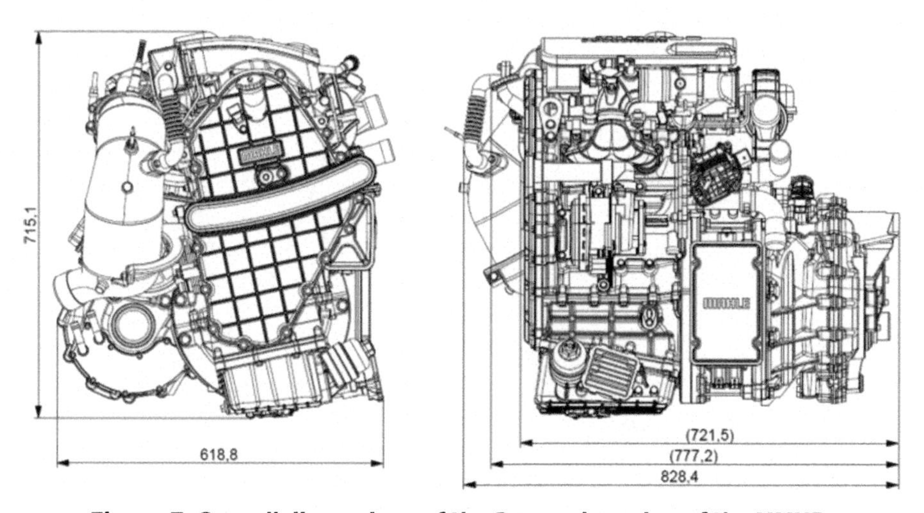

Figure 7. Overall dimensions of the 2-speed version of the MMHP.

4 MULE ENGINE TEST RESULTS

Initial validation of the combined use of passive MJI system, high geometric compression ratio and a high degree of Miller-cycle operation has been conducted using a version of the Di3 engine with a cylinder head that has been adapted for MJI operation. Figure 8 shows a comparison of the combustion events, at the target peak power operating point of 4000 rev/min at an engine load of 18 bar BMEP, between the passive MJI system and a conventional central spark plug (CSP). The pre-chamber combustion event can clearly be seen in the pre-chamber pressure trace just prior to top-dead-centre firing (0° crank angle). This in-turn initiates the rapid combustion in the main chamber. Table 5 summarises some of the key combustion metrics for the two cases.

At this test condition the 10 to 90 % burn duration is reduced by 40 %, from 25.9 °CA to 15.4 °CA, through use of MJI which then enables the combustion phasing to be advanced by almost 9° CA before the onset of knock. As a consequence of the faster and more advanced combustion, higher levels of maximum cylinder pressure and the maximum rate of pressure rise are seen. The MJI system provides an improvement in combustion stability at this operating condition, with an improvement in the CovNMEP of 2.54 % down to 0.85 %.

Using the Di3 engine, adapted for passive MJI, operating with a very high geometric compression ratio (the same as that intended for the DHICE of the MMHP) in conjunction with a PFI system, Miller-cycle intake camshaft (140 °CA duration at 1 mm lift), and an externally cooled EGR system, a whole operating map for the DHICE was generated. This is shown in Figure 9.

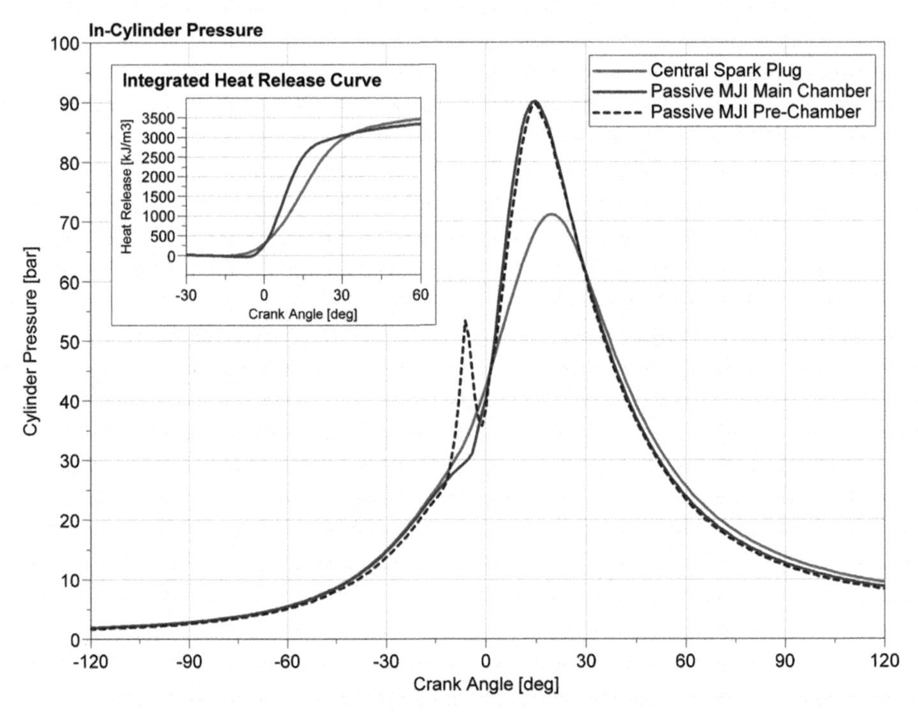

Figure 8. Comparison of passive MJI combustion to a conventional central spark plug at 4000 rev/min and 18 bar BMEP.

Table 5. Comparison of combustion metrics at 4000 rev/min and 18 bar BMEP between MJI and CSP (results without EGR).

Parameter	Units	MJI	CSP
Combustion Phasing (50% Mass Fraction Burned)	°ATDFC	8.6	17.4
Burn Duration (10-90%)	°CA	15.4	25.9
Maximum Cylinder Pressure	bar	90.9	71.2
Maximum Rate of Pressure Rise	bar/°CA	5.2	1.98
Combustion Stability - Coefficient of variance of NMEP (CovNMEP)	%	0.85	2.54

The initial status map shown in Figure 9 gives a promising indication of the potential of the combustion system to achieve the challenging BTE targets. This testing demonstrated that a maximum BTE level of over 40 % is achievable at an engine speed of 3000 rev/min and a load of 14 bar BMEP as well as a wide area of operation above a BTE of 35%. The 20 kW continuous operating line of the series hybrid generator is shown for reference. The current maximum BTE value that has been achieved is limited by the boosting system fitted to the engine. Further optimisation of the boosting system is expected to enable higher rates of external EGR to be delivered, which are in turn anticipated to yield further improvement in BTE. Figure 10 shows the results of an EGR sweep at the highest efficiency operating point.

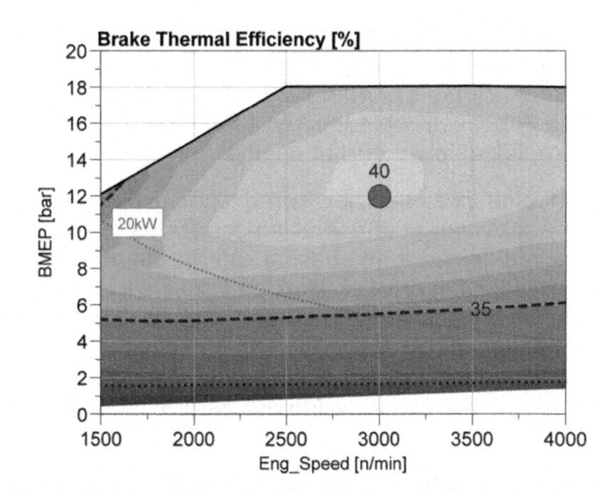

Figure 9. DHICE brake thermal efficiency map based on mule engine testing.

Figure 10 shows the BTE improvement that can be made through the addition of EGR. Approximately 13 % is the maximum amount of external EGR that can be provided to the engine at this operating point, with the current boosting system, using the optimum phasing for the Miller-cycle inlet cam. At this level of EGR the combustion stability is well below the typical limit of 3 % COV NMEP and the 10-90 % burn duration, whilst slowed by the addition of EGR, is still significantly faster than that achieved by a CSP without any EGR. An additional benefit of the addition of EGR and the associated slowing of the burn duration is the reduction in the maximum rate of pressure rise.

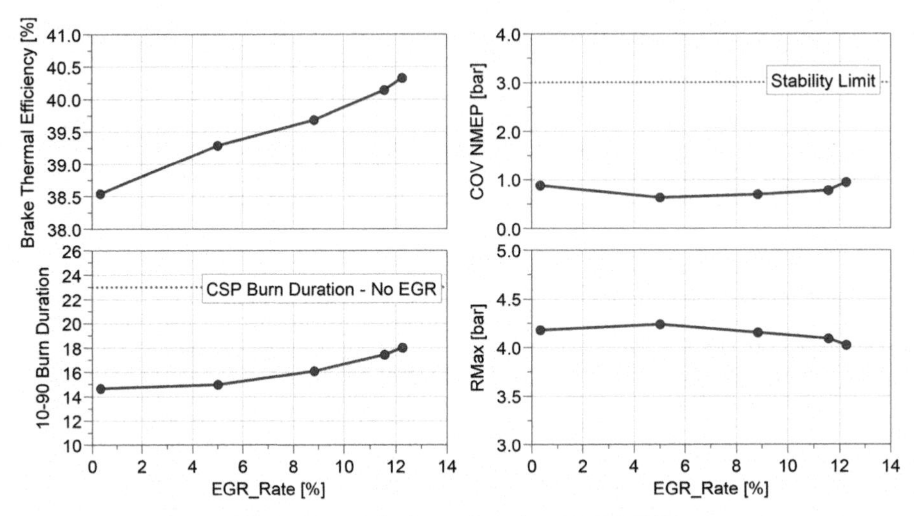

Figure 10. 3000 rev/min and 14 bar BMEP EGR sweep.

Testing is currently underway, targeting >41% BTE through the re-optimisation of the boosting and EGR systems. The Di3 engine used for this testing is a research engine that has been shown to be capable of operating reliably at up to 35 bar BMEP and 160 kW/l (11), further benefits would be expected for a DHICE due to the lower friction anticipated for this, simplified, lower output engine.

A challenge for developing pre-chamber based combustion systems is to achieve stable operation over a wide range of operating conditions, especially under low load conditions or with delayed combustion phasing as typically employed during the cold start catalyst heating phase of operation. Figure 11 shows a comparison of the results from a steady-state chilled fluids (25 °C oil and coolant) spark-retard test used to assess catalyst heating capability, for the Di3 engine with a conventional CSP against our initial passive pre-chamber concept and the developed solution. The developed MJI pre-chamber concept is the same design as used to generate the data presented in Figures 9 and 10.

With the initial pre-chamber design the combustion stability degraded rapidly once the combustion phasing was retarded beyond 20° ATDCF, significantly less than the 70° ATDCF phasing attained by the CSP. This degradation was caused by the onset of sporadic poor combustion events or misfires within the pre-chamber which leads to poor, or no, jet formation and the subsequently very poor combustion, or even total misfire, in the main chamber.

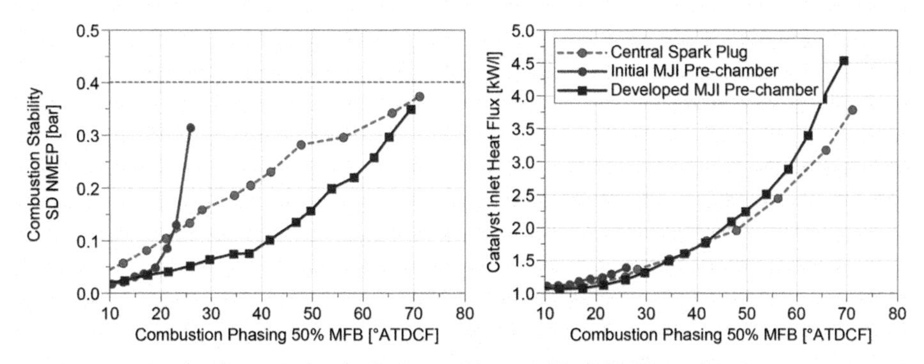

Figure 11. 1500 rev/min and 2 bar NMEP chilled fluid spark retard sweep.

With the developed MJI pre-chamber it was possible to eliminate these sporadic poor combustion events within the pre-chamber enabling the combustion phasing to be retarded to 70 °ATDCF matching the capability of the CSP at a very similar and acceptable level of combustion stability. As can be seen in Figure 11, the combustion stability throughout the spark sweep is significantly better with the phasing retarded up to 60 ° ATDCF. Figure 11 also shows the catalyst inlet heat flux (specific heating power) available to heat the catalyst. The results presented in Figure 11 show that the developed solution is capable of providing three times the heat flux of the initial design and can achieve a comparable, or higher, level than the CSP throughout the combustion phasing sweep conducted. The engine out gaseous and particulate emissions measured for the MJI equipped engine during this test, and also for the whole area operating map testing, are comparable to those measured with the CSP.

Testing has also been completed to assess the ability of the combustion system under hot idle conditions. This testing demonstrated the ability to operate with stable combustion below 750 rev/min with adequate spark authority (+/- 10 °CA) for fast idle speed control and load compensation.

5 CONCLUSIONS

The MMHP has been developed to meet the legislative requirements for passenger cars in the 2030 timeframe, and beyond. It has been based around a high-voltage PHEV architecture, with a dedicated hybrid internal combustion engine (DHICE) integrated with a dual-mode hybrid electric drive, featuring two electric machines. The traction motor provides full vehicle dynamic performance and vehicle maximum speed without assistance from the DHICE. This enables improved emissions and reduced after treatment complexity. Additionally, the direct drive arrangement enables seamless torque delivery, enabling the use of a simple automated manual transmission. The design of the unit has targeted reduced cost, complexity, package size and weight compared to current hybrid powertrains. The concept is scalable across multiple vehicle applications with, 1, 2 or 4 speed transmission and 2 or 3-cylinder engine options available.

Results of initial testing, undertaken to validate a pre-chamber combustion concept proposed for the DHICE, have been presented. The combination of the pre-chamber based combustion layout, together with high geometric compression ratio, external cooled exhaust gas recirculation (EGR) and aggressive Miller-cycle operation enable extremely high brake thermal efficiency levels, up to <40% BTE, to be achieved with very modest technology requirements for the engine.

Abbreviations

AES	Auxiliary emissions strategy
BES	Base emissions strategy
BEV	Battery electric vehicle
BMEP	Brake mean effective pressure
BTE	Brake thermal efficiency
CovNMEP	Coefficient of variance of NMEP
CR	Compression ratio
CSP	Central spark plug
DHICE	Dedicated hybrid ICE
Di3	Downsized, inline, 3-cylinder
EGR	Exhaust gas recirculation
GDI	Gasoline direct injection
GVW	Gross vehicle weight
ICE	Internal combustion engine
LHV	Lower heating value
MG1	Motor-generator 1
MG2	Motor-generator 2
MJI	MAHLE Jet Ignition®
MMHP	MAHLE modular hybrid powertrain
NEDC	new European drive cycle
NMEP	Net mean effective pressure
NVH	Noise, vibration and harshness
PFI	Port fuel injection
PHEV	Plug-in hybrid electric vehicle
RDE	Real drive emissions
REEV	Range extended electric vehicle
REx	Range extender
SOC	State of charge
SUV	Sports utility vehicle
WLTP	World-harmonised light-duty test proceedure

ACKNOWLEDGEMENTS

The authors would like to acknowledge the support of Ambixtra, during the engine testing phase of this project, by the loan of a variable spark ignition (VSI) system. The authors would like to also acknowledge the contribution contributions made to this work by their colleagues both from MAHLE Powertrain, MAHLE ZG and from within the wider MAHLE Group. In particular David Hancock, Simon Duke, Steven Simmonds, Jonathan Gardner, Dr. Greg Taylor, Gery Fossaert, Anthony Harrington, Michael Bunce, Dr. Lovrenc Gasparin, Stefan Jauss, Thomas Strauss, Roberto Almeida E Silva, Dr. Christian Wirth, Sebastian Stoll and Dr. Otmar Scharrer.

REFERENCES

[1] Hayes, J., Department for Transport, Explanatory Memorandum on European Union Legislation, 23rd January 2017: http://europeanmemoranda.cabinetoffice.gov.uk/files/2017/01/170118_-_Real_Driving_Emissions_(1).pdf (accessed 3rd June 2019).

[2] C(2017) 352 final, Guidance on the evaluation of Auxiliary Emission Strategies and the presence of Defeat Devices with regard to the application of Regulation (EC) No 715/2007 on type approval of motor vehicles with respect to emissions from light passenger and commercial vehicles (Euro 5 and Euro 6), European Commission, Brussels, 26th January 2017. https://www.greensefa.eu/files/doc/docs/e1d4fd81911a47c76fcbd4458d6bfdd5.pdf (last accessed 13th October 2017).

[3] Bassett, M., Hall, J., and Warth, M., "Development of a dedicated range extender unit and demonstration vehicle", EVS27, Barcelona, Spain, November 17-20th, 2013.

[4] Wirth, M., "Ford's EcoBoost Technology: A Central Element of a Sustainable CO_2 and Fuel Economy Strategy with Affordable Products", Automotive Summit, Brussels, 9-10th November 2010.

[5] Hancock, D., Fraser, N., Jeremy, M., Sykes, R., Blaxill, H., "A New 3 Cylinder 1.2l Advanced Downsizing Technology Demonstrator Engine," SAE Technical Paper 2008-01-0611, 2008, https://doi.org/10.4271/2008-01-0611.

[6] Lumsden, G., OudeNijeweme, D., Fraser, N., Blaxill, H., "Development of a Turbocharged Direct Injection Downsizing Demonstrator Engine", SAE Paper 2009-01-1503, 2009. https://doi.org/10.4271/2009-01-1503.

[7] Korte, V., Rueckauf J., Harms K, Miersch J., Brandt M. Muenz S. and Rauscher M., "MAHLE-Bosch Demonstrator Vehicle for Advanced Downsizing", 19th Aachen Colloquium Automobile and Engine Technology, 2010.

[8] Bassett, M., Hall, J., Hibberd, B., Borman, S., Reader, S., Gray, K. and Richards, B., "Heavily Downsized Gasoline Demonstrator," SAE Int. J. Engines 9(2):729-738, 2016, https://doi.org/10.4271/2016-01-0663.

[9] Bassett, M., Hall, J., Cains, T., Reader, S., Wall, R., "Dynamically Downsized Gasoline Demonstrator Vehicle," IMechE Internal Combustion Engines Conference, Birmingham, UK, 6-7 December, 2017.

[10] Bassett, M., Vogler, C., Taylor, J., Hall, J., Gray, K., "Analysis of a Highly Downsized Gasoline Operating at Lambda 1," IMechE 13th International Conference on Turbochargers and Turbocharging, London, UK, 16-17 May, 2018.

[11] Bassett, M., Hall, J., Vogler, C., Preece, D., Cooper, A. and Berger, M., "Analysis and Design of a High-power Module for 48V Applications", EVS31, Kobe, Japan, 1-3 October, 2018.

[12] Cooper, A., Stodart, A., Hancock, D., Duke, S. et al., "Development of Two New High Specific Output 3 Cylinder Engines for the Global Market with Capacities of 1.2l and 1.5l," SAE Technical Paper 2019-01-1193, 2019, https://doi.org/10.4271/2019-01-1193.

[13] Bunce, M. and Blaxill, H., "Sub-200 g/kWh BSFC on a Light Duty Gasoline Engine," SAE Technical Paper 2016-01-0709, 2016. https://doi.org/10.4271/2016-01-0709.

[14] Bunce, M., Blaxill, H., Kulatilaka, W., and Jiang, N., "The Effects of Turbulent Jet Characteristics on Engine Performance Using a Pre-Chamber Combustor," SAE Technical Paper 2014-01-1195, 2014, https://doi.org/10.4271/2014-01-1195.

[15] Attard, W. and Blaxill, H., "A Lean Burn Gasoline Fueled Pre-Chamber Jet Ignition Combustion System Achieving High Efficiency and Low NOx at Part Load," SAE Technical Paper 2012-01-1146, 2012, https://doi.org/10.4271/2012-01-1146.

[16] UNECE Addendum 100: Regulation No. 101, Revision 3 (E/ECE/324/Rev.2/Add.100/Rev.3), 16th October 1995. http://www.unece.org/fileadmin/DAM/trans/main/wp29/wp29regs/2015/R101r3e.pdf, (accessed 3rd June 2019).
[17] Bassett, M., Cooper, C., Hall, J., Reader, S. and Berger, M., "Hybrid Powertrain technology Roadmap", EVS 32, Lyon, France, 20-22 May 2019.
[18] Cooper, A., Bassett, M., Hall, J., Harrington, A. *et al.*, "HyPACE - Hybrid Petrol Advance Combustion Engine - Advanced Boosting System for Extended Stoichiometric Operation and Improved Dynamic Response," SAE Technical Paper 2019-01-0325, 2019, https://doi.org/10.4271/2019-01-0325.

SESSION 3: EMISSIONS AND AFTER-TREATMENT

Modeling of three-way catalyst dynamics during fast transient operation for a CNG heavy-duty engine

Dario Di Maio[1], Carlo Beatrice[2], Valentina Fraioli[2], Stefano Golini[3], Francesco Giovanni Rutigliano[3], Marco Riccardi[4]

[1]Istituto Motori CNR, Via G. Marconi, Naples, Parthenope
[2]Istituto Motori CNR
[3]FPT Industrial SpA
[4]Università degli Studi di Napoli Federico II

ABSTRACT

In this work, a numerical "quasi-steady" model was developed to simulate the chemical and transport phenomena of a specific Three-Way Catalyst (TWC) for a natural-gas heavy-duty engine.

Goal of the present research activity was to investigate the effect of very fast composition transitions of the engine exhaust typical of real world driving operating conditions, as fuel cutoff phases or engine misfire, which produce strong deviations from stoichiometric in Air-to-Fuel (A/F) ratio and characterize catalytic converter efficiency. In fact, according to the literature it is confirmed that the catalyst dynamic behavior differs from the steady-state one due to oxygen storage phenomena.

A dedicated experimental campaign has been performed in order to evaluate the catalyst response to a defined λ variation pattern of the engine exhaust stream, thus providing the data necessary for the numerical model validation. A surface reactions kinetic mechanism, concerning CH_4, CO, H_2 oxidation and NO reduction, has been appropriately calibrated, with a step-by-step procedure, both in steady-state conditions of the engine work plan, at different A/F ratios, and during transient conditions, through cyclical and consecutive transitions of variable frequency between rich and lean phases.

The activity also includes a proper calibration of the reactions involving Cerium inside the catalyst, in order to reproduce oxygen storage and release dynamics. Sensitivity analysis and a reactions rate continuous control allowed evaluating the impact of each of them on the exhaust composition in several operating conditions.

The proposed model predicts tailpipe conversion/formation of the main chemical species, starting from experimental engine-out data and provides a useful tool for evaluation of the catalyst performance.

1 INTRODUCTION

Given the need to achieve zero or near-zero emissions in the field of transport and sustainable mobility, the interest in heavy-duty engines powered with CNG is continuously growing. As a result, the development of reliable and predictive numerical models for virtual design of these vehicles has become necessary, offering a fundamental contribution in time to market reduction. This type of activity is placed in an industrial context that foresees the use of ever faster and more reliable tools in anticipation of multiple operating conditions through simulations of the entire layout of future powertrains. The automotive industries are dealing with significant advancements on pollutants control strategies to comply with always more stringent emissions standard [1]. Especially in heavy-duty truck and bus engines, NG became more and more attractive as alternative fuel in terms of emissions and performance in comparison to traditional fuels, reducing adverse health effects and social costs of air pollution [2].

For SI stoichiometric CNG engines, the most suitable pollutant abatement system is the Three-Way catalytic converter (TWC). Similarly to SI gasoline engines, this device permits to control NOx, unburned methane (CH_4) and other pollutant emissions (CO, NMHC). The simultaneous conversion of these species is possible exclusively in a very tiny range of inlet stream composition around stoichiometric condition [3]. Compared to gasoline engine, this optimum operating point is further reduced, due to a sudden drop in NOx conversion efficiency as soon as a slightly lean λ value is achieved and to a non-complete conversion of THCs both in lean and in rich conditions [4].

Previous studies have widely demonstrated the differences in conversion efficiency between a steady-state test and a dynamic condition [5,6]. Inside the washcoat there are some species that have the ability to be oxidized or reduced according to the exhaust gas composition. The most important of these components is cerium, which is added in the washcoat as a stabilizer and a medium for oxygen storage component. These effects have a considerable impact on the lambda value inside the catalyst which must be appropriately investigated in order to better manage TWC behavior. Under real world driving conditions, several deviations of the AFR from the stoichiometric value take place, due to fuel cutoff phases, engine misfire and response lag of fuel-control system [7].

In this scenario, the numerous tasks and the mentioned working issues typical of this aftertreatment device have required, since its introduction, a development of mathematical models capable of analyzing specific operating conditions.

Modelling approaches are generally categorized as 0D, 1D and 3D, with increasing accuracy and complexity levels. A zero dimensional approach is solely used for Steady-State conditions, as only the mean exhaust gas mass flow is considered, thus reducing the reacting device to an element which performs the chemical conversion according to kinetic parameters given as input. More complex is the 3D approach, used when the characterization of flow distribution inside the reactor is required. The results accuracy is certainly improved because it permits to identify radial diffusion effects of the flow inside the catalyst; the real limitation is represented by high computational efforts, which do not allow an investigation in a very wide range of operating conditions, such as the variables of a dynamic cycle, in time compatible with project targets. A good compromise is represented by 1D models which allow to simulate a thermo-fluid dynamic behavior of the whole exhaust system during both Steady-State and Transient condition with a reduced computational effort [8].

Calibration of the reactions kinetic model inside the catalyst is certainly one of the most challenging topics. Several studies are available in literature, mostly concerning traditional gasoline spark ignition (SI) engines [9,10,11].

An interesting modeling approach, based on 104 reaction steps, of a TWC kinetic scheme with exhaust mixture from natural gas-fueled engines was proposed by Zeng et al. However, in this study only SS conditions were analyzed, thus Oxygen Storage phenomena and perturbations in AFR were not considered [12]. Very little information is available in literature on the specific features of TWC systems applied to natural gas vehicles. In this respect, an important contribution is provided by Tsinoglou et al. through a comparison between honeycomb and ceramic foam catalysts [13].

With a favorable ratio between accuracy and calculation time, in the present work a "quasi-steady" model, equipped with comprehensive oxygen storage and release submodels, is setup to analyze the effect of cyclical perturbations in the exhaust gas of a NG heavy-duty engine on TWC efficiency in different load conditions. Catalyst performances under fast transient AFR dynamics, from lean to very rich conditions, are investigated.

2 EXPERIMENTAL LAYOUT AND AFTERTREATMENT

This section is divided in two paragraphs. The first one comprises a description of the layout, including the CNG 6-cylinder engine, adopted to perform the experimental activity. In the subsequent paragraph are then illustrated the TWC catalyst aspects for the numerical modeling.

2.1 NG 6-cylinder engine

The experimental activities were conducted on compression ignition (CI) Heavy Duty Natural Gas production engine, with a combustion system design compliant with Euro VI regulations. The PFI NG injectors are fed by a separate NG low pressure line, operating at a pressure of about 10 barG. The NG consumption is measured by means of an Emerson Coriolis effect device; the air flow rate is measured by means of an Air Mass Flow meter. The experimental layout is reported in Figure 1, while Table 1 describes the main characteristics of the engine.

Table 1. Main features of the natural gas 6-cylidnder engine.

Displaced volume	12.8 L
Stroke	150 mm
Bore	135 mm
Compression ratio	12:1
Number of Valves	4
Rated Power	338 kW @ 2000 rpm
Torque	2000 Nm @ 1100-1620 rpm
PFI Injector	Natural gas

The chemical composition of the adopted fuel is summarized in Table 2.

2.2 Three-way catalytic converter

The catalytic converters used in automotive applications are commonly equipped with multiple parallel channels of small square or honeycomb cross-section over a large surface area, in order to obtain a laminar flow field. All pollutant emissions are measured by AVL i60 devices: THC and CH_4 by FID, NO_x by CLD, CO, CO_2 by IRD, O_2 by PMD, NH_3 by LDD.

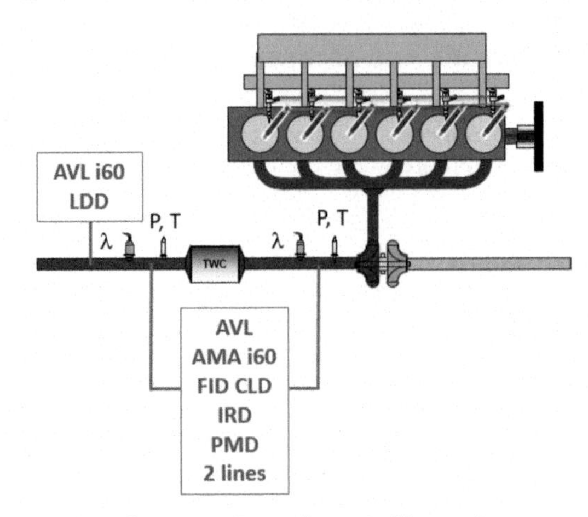

Figure 1. Experimental layout.

Table 2. CNG fuel mixture used during experimental tests.

Fuel composition		
Methane	CH_4	84,78%
Ethane	C_2H_6	8,88%
Propane	C_3H_8	1,88%
N-Butane	C_4H_{10}	0,50%
N-Pentane	C_5H_{12}	0,08%
N-Hexane	C_6H_{14}	0,04%
Nitrogen	N_2	1,90%
Carbon Dioxide	CO_2	1,87%
Helium	He	0,07%

3 EXPERIMENTAL TEST

The experimental tests were performed in two ways. For each test the AFR ECU control is disabled and the AFR is superimposed by means of injection quantity at a fixed air mass flow. The experimental tests in SS conditions (15 engine operative points) were carried out with λ sweep from 0.9 to 1.1. However, the real value of AFR can be slightly different from the value that derives from ECU, taking into account the variable measurement dynamics of the chemical species analyzers and the behavior of the installed λ sensors.

Several analyses have been carried out in order to identify a reference λ value to describe discrepancies in exhaust gas composition. In fact, the engine is equipped with two types of sensors that provide an AFR measurement. The first is the Smart NOx Sensor (SNS) 120 by Continental™, consisting of a ceramic sensor element made of zirconia (ZrO_2) and an electronic control unit, while the second is the Universal

Lambda Sensor (ZFAS-U) by NGK™, made of two zirconia (ZrO_2) substrate elements, one is the O_2 pumping cell (Ip cell), the other is the O_2 detecting cell (Vs cell) heated by a ceramic heater which is supplied by a very small current. The SNS sensor has a very rapid response at composition changing and is very precise at stoichiometric conditions, but suffers of higher errors when the exhaust gas is in rich or lean conditions, due to linear correlation between oxygen concentration of residual gas and AFR. On the other hand, for the ZFAS-U sensor, very precise at stoichiometric conditions, even a slight variation in the exhaust gas composition, typical of NG engines, can considerably affect the sensed AFR value, determining oscillating values. For these reasons it is not reliable in transient operations but remains suitable for SS conditions. Therefore, the most reliable method to calculate λ for SS conditions was identified as an analytical relation present in Regulation 49 [14] from test best bench data:

$$\lambda_i = \frac{\left(100 - \frac{c_{COd} \cdot 10^{-4}}{2} - c_{HCw} \cdot 10^{-4}\right) + \left(\frac{\alpha}{4} \cdot \frac{1 - \frac{2 \cdot c_{COd} \cdot 10^{-4}}{3.5 \cdot C_{CO2d}}}{1 + \frac{c_{COd} \cdot 10^{-4}}{3.5 \cdot c_{CO2d}}} - \frac{\varepsilon}{2} - \frac{\delta}{2}\right) * \left(c_{CO2d} + c_{COd} \cdot 10^{-4}\right)}{4.764 * \left(1 + \frac{\alpha}{4} - \frac{\varepsilon}{2} + \gamma\right) * \left(c_{CO2d} + c_{COd} \cdot 10^{-4} + c_{HCw} \cdot 10^{-4}\right)}$$

where:

λ_i	is the instantaneous "relative Air to Fuel Ratio"
c_{CO2d}	is the dry CO_2 concentration, in percentage
c_{COd}	is the dry CO concentration, ppm
c_{HCw}	is the wet HC concentration, ppm
α	is the molar hydrogen ratio (H/C)
β	is the molar carbon ratio (C/C)
γ	is the molar sulphur ratio (S/C)
δ	is the molar nitrogen ratio (N/C)
ε	is the molar oxygen ratio (O/C)

The usage of this formula for the λ calculation permits to neglect the mentioned uncertainties linked to data provided by the sensors, as this calculation is directly based on the species concentrations actually measured at the engine exhaust.

Finally, hydrogen concentration values were estimated: such hypothesis, despite not significantly affecting λ values, can certainly influence the conversion of pollutants, especially NOx, and remains one of the points to be explored in further studies.

The experimental tests in dynamic conditions were carried out with-a schematic pattern of the λ target reported in Figure 2. Usually, in the evaluation of the oxygen storage phenomena, and in the characterization of the change of behavior of the catalyst during the transitions from rich to lean and vice versa, wide duration tests, similar to stat are used for AFR scan [6] or as an alternative similar conditions are investigated with the introduction of a synthetic gas [15]. The innovative design of the experiments provides three consecutive transitions made through an AFR control system that allows the exploration of the emissions of a real engine during these rather complex phases. As shown in Figure 2, target value of λ varied between 0.90 in rich conditions and 1.10 in lean conditions. The maximum duration of a single step was set equal to 10s, as in this timeframe the exhaust gas, reaching a stationary composition, fully oxidizes the cerium contained within the catalyst. On the contrary, the highest achievable frequency was set at 1 Hz: below this value the catalytic converter cannot follow the input dynamics, giving rise to an unaffected efficiency with faster oscillations. In fact, especially at low engine load, the presence of an empty volume upstream the active catalytic zone can damp the temporal evolution of the species concentration.

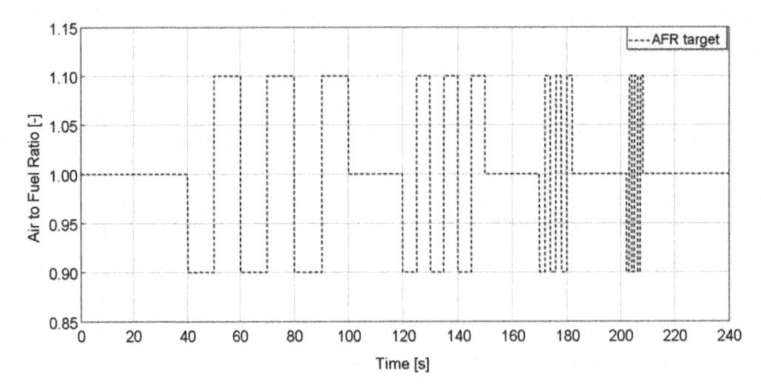

Figure 2. λ target in dynamic experimental test.

The concentration of the main exhaust gas components was measured before and after TWC. These values were determined by analyzing a fully dried sample stream for CO_2, CO, O_2 and NOx and a fully wet stream with a Flame Ionization Detector (FID) for unburned hydrocarbons. Based on these experimental data, water concentration was calculated according the following formula:

$$\tilde{x}_{H_2O} = \frac{m}{2n} \frac{\tilde{x}^*_{CO} + \tilde{x}^*_{CO_2}}{\left[1 + \tilde{x}^*_{CO}/\left(K\tilde{x}^*_{CO_2}\right) + (m/2n)\left(\tilde{x}^*_{CO} + \tilde{x}^*_{CO_2}\right)\right]}$$

In this formula, K is a constant equal to 3,65 while m,n are typical values obtainable from the global chemical formula $C_nH_mO_r$ representing the employed NG mixture reported in Table 2, and $\tilde{x}^*$ denotes the dry mole fraction of the species in sub-script [16].

4 ASSUMPTIONS

The main hypotheses commonly adopted for catalyst 1D models are still valid also for this application because, as mentioned in the introduction, the TWC "quasi-steady" mathematical model, here extended to CNG engines, was based on previous validated works for traditional gasoline engines. Along the catalytic converter changes in potential and kinetic energy are neglected, as well as heat losses to the surroundings. In order to analyze the TWC performance when exposed to real concentrations values of the pollutants in the exhaust stream, it is necessary to measure, at the reactor inlet, the exhaust mass flow rate and temperature. These values are constant in SS tests, while vary during dynamic λ sweep tests. However, considering that tests are performed at constant load and at fixed engine speed, the fluctuations are due to the acquisition dynamics. Constants pressure is assumed along the systems. Radial diffusion is not considered. Wet concentration of main species present in exhaust gas as CO_2, CO, H_2O, NO, NO_2, H_2, CH_4, C_3H_8, O_2 and N_2 (evaluated as complement to unity of total mass flow) are imposed at the inlet of the TWC. As known, these categories of engines, given the high H/C ratio, produce a conspicuous quantity of hydrogen, which has an appreciable impact on catalyst reactions, especially on Oxygen Storage Capacity (OSC). Since hydrogen concentration measurements in the exhaust gas were not available, an assumption had to be made, starting from the empirical correlation from [17], which relates its values to CO concentration ones, for every AFR, as graphically represented in Figure 3. This characteristic has been extended to values with lean mixture conditions.

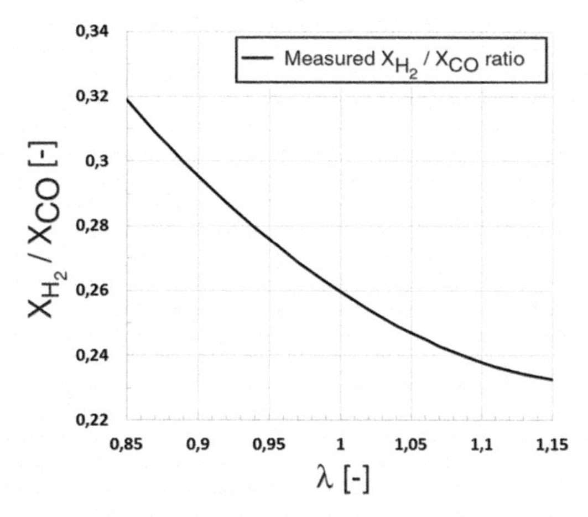

Figure 3. H_2/CO ratio. Empirical correlation from Holder *et al.* [17].

It is clear that a measure of hydrogen both at the inlet and outlet of the TWC would allow a better calibration for the nitrogen and carbon oxides conversion. In addition, its measurement could be also interesting to evaluate possible interactions with NO in the direct production of NH_3 and N_2O.

It is worth to mention that the residual oxygen concentration in exhaust gases, especially in stoichiometric and in rich conditions, represents a critical issue: indeed, its concentration, measured by the analyzer, is approximately equal to 0,2% even in the conditions at lowest AFR. As a result, oxygen analyzer measurement turned out to be an important element for the simulations setup, influencing, as known, the oxidation/ reduction of the pollutant species.

A NO/NO_2 ratio equal to 90/10 was used, in line with the assumptions generally made for traditional SI engines.

5 MODEL VALIDATION

The present paragraph briefly describes the procedure adopted to validate the kinetic conversion model of main gaseous pollutants and the oxygen storage phenomena. First of all, a default kinetic mechanism, already implemented in GT-Power code, with a reaction scheme from [17], was the starting reference model. This mechanism, comprising 13 chemical species and 21 reactions, was specifically developed to summarize the most relevant reaction paths taking place in a TWC for exhaust streams from a gasoline engine.

Applying this initial model to simulations of a NG engine, it is not possible to obtain a correct prediction of the main pollutant species concentration at the tailpipe, even though the injection mode is similar to a traditional Gasoline SI Engine. As an example, Figure 4 shows engine out and tailpipe concentration of the main converted species obtained through the default kinetic scheme [17], available in the original version of the employed software. For sake of brevity only the model answer in the most favorable conditions at high load in SS is proposed. The wide deviations between the pink curve and the green curve are significant, demonstrating the inaccuracy of the starting model.

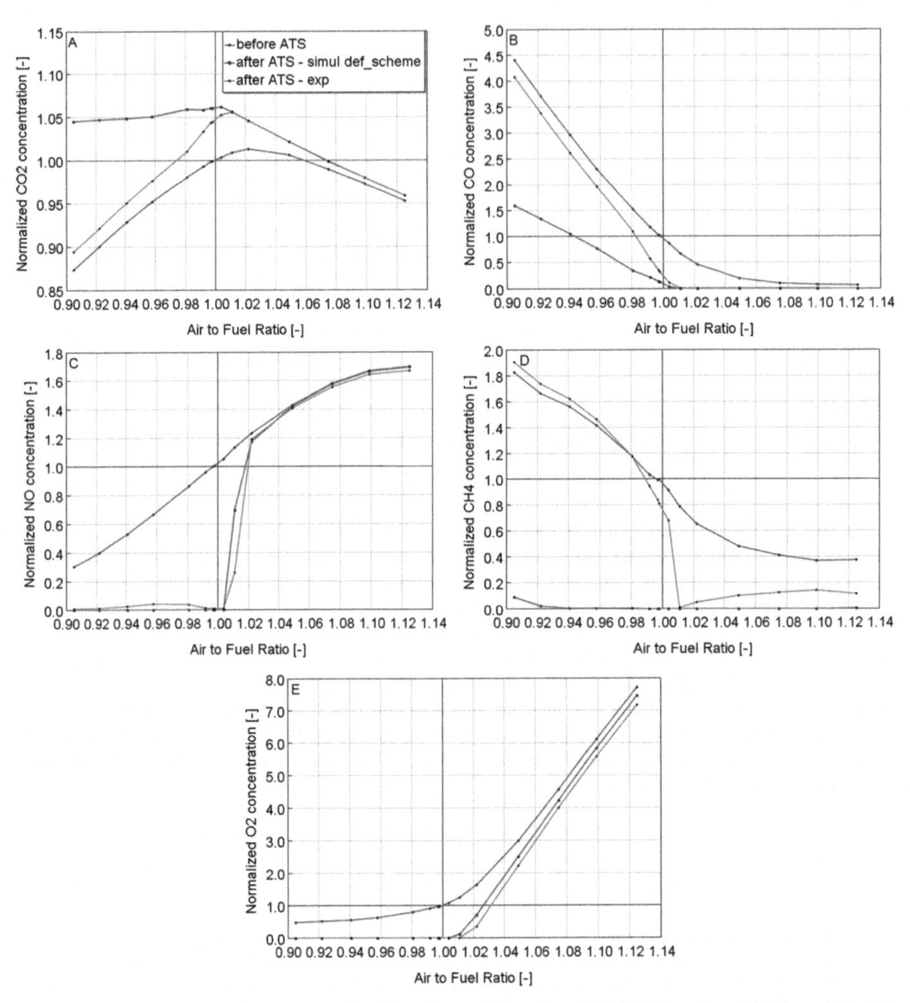

Figure 4. Main results at steady-state conditions at 100% load with the default GT-Suite reactions kinetic scheme for TWC. [17] gas temperature ≈ 760°C.

In order to properly calibrate the chemical kinetics of the main reactions occurring in the TWC, a step-by-step procedure, or also try-and-error approach, was adopted. To this aim, a sensitivity analysis was performed for the pre-exponential factor **A** and the activation energy $\mathbf{E_a}$, which characterize the rate expression of each reaction. The global reaction rate are generally of the form $\omega = k\prod_k X_k^{a_{kr}} \prod_m \Gamma_l \theta_m^{\beta_{mr}} / \prod_k F_j$, with the Arrhenius terms $k(kmol/m^3) = AT_s^{\beta}\exp[-E_a/RT_s]$. Calibration procedure began with Steady-State cases, monitoring the rate of each reaction. At this point, all the reactions were deactivated, thus starting from a tailpipe composition equal to the engine-out exhaust stream. Then an algorithm based on the order of activation of each reaction of the kinetic scheme was implemented. Activation order is dictated by the species that have higher concentrations in exhaust. The PGM chemical reaction calibration procedure can be summarized in the following steps:

1. Deactivation of all reactions.
2. Activation of reactions with CO and CO_2 in reagents and products.
3. Run simulation and reaction rates monitoring.
4. Further deactivation of all reactions unless the one with higher rate.
5. Parameterization of this reaction to get the result as close as possible to experiments.
6. Activation and calibration of each reaction based on the highest rate.
7. Repeat steps 2 to 6 for reactions with CH_4.
8. Check and recalibration of reactions with CO and CO_2
9. Repeat steps 2 to 6 for reactions with NO.
10. Check and recalibration of reactions with CO, CO_2 and CH_4.

The calibration of the kinetic scheme was carried out under the operating conditions 1900 RPM - 100% both in Steady-State and in transient; the reaction model was then verified also at partial loads in order to validate its consistency.

5.1 Oxygen storage submodel

The same calibration protocol was used for the reactions involving Cerium, which, as mentioned, is mainly responsible for the accumulation/release of oxygen (OSC) Cerium is normally present in high quantities in the Washcoat (around 30% by weight, i.e., 1000 g/ft^3 or 35.31 kg/m^3). Cerium stabilizes the washcoat layer, enhances precious metal activity and improves thermal resistance. Oxygen storage is due to the cerium's ability to form 3- and 4- valent oxides. [7] The following reaction represents cerium oxidation and the corresponding oxygen capture:

$$\frac{1}{2}O_2 + Ce_2O_3 \rightarrow 2CeO_2$$

This reaction stands for the storage of an oxygen atom by increasing the cerium oxidation state. Due to the significant presence of carbon oxides and hydrogen in the exhaust gas, additional important pathways are:

$$CO + 2CeO_2 \leftrightarrow CO_2 + Ce_2O_3$$

$$H_2 + 2CeO_2 \leftrightarrow H_2O + Ce_2O_3$$

As demonstrated in previous works [7,11], cerium also interacts with nitrogen oxides, according to the following reaction:

$$NO + Ce_2O_3 \rightarrow \frac{1}{2}N_2 + 2CeO_2$$

which has been added to the initial kinetic scheme to better characterize the dynamics of TWC during fast transient λ sweep. It also provides a mechanism for NO reduction under lean conditions, which can be particularly important under real world driving conditions. Finally, the importance of this pathway is confirmed by the typical delay in the NO release at the TWC outlet, also detected in the present experimental activity.

6 STEADY-STATE RESULTS

This section shows the main results obtained through the kinetic scheme calibration. The following graphs show the concentration of significant chemical species, normalized with respect to stoichiometric emission values (to keep OEM confidential data) and measured upstream and downstream the catalyst. The figures in the present section thus report CO_2, CO, CH_4 and NO concentrations, combined with the AFR calculated through the R49 formula, as reported before. Looking at the x-axis of Figures 5-6-7, it is possible to notice that the λ variation step is not uniformly spaced between lean and rich limits, but a higher number of data were collected close to the stoichiometric value, in order to better describe the variation of the TWC efficiency in this specific critical range. As previously seen in Figure 4, the same letters from A to E have been used in the following graphs in order to simplify the comparison between numerical and experimental data.

6.1 Carbon oxides

The main reactions involving CO conversion in the TWC are essentially the direct oxidation via oxygen and the Water Gas Shift (WGS) reported in the Appendix as reaction 1 and reaction 6:

1. $CO + \frac{1}{2}O_2 \rightarrow CO_2$

6. $CO + H_2O \rightarrow CO_2 + H_2$

The first reaction has a high rate in lean and stoichiometric conditions, when a considerable amount of oxygen is present in the exhaust gas, while it has a weak influence in extremely rich conditions. The higher H_2O concentration in the exhaust, compared to a traditional gasoline engine, makes WGS very influential among carbon oxides reactions. For this reason, one of the first calibration steps foresaw to properly balance these reactions, to obtain a good match in tailpipe CO/CO_2 concentration.

Figures 5A-B report the results for the full engine load, clearly indicating an adequate response of the model on carbon oxides conversion. It is possible to notice that CO concentration is slightly underestimated around the stoichiometric and overpredicted as the mixture tends to very rich values. Such behaviour, correspondingly affecting CO_2 conversion, was obtained in all the operating conditions, even at partial engine load. Nonetheless, it has to be specified that the discrepancies are contained, in all the cases, within a range of 0.2%, giving rise to a reasonably consistent agreement between numerical and experimental results.

6.2 Nitrogen oxides

Nitrogen oxides reduction occurs if adequate concentration values of CO and H_2 are achieved inside the catalyst. The global reactions describing this process in the TWC are:

10. $CO + NO \rightarrow CO_2 + \frac{1}{2}N_2$

15. $CO + 2NO \rightarrow CO_2 + N_2$

11. $H_2 + NO \rightarrow H_2O + \frac{1}{2}N_2$

13. $H_2 + 2NO \rightarrow H_2O + N_2O$

As displayed in Figures 5C, 6C and 7C, at all the engine load values, nitrogen oxides conversion efficiency quickly drops as soon as λ exceeds unity. As the mixture approaches even slightly lean conditions, TWC conversion efficiency is approximately zero and NO tailpipe concentration remains identical to engine-out values. The predicted values suitably reproduce the detected behaviour, with some minor inaccuracies at AFR values in the range between 1.01 and 1.04. Taking into account the extreme sensitivity of the converter with respect to these species concentration and the related experimental uncertainties, possible improvements to the model could be achieved collecting a higher number of experimental data within the indicated λ range.

Certainly, as mentioned, hydrogen participation in the reaction mechanism should be further investigated, because the assumption of Figure 3 may not be respected in all the temperature ranges, especially in specific operating conditions, like dynamic ones. In spite of these considerations, the kinetic scheme, firstly calibrated at full engine load conditions, generally provides a good numerical/experimental agreement even at partial loads.

Finally, around $\lambda \approx 0.94 - 0.96$ measured emissions show a small NOx spike of a few ppm: the numerical model was not capable of capturing such phenomenon, which should be better clarified through additional investigations.

6.3 ***Methane***

Methane conversion model involves the use of the following reactions:

5. $CH_4 + 2O_2 \rightarrow 3CO_2 + 2H_2O$

9. $CH_4 + H_2O \rightarrow CO + 3H_2$

The calibration of these reactions, always carried out at full load and with a continuous check on their influences on the previous ones, allowed to reach an adequate response of the model in the medium-high load conditions.

Looking at the measured trends in Figures 5D, 6D and 7D (red and green curves), it is possible to notice that, for all the engine load levels, methane was fully converted only at the stoichiometric condition. As known, the different state of the catalyst surface under lean and rich conditions affects methane conversion in the reactor. In the traditional TWC used for gasoline engines, methane is completely converted in lean conditions. In fact its diffusion is the limiting step and hydrogen species which form on the surface as a result of methane dissociation are oxidized to CO_2 and H_2O by residual oxygen [18]. On the contrary, for the present CNG engine, an efficiency loss in lean conditions was observable, more significant decreasing the engine load. One of the main reasons could be related to a different response of the Pd storage reactions on the catalyst surface. In order to better understand this phenomenon, dedicated tests could be useful with a linear increase in temperature and constant methane concentration, possibly with the use of a Synthetic Gas Bench (SGB). The model is not able to predict the loss of efficiency that occurs in extremely lean conditions, especially at partial loads where the contribution of water vapour is directly proportional to the temperature decrease. This effect could also be due to an excessive TWC aging, in addition to a high presence of CH_4 at the exhaust compared to gasoline engines [19].

Moving to tests at AFR lower than unity, the catalyst turned out to be completely unable to reduce methane concentration, which resulted equal to the inlet one. In these rich conditions, oxygen is the limiting reactant; its concentration is near to zero and the surface is covered by the partial oxidation species, CO and H in particular. The TWC model substantially reproduces this trend at all the engine load values, with a loss of accuracy at an engine load of 40%, as visible in Figure 7.

To sum up, a reasonable agreement between measured and calculated trends was reached, with some discrepancies around stoichiometric and lean AFR values at partial load (Figures 6 and 7), indicating the necessity to improve its accuracy in such operating conditions.

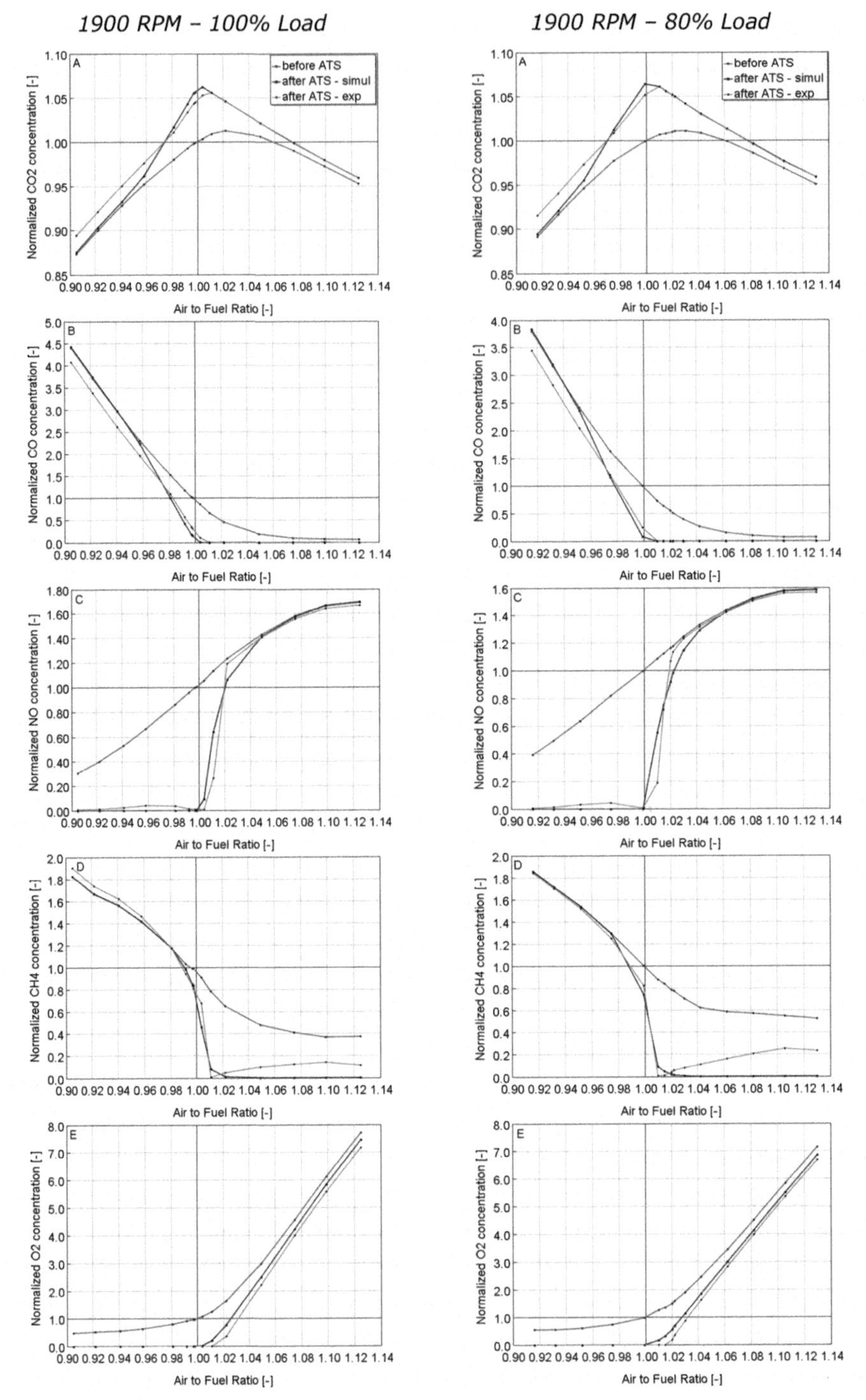

Figure 5. SS tests at 100% load.gas temp. ≈ 760°C.

Figure 6. SS tests at 80% load.gas temp. ≈ 750°C.

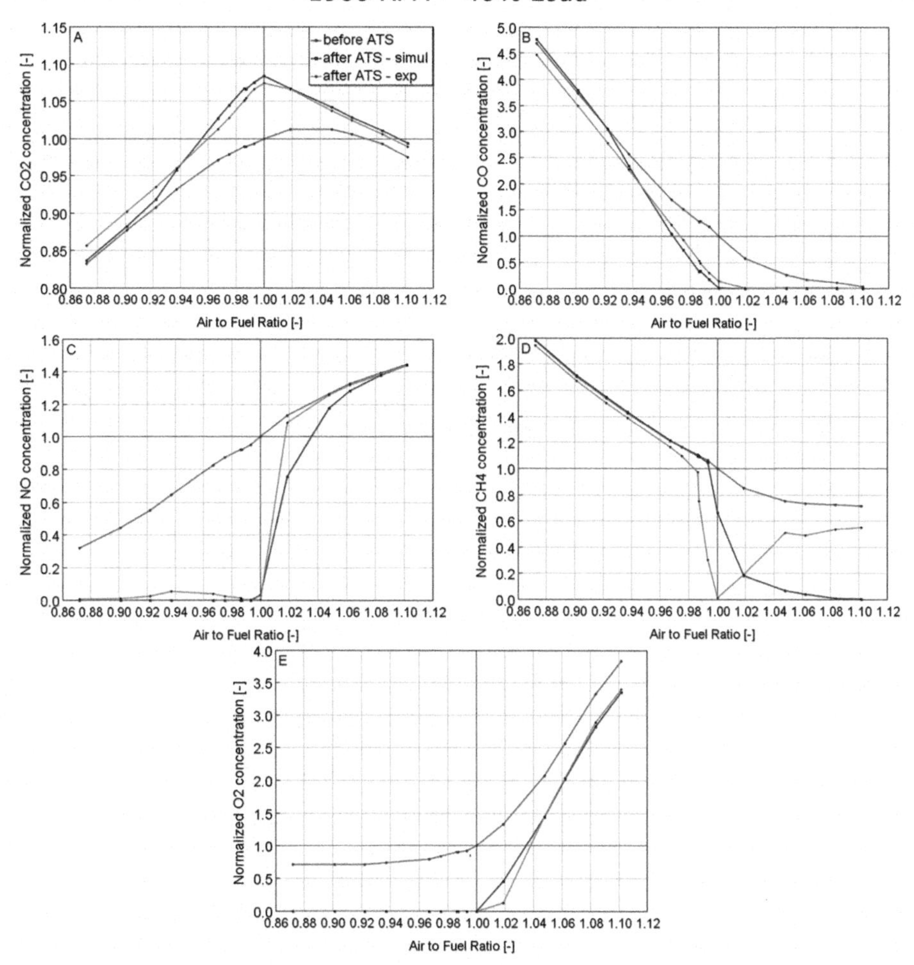

Figure 7. SS tests at 40% load.gas temp. ≈ 620°C.

7 TRANSIENT RESULTS

As mentioned, the oxygen storage phenomenon consists in the formation of different cerium oxides which affect local concentration values of the main pollutants and significantly modify local AFR values in lean to rich mixture transitions and vice versa. Conversion of each analyzed species takes place in variable times, depending on the concentration and temperature -as in steady-state cases, but also on the aforementioned cerium oxides, which react simultaneously with all the other species. It should be emphasized that also in this case, the value of AFR is not exactly what is designated by the ECU control. In fact, there are situations - further described in the next - in which the catalyst does not exhibit maximum conversion efficiency even though it nominally operates at the stoichiometric. It is important to recall that, given the extreme sensitivity of the catalyst to the inlet gas composition, the uncertainty in the knowledge of the actual istantaneous value of lambda represents an open critical issue. In fact, the measured lambda temporal evolution -to be used as reference to validate numerical results- can display a considerable variability, according to the adopted measurement technique. To illustrate this point, Figure 8 (top) reports the AFR profile obtained by means of the LinearLambdaNOx sensor, compared to the imposed target, upstream the TWC. It is possible to notice that the phasing of the measured profiles is correctly reproduced, but detected maximum and minimum lambda values display a discrepancy with respect to imposed target values (required to be equal to 1.10 and 0.9 respectively). Such discrepancy increases at decreasing the engine load levels. Figure 8 (bottom) also shows the comparison, for the full engine load case, of the lambda profile detected upstream and downstream the TWC. Thanks to its fast measurement dynamics, oxygen storage phenomena are captured by the LinearLambdaNOx probe, with storage and release phases determining the observed differences in the two lambda profiles. On the other hand, the reduced sensor accuracy at the leanest and richest phases shown in Figure does not permit to fully rely on these data for these transient conditions.

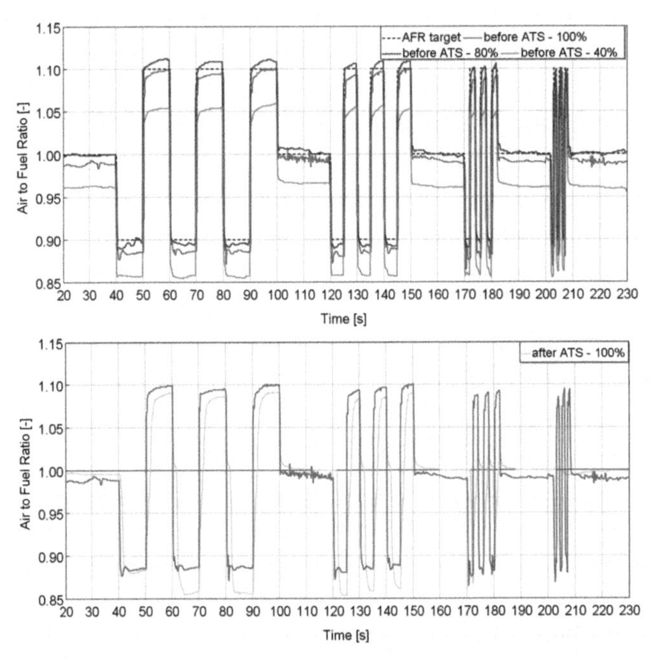

Figure 8. Lambda profiles measured by linearLambda NOx Sensor upstream the TWC, during dynamic sweeps at different engine load values (top). comparison between measurements upstream and downstream the TWC for the full load case (bottom).

To sum up, when looking at the results reported in the following paragraphs, it should be taken into account that the AFR dynamics at the catalyst inlet is known within the limits of the discussed uncertainties, with the subsequent impact on the initial conditions for the simulations. The following graphs are made with the same logic used in the previous chapter and the reference value used to normalize the emissions is the same of the respective case at equal torque of the steady-state conditions. As for the steady-state conditions, the submodel- of the reactions describing the oxygen storage was calibrated at high load and additional tests were then performed at lower load. Figure 9 reports CO_2, CO, NOx and CH_4 concentration histories measured upstream and downstream the TWC, compared with the corresponding calculated profiles at full engine load. Subsequent Figures 10 and 11 display the same curves measured and predicted for engine load values equal to 80% and 40% respectively.

7.1 Carbon oxides

Generally speaking, as regards the measured profiles, it is worth specifying that the first AFR step of each sequence of 3 steps (rich-lean) should be considered as TWC pre-conditioning cycles and not considered for results repeatability. However in the subsequent two steps the conversion of the species can be sufficiently characterized. After the lean to rich transitions, there are different phases that characterize the oxidation of carbon monoxide. In the 10s steps there is a first part, lasting 2-3 seconds for full load conditions and 4-5 seconds for medium-low load conditions, where the TWC converts all CO into CO_2, with a consequent concentration peak clearly visible in Figures 9A, 10A, 11A. After this initial phase, the conversion efficiency of the TWC is approximately 50%.

The main reason for this change in CO oxidation is due to the fact that, after a lean mixture treatment, the TWC is crossed by a higher concentration of O_2 that is stored in the catalyst in the form of CeO_2. Thus, given the abundant availability of this species, the CO oxidation via reaction 19 prevails with respect to reaction 20 for a certain time interval, which can be suitably calibrated by acting on the parameters that involve these reactions. At high load, as shown in Figures 9 and 10 at 100% and 80% torque levels, the model is able to predict very accurately the dynamics involving the conversion of CO and CO_2 (with the exception of the first step).

It is interesting to note that there is a sort of adjustment on this efficiency value in the 10-5 s amplitude transitions, demonstrating the achieved equilibrium conditions of the system after the first phase with maximum conversion efficiency. In the 2s transitions, instead, the timing is not sufficient to ensure that the system returns to balance. Thus the change in behavior of the catalyst and, therefore, the change in the rate of reactions involving carbon oxides is interrupted, with a conversion efficiency of approximately 80%. Experimentally, the transitions of 1Hz are too fast to verify the influence of the AFR sweeps on the high conversion efficiency due to dilution effects along the analyzer's measurement line, which are gradually greater as the load decreases. Finally, it is interesting to notice that the model is also able to capture the variation of the CO profile slope when the engine load is reduced, reproducing the steep transitions of full load and the slower dynamics at partial load.

7.2 Nitrogen oxides

Looking at Figures 9C, 10C and 11C, as described for carbon oxides, after the lean to rich transition there is a first phase in which the conversion efficiency is maximum, while in a subsequent phase there is no impact in the reduction of nitrogen oxides. As said, with respect to the starting model included in the simulation software, the addition of reaction 22 allows a more immediate calibration, since it permits to act directly on the NO species, which would otherwise be managed exclusively by an indirect calibration on carbon oxides reactions.

Generally speaking, the obtained model accurately reproduces the conversion within the catalyst during these transient phases at all the considered engine load levels. Some minor discrepancies could be observed, similarly to what occurred in the steady-state conditions (see Figures 5C, 6C, 7C), around the stoichiometric λ value, giving rise, as an example, to a slightly slower NO decrement between 100-120s.

7.3 *Methane*

As already discussed, in all the considered steady-state conditions, methane was fully converted only at the stoichiometric condition, while an efficiency loss was observed in lean cases and a no conversion occurred in rich ones. On the contrary, as visible in Figure 9, during the dynamic λ transitions, CH_4 tailpipe concentration displayed a different behaviour, resulting substantially removed regardless of all the inlet spikes. The only exception in Figure 9 is represented by the first spike, but, as already mentioned, the first step could be neglected as TWC preconditioning. A similar response was obtained in the experiments at lower engine loads reported in Figures 10 and 11, where methane profiles at the catalyst outlet appear nearly unaffected by the inlet concentration dynamics. In order to try to capture the observed TWC behaviour, the reaction mechanism was modified, including some steps involving Palladium. In fact, it is known that methane can interact with the oxides of noble metals, such as Palladium [18]. For this reason, the following two reactions have been added to the kinetic scheme, identified in the Appendix with the numbers 23 and 24.

23. $\boldsymbol{Pd + O_2 \rightarrow PdO_2}$

24. $\boldsymbol{PdO_2 + CO \rightarrow Pd + 2CO_2}$

The link between methane and palladium oxide was created modifying the rate expression of the methane oxidation reaction (reaction 5 in the App.), introducing an inhibition term Ω, defined as:

$$\Omega = \frac{mol\ of\ PdO_2}{2 \cdot mol\ of\ Pd + mol\ of\ PdO_2}$$

The kinetic parameters of the added/modified reactions were only calibrated at full engine load, then applied to the other test cases. As visible in Figure 9D, thanks to this modification, the model could reproduce a consistent methane conversion during the AFR sweep, especially in rich phases, despite not completely abated as shown in the measured profile. At lower engine load values (Figures 10D and 11D), the agreement between experimental and predicted methane histories is less satisfying, but it is likely that a further calibration work, possibly combined with additional experiments, could significantly improve these results.

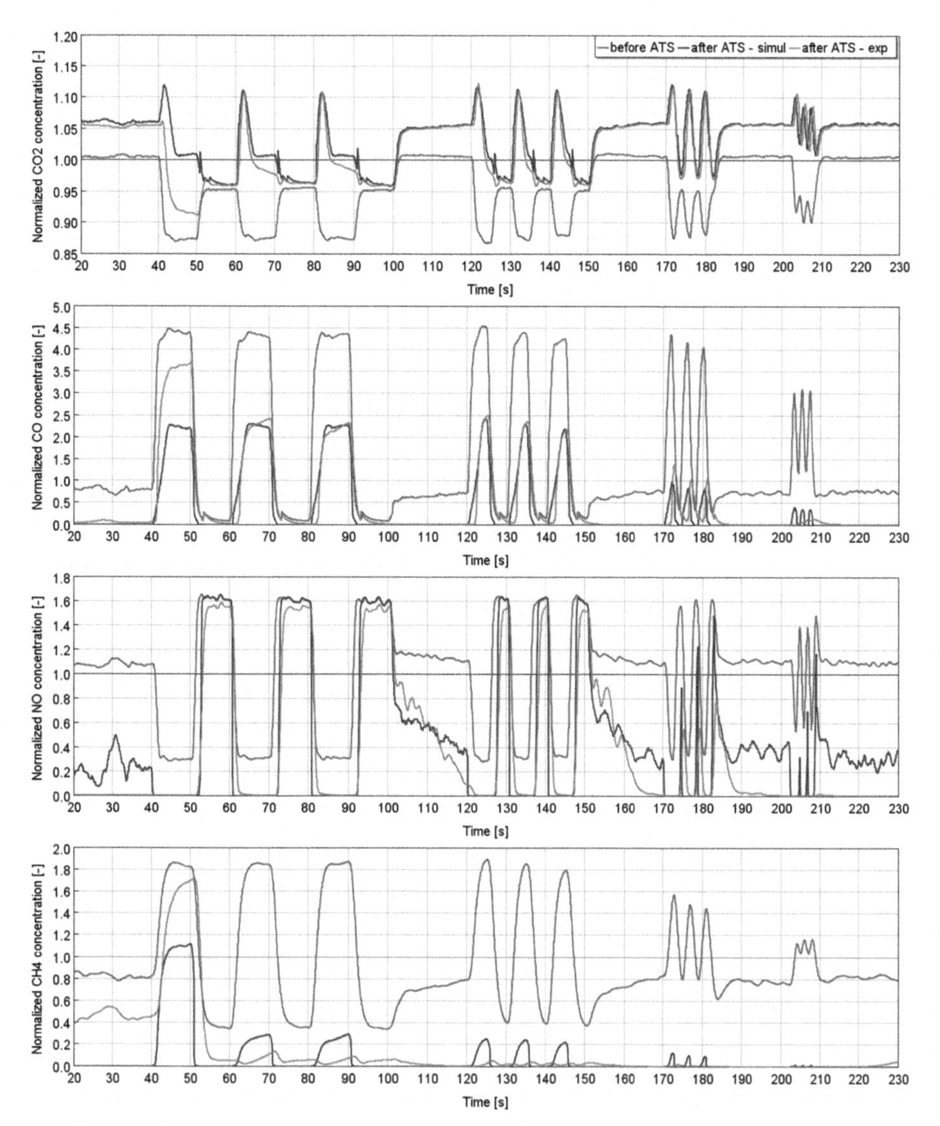

Figure 9. Transient conditions at 100% load. gas temperature ≈ 750°C A) Num/Exp. CO_2 engine out and tailpipe concentration. B) Num/Exp. CO engine out and tailpipe concentration. C) Num/Exp.NO engine out and tailpipe concentration D) Num/Exp.CH_4 engine out and tailpipe concentration.

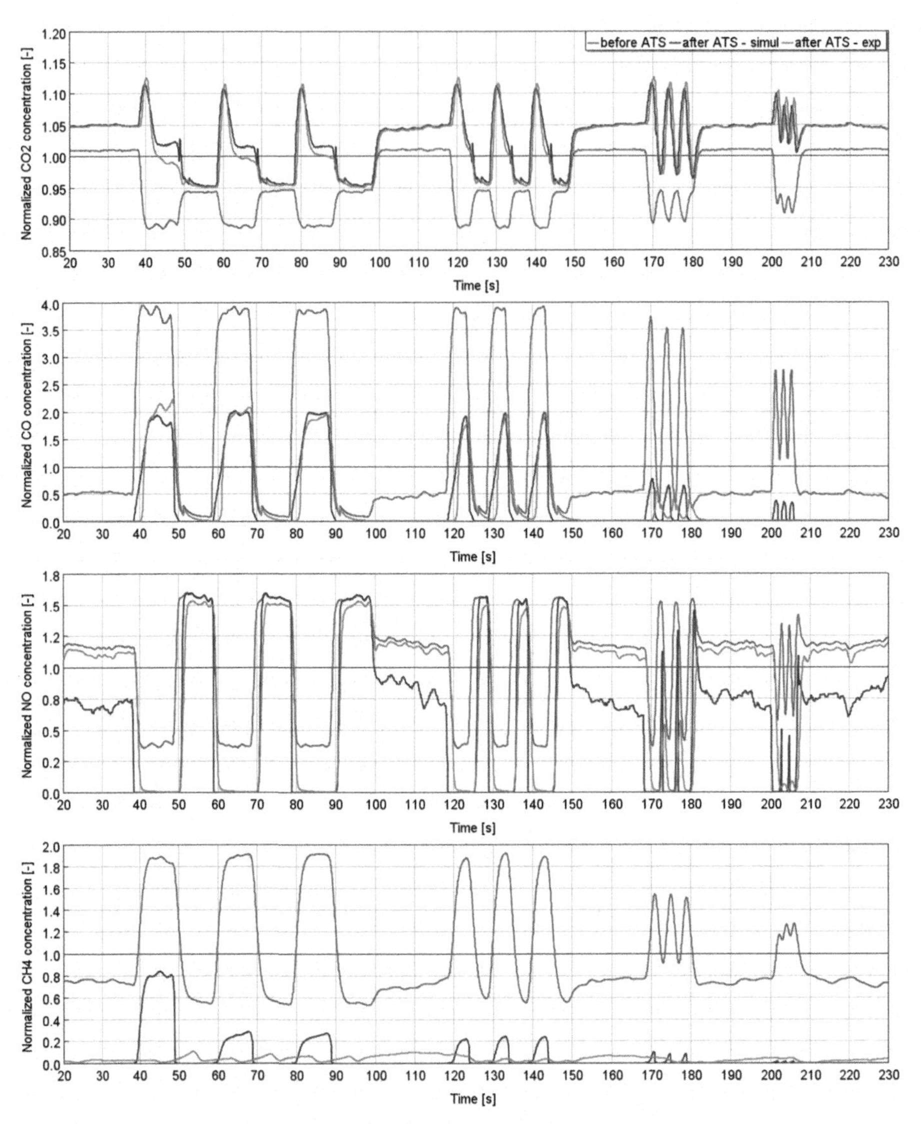

Figure 10. Transient conditions at 80% load. gas temperature ≈ 740°C A) Num/Exp. CO_2 engine out and tailpipe concentration. B) Num/Exp. CO engine out and tailpipe concentration. C) Num/Exp.NO engine out and tailpipe concentration D) Num/Exp.CH_4 engine out and tailpipe concentration.

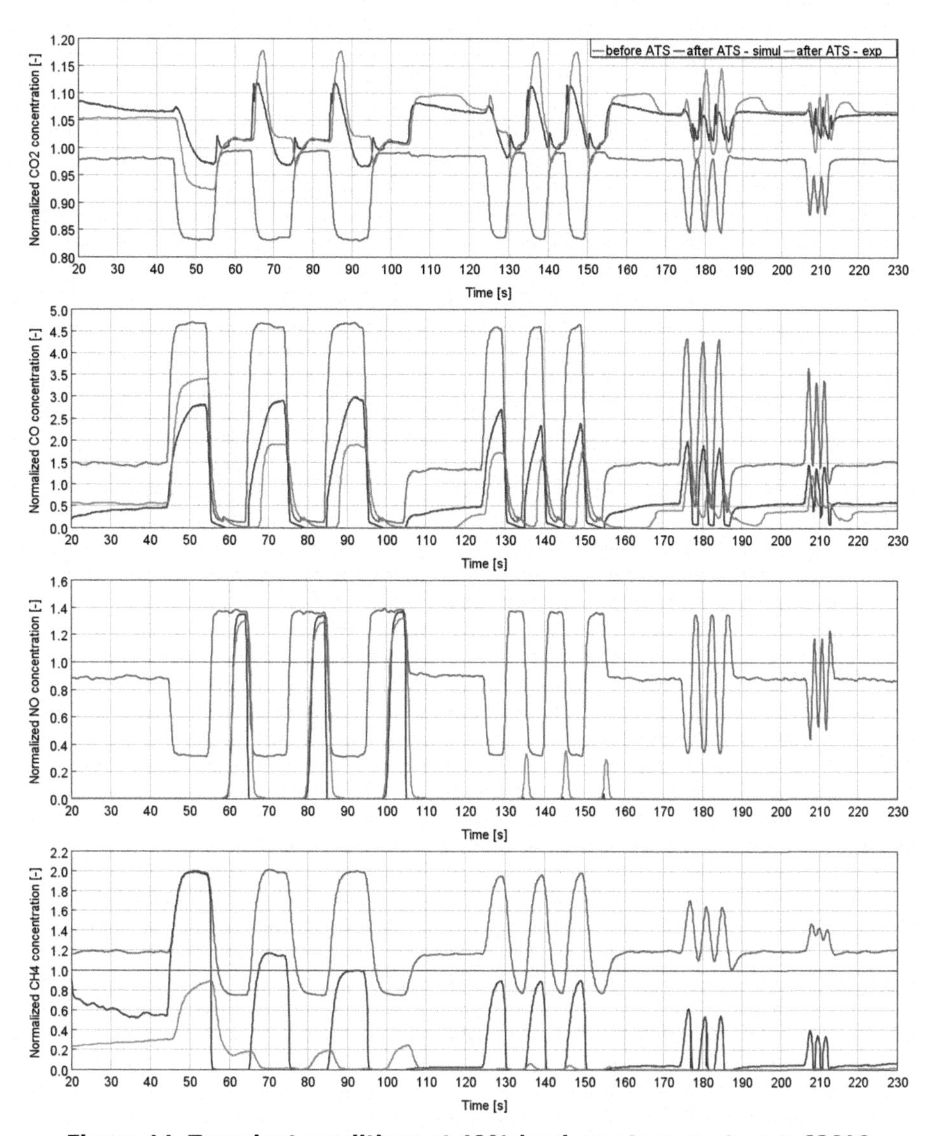

Figure 11. Transient conditions at 40% load. gas temperature ≈ 620°C A) Num/Exp. CO_2 engine out and tailpipe concentration. B) Num/Exp. CO engine out and tailpipe concentration. C) Num/Exp.NO engine out and tailpipe concentration D) Num/Exp.CH_4 engine out and tailpipe concentration.

8 CONCLUSIONS

A predictive model of catalyst behavior with oxygen storage and release of a heavy-duty engine powered with NG was developed. The need to re-adapt the existing reaction scheme for TWC of gasoline powered engines in order to make them suitable for NG has been demonstrated. A protocol for the calibration of the main reactions was identified through an iterative try-and-error approach. The calibration of the reactions kinetic scheme was specifically carried out starting from the emissions at full engine load during a wide AFR sweep, both in Steady-State and transient conditions. It represents a good starting point and a first goal in the use of a simplified scheme to manage the complex phenomena occurring in a TWC for the aftertreatment of emissions by NG engines. Indeed, the response of the model in terms of emissions is adequate in high load conditions and is still acceptable under medium-low load conditions, considering also the higher measurement uncertainties from the analyzer devices, present in these circumstances. CH_4 oxidation represents the major open point of the current scheme in lean conditions and during the dynamic lambda scan, characterized by different phenomena with respect to similar stationary conditions. Generally speaking, a reasonable predictivity of the model is obtained, resulting in a sufficiently adequate representation of the catalyst reactivity in dynamic conditions. Cold start phases before light-off temperature and the analysis in driving cycles representative of the real transient ATS working conditions, such as the WHTC, need further investigation, as well as model validation with different NG engines.

Definitions/Abbreviations

AFR	Air to Fuel Ratio
CI	Compression Ignition
CLD	Chemi-luminescence detector
ECU	Engine Control Unit
FID	Flame Ionization Detector
IRD	Infra-Red Detector
NG	Natural Gas
NMHC	Non-Methane Hydrocarbons
OEM	Original Equipment Manufacturer
OSCf	Oxygen Storage Capacity
PFI	Port Fuel Injection
PMD	Paramagnetic Detector
SI	Spark Ignition
SS	Steady State
SGB	Synthetic Gas Bench
SNS	Smart NOx Sensor
THC	Total hydrocarbons
TWC	Three-Way Catalyst
WGS	Water-Gas Shift
ZFAS-U	Universal Lambda Sensor

Appendix

Chemical reactions comprised in the final mechanism:

1. $CO + \frac{1}{2}O_2 \rightarrow CO_2$
2. $H_2 + \frac{1}{2}O_2 \rightarrow H_2O$
3. $C_3H_8 + 5O_2 \rightarrow 3CO_2 + 4H_2O$
4. $C_3H_6 + \frac{9}{2}O_2 \rightarrow 3CO_2 + 3H_2O$
5. $CH_4 + 2O_2 \rightarrow 3CO_2 + 2H_2O$
6. $CO + H_2O \rightarrow CO_2 + H_2$
7. $C_3H_8 + 3H_2O \rightarrow 3CO + 7H_2$
8. $\mathbf{C_3H_6 + 3H_2O \rightarrow 3CO + 6H_2}$
9. $\mathbf{CH_4 + H_2O \rightarrow CO + 3H_2}$
10. $\mathbf{CO + NO \rightarrow CO_2 + \frac{1}{2}N_2}$
11. $\mathbf{H_2 + NO \rightarrow H_2O + \frac{1}{2}N_2}$
12. $\mathbf{C_3H_6 + 9NO \rightarrow 3H_2O + 3CO_2 + \frac{9}{2}N_2}$
13. $\mathbf{H_2 + 2NO \rightarrow H_2O + N_2O}$
14. $\mathbf{N_2O + H_2 \rightarrow H_2O + N_2}$
15. $\mathbf{CO + 2NO \rightarrow CO_2 + N_2O}$
16. $\mathbf{N_2O + CO \rightarrow CO_2 + N_2}$
17. $\mathbf{H_2 + 2CeO_2 \rightarrow H_2O + Ce_2O_3}$
18. $\mathbf{H_2O + Ce_2O_3 \rightarrow H_2 + 2CeO_2}$
19. $\mathbf{CO + 2CeO_2 \rightarrow CO_2 + Ce_2O_3}$
20. $\mathbf{CO_2 + Ce_2O_3 \rightarrow CO + 2CeO_2}$
21. $\mathbf{\frac{1}{2}O_2 + Ce_2O_3 \rightarrow 2CeO_2}$
22. $\mathbf{NO + Ce_2O_3 \rightarrow \frac{1}{2}N_2 + 2CeO_2}$
23. $\mathbf{Pd + O_2 \rightarrow PdO_2}$
24. $\mathbf{PdO_2 + CO \rightarrow Pd + 2CO_2}$

REFERENCES

[1] Thiruvengadam, A., Besch, M., Padmanaban, V., Pradhan, S. et al., "Natural gas vehicles in heavy-duty transportation-A review," Energy Policy, 122:253-259, 2018, doi.org/10.1016/j.enpol.2018.07.052.

[2] Turrio-Baldassarri, L., Battistelli, C. L., Conti, L., Crebelli, R.,et al. "Evaluation of emission toxicity of urban bus engines: Compressed natural gas and comparison with liquid fuels," Science of the Total Environment, 355(1–3):64-77, 2006, doi.org/10.1016/j.scitotenv.2005.02.037.

[3] Wahbi, A., Tsolakis, A., and Herreros, J. "Emissions Control Technologies for Natural Gas Engines". Natural Gas Engines, 359–379, 2018, doi:10.1007/978-981-13-3307-1_13.

[4] Raj, B. A. "A review of mobile and stationary source emissions abatement technologies for natural gas engines," Johnson Matthey Tech, 60:228-235, 2016, doi:10.1595/205651316x692554.

[5] Herz, R. K.; Klela, J. B.; Sell, J. A. "Dynamic Behavior of Automotive Catalysts. 2. Carbon Monoxide Conversion under Transient Air/Fuel Ratio Conditions," Ind. Eng. Chem. Prod. Res. Dev., 22:387-396, 1983 doi: 10.1021/i300011a002.

[6] Koltsakis, G., Kandylas, I. and Stamatelos, T. "Three-Way Catalytic Converter Modeling and Applications," Chemical Engineering Communications. 164:153-189, 1998, doi:10.1080/00986449808912363.

[7] Tsinoglou, D. N., Koltsakis, G. C., and Peyton Jones, J. C., "Oxygen storage modeling in three-way catalytic converters," Industrial & engineering chemistry research, 41(5):1152–1165,2002, doi:10.1021/ie010576c.

[8] Montenegro, G. and Onorati, A., "1D Thermo-Fluid Dynamic Modeling of Reacting Flows inside Three-Way Catalytic Converters," SAE Int. J. Engines 2 (1):1444–1459, 2009, doi:10.4271/2009-01-1510.

[9] Oh, S. and Cavendish, J., "Transients of Monolithic Catalytic Converters: Response to Step Changes in Feedstream Temperature as Related to Controlling Automobile Emissions," Ind. Eng. Chem. Prod. Res. DeV. 21:29–37, 1982, doi:10.1021/i300005a006.
[10] Koltsakis, G. C., Konstantinidis, P. A., and Stamatelos, A. M. "Development and application range of mathematical models for 3-way catalytic converters," Applied Catalysis B: Environmental, 12(2–3):161-191,1997, doi:10.1016/S0926-3373(96)00073-2.
[11] Tsinoglou, D. N., and Weilenmann, M. "A simplified three-way catalyst model for transient hot-mode driving cycles," Industrial & Engineering Chemistry Research, 48(4):1772–1785,2009, doi: 10.1021/ie8010325.
[12] Zeng, F., and Hohn, K. L., "Modeling of three-way catalytic converter performance with exhaust mixture from natural gas-fueled engines," Applied Catalysis B: Environmental, 182:570-579,2016, doi: 10.1016/j.apcatb.2015.10.004.
[13] Tsinoglou, D. N., Eggenschwiler, P. D., Thurnheer, T. and Hofer, P. "A simplified model for natural-gas vehicle catalysts with honeycomb and foam substrates," Proceedings of the Institution of Mechanical Engineers, Part D: Journal of Automobile Engineering, 223(6), 819–834, 2009. doi:10.1243/09544070JAUTO1095.
[14] Regulation 49, "Uniform provisions concerning the measures to be taken against the emission of gaseous and particulate pollutants from compression-ignition engines for use in vehicles, and the emission of gaseous pollutants from positive-ignition engines fuelled with natural gas or liquefied petroleum gas for use in vehicles", E/ECE/324/Rev.1/Add.48/Rev.5.
[15] Ferri, D., Elsener, M., and Kröcher, O., "Methane oxidation over a honeycomb Pd-only three-way catalyst under static and periodic operation," Applied Catalysis B: Environmental, 220:67-77,2018, doi: 10.1016/j.apcatb.2017.07.070.
[16] Heywood, John B. "Internal combustion engine fundamentals." (1988).
[17] Holder, R., Bollig, M., Anderson, D. R. and Hochmuth, J. K., "A discussion on transport phenomena and three-way kinetics of monolithic converters," Chemical Engineering Science. 61:8010-8027, 2006. doi:10.1016/j.ces.2006.09.030.
[18] Lyubovsky, M., Smith, L. L., Castaldi, M., Karim, H., et al. "Catalytic combustion over platinum group catalysts: fuel-lean versus fuel-rich operation." Catalysis Today, 83(1–4):71-84,2003, doi: 10.1016/S0920-5861(03)00217–7.
[19] Martı´n, L., Arranz, J. L., Prieto, O., Trujillano, R.,et al. "Simulation three-way catalyst ageing: analysis of two conventional catalyst," Appl. Catalysis B, 44 (1):41–52,2003, doi:10.1016/S0926-3373(03)00008-0.

Internal Combustion Engines and Powertrain Systems for Future Transport 2019 – Institute of Mechanical Engineers, ISBN 978-0-367-90356-5

Use of cryogenic fluids for zero toxic emission hybrid engines

M. Jaya Vignesh[1], S. Harvey[1,2], A. Atkins[2], P. Atkins[1], G. De Sercey[1], M. Heikal, R. Morgan[1], K. Vogiatzaki[1]

[1]Centre for Automotive Engineering, University of Brighton, UK
[2]Ricardo Innovations, Shoreham Technical Centre, UK

ABSTRACT

In this paper we present the basic concepts of operations of a hybrid liquid nitrogen and internal combustion engine currently under development by Ricardo Innovations, Dolphin N2 and the University of Brighton, the CryoPower recuperated split cycle engine (RSCE). The engine is based on a new split-cycle combustion concept utilising isothermal compression via cryogenic injection to maximise the efficiency of the engine, while also providing near zero polluting emissions from the dilution effects. Combined experimental and numerical findings will be presented and the effect of evaporation dynamics of the LN_2 are explored. This study aims to improve the understanding of the spray process evolution in order to achieve optimal isothermal compression.

1 INTRODUCTION

Cryogenic liquids are gases which have been converted to liquids by cooling them to very low temperatures. It is generally agreed that a liquid is cryogenic if it exists in its liquid state at temperatures below 122K. Common cryogenic liquids which have found applications in several technologies are Liquid Helium (LHe), Liquid Oxygen (LOX), Liquid Hydrogen (LH_2) and Liquid Nitrogen (LN_2). LHe is used to achieve very low temperatures and is used in Magnetic Resonance Imaging (MRI) scanners and other superconductor applications. LO_X and LH_2 are used as oxidizer and fuel combinations respectively in cryogenic rocket engines. LN_2 is currently mostly used for refrigeration and cryo-preservation due to its cheap and abundant availability.

Recently there has been an increase in interest for research relevant to the use of cryogenic fluids in transportation applications. Examples of such systems are the Dearman engine, which is a zero-emission engine featuring no combustion, and the CryoPower RSCE, a hybrid high efficiency engine aimed at heavy duty applications. If the production of the cryogenic fluid is obtained from renewable energy resources, or wasted heat such as wasted cold during regasification of LNG, these engines have the potential to be game changing energy systems in terms of environmental impact. Nitrogen is already a by-product of industrial oxygen production which can be readily re-liquefied for use in transport. Many industrial cryogenic production facilities are operated intermittently using off peak (currently overnight) energy. In the future, surplus renewable energy could be used for cryogenic fluid production, offering a way of absorbing and effectively storing renewable energy. There is already an extensive distribution infrastructure for liquid nitrogen fluids (unlike hydrogen) and refuelling technologies developed for CNG can be readily utilised.

The Dearman engine works on the principle that cryogenic liquids expand around 700 times in volume when they undergo the transition from liquid to an atmospheric gaseous state. This expansion is used to drive a reciprocating engine, relying on the increase in pressure to drive a piston to create torque in a same way as an Internal Combustion Engine (ICE), but without emitting polluting gases since no chemical reaction takes place inside the engine. On the other hand, the CryoPower engine is a split cycle engine where compression and expansion (though combustion) happen in separate chambers. This allows the compression and expansion cylinder processes to be individually optimised, which is not possible in conventional ICE architectures. In the RSCE a further increase in efficiency is achieved by injecting small amounts of coolant liquid during the compression process, in order to maintain isothermal conditions. This significantly reduces the compression work and enables the recovery of waste heat from the engine exhaust heat to the working fluid before combustion. Several variations of the RSCE are underdevelopment, Dry ThermoPower, Wet ThermoPower and CryoPower. The Dry ThermoPower RSCE does not have any coolant injection during the compression process while wet ThermoPower uses water. The CryoPower RSCE represents the largest technological challenge due to the unique process of using LN_2 injection as a coolant during compression. However, this concept shows the most promise with research to date suggesting that the CryoPower concept has the potential to achieve up to and possibly over 60% efficiency and a significant reduction in emissions.

In the following sections, the basic principles of the operation of the RSCE will initially be presented, which is currently under development by the University of Brighton (UoB) in collaboration with Dolphin N2 and Ricardo. The focus will then be turned to the examination of the thermodynamic processes taking place in the compression chamber. After which experimental evidences for the dynamics of cryogenic liquids based on extensive review of existent experiments as well as new ones currently under development at the UoB will be presented. Next a comprehensive analysis of cryogens from thermodynamic point of view will be performed along with an investigation of the challenges of moving from water to LN_2 as coolant. Finally, the complications in numerical modelling of cryogenic liquid properties are presented.

2 RECUPERATED SPLIT CYCLE ENGINE

The idea of a split cycle engine was first proposed by George Brayton in 1876 but the first commercial split cycle engine was the Dolphin engine by Sir Harry Ricardo in 1908. Interest has renewed in split cycle concepts over the last couple of decades with the pressing need to primarily (a) increase engine efficiency (b) decrease CO_2 emissions and (b) reduce toxic emissions. Figure 1 presents a schematic of a RSCE. It can be seen that the compression chamber ingests ambient air from atmosphere, which is then compressed and simultaneously cooled by the induction of a coolant such as water or liquid nitrogen. This coolant should maintain a near constant temperature during compression, towards isothermal compression conditions. This cool compressed air is then passed through a recuperator which absorbs some of the otherwise wasted heat from the exhaust. The preheated air is then passed into the combustion chamber where combustion with fuel takes place. Some of the wasted heat from exhaust is re-utilised in the recuperator to heat the cool compressed air.

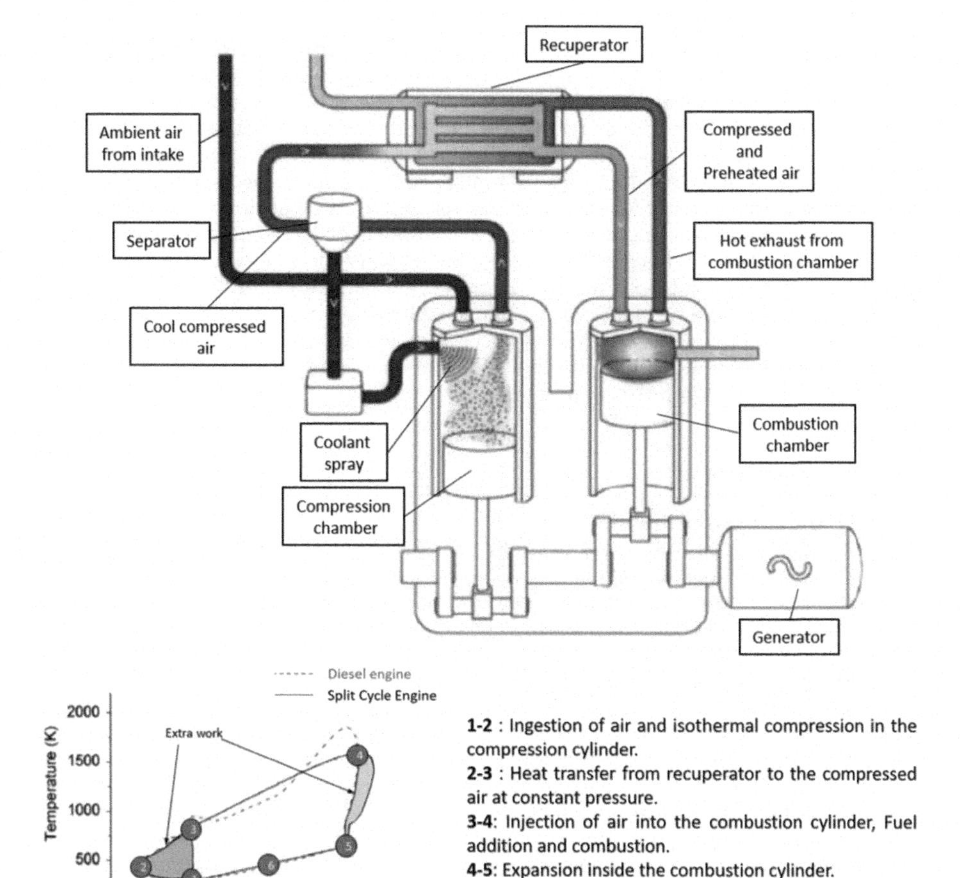

Figure 1. Schematic of a Split Cycle Engine by Ricardo[6] with T-S diagram of a RCSE compared to a diesel engine by Morgan[1]. The numbers referring the positions in the cycle are described on the right.

One important characteristic of this engine is that although 55 % efficiency is the max upper limit for conventional diesel engines with large 2 stroke low speed engines approaching this, the efficiency of RCSE at its limit is calculated to be over 55 % and up to 60% [1]. The factors upon which primarily the higher efficiency is dependent are, optimisation of compression and expansion cylinder processes, achieving near isothermal compression and the recuperation efficiency of the wasted heat [2]. In order to explain this further, the thermodynamic cycle of a RSCE compared to a Diesel engine by Morgan et al [1] is shown in Figure 1 (graph on the bottom). This clearly shows the additional work extracted by the cycle and thereby increasing the efficiency, due to the above mentioned factors.

Quasi-isothermal compression alone can achieve 5% thermal efficiency in the RSCE as calculated by Dong et al [3]. Maintaining quasi-isothermal compression is linked to the way the coolant will atomise. The rate at which the air will heat up because of the compression should be balanced by the rate by which the coolant is capable of absorbing heat and vaporises. Past experiments used water as a coolant and have achieved predicted overall efficiencies of 60% [4] in 2004, but due to the technological constraints of that time, stable, efficient combustion could not be achieved in the combustion chamber. Nevertheless, the isothermal compression concept and overall cycle were demonstrated [5]. Recent R&D developments were focused on using liquid nitrogen as the coolant for the isothermal compression. Unlike in the previous work by Coney et al [4], stable efficient combustion was achieved at remarkably low levels of NOx emissions. This was attributed to a combination of improved mixing, low temperature retarded combustion phasing and dilution of the charge air with nitrogen from the isothermal compression process.

3 EXPERIMENTS

3.1 Previous experiments

If cryogenic fluids are to be used extensively in future energy and transport applications, their thermofluids behaviour must be understood. One of the challenges is that the fundamental properties and dynamics of cryogenic fluids have not been extensively studied at the conditions present in an engine or compressor. After performing a critical review on the existent available data, Figure 2 has been produced and shows an overview of the existing experimental data of cryogenic liquids compared to the operational regime of the compression chamber of a current RSCE prototype on a phase diagram chart of nitrogen. For reference, a split cycle engine is expected to operate at a maximum of 17 MPa compression chamber pressure [4]. The current experimental work on the CryoPower engine is expected to operate at 7 MPa. Two of the important conclusions of the figure are: a) Existent experiments are not representative of the operating conditions of the RSCE and b) The maximum pressure in the compression chamber of all the demonstrator prototypes are above nitrogen's critical pressure of 3.39 MPa (see also Table 1 and Figure 3), which means the cryogenic fluid will transition from sub to super-critical conditions in the compression cylinder. The behaviour even of common fluids, let alone cryogens, is not well understood under these conditions.

Experiments done by Mayer [7]–[9], Chehroudi [10] and Oschwald [11] provide valuable information in the quest to understand cryogenic jets in their supercritical state. The observation from these experiments can be summarised as:

1. Starting from a sub-critical pressure, with increase in pressure the cryogenic fluids reach a trans-critical state where surface tension reduces considerably, and the fluids lose their capability to produce observable droplets. This is not specific to cryogenic fluids, but it also happens in any fluid that reaches these conditions. What is particular for cryogenics is that the super-critical state is reached at conditions much closer to atmospheric than other fluids/fuels.
2. As a result of the reduced surface tension, with increase in super-critical pressure the mixing of cryogenic fluid with ambient fluid becomes more rapid.
3. The pressure at which the transition of the fluid properties from liquid to gas happens, does not depend only on the critical pressure of the injected fluid but on the ambient fluid as well.

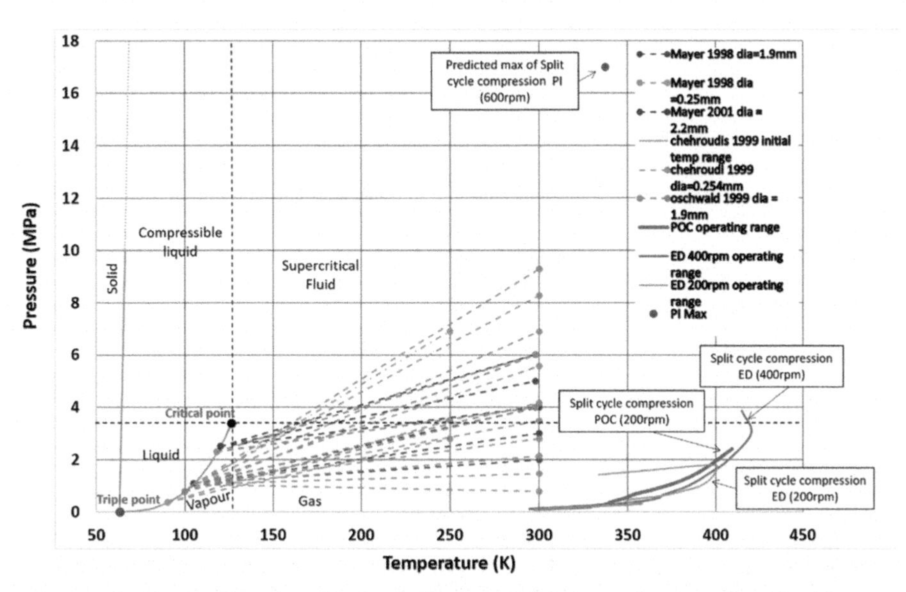

Figure 2. Phase diagram chart of nitrogen with experiments done by Mayer, Chehroudi and Oschwald. The operating conditions range of a split cycle compression chamber is also displayed for comparison. The initial and final conditions of the cryogenic experiments are represented by solid line and solid dots. The line linking initial conditions to final conditions is displayed as dashed line. 'dia' represents injector diameter. POC represents Proof of Concept and ED represents Engineering Demonstrator split cycle engine prototypes.

Table 1. Operating conditions of compression chamber in various split cycle prototypes.

Split cycle prototype	Compression chamber				Coolant	Efficiency	
	Cylinder (No x Bore x Stroke) (mm)	Max Pressure (MPa)	Max Temp (K)	Crank-shaft Speed (rpm)		Work Saving	Total efficiency
Proof of Concept (POC)	1 x 300 x 200	3.25	413	50 - 200	Water	28%	
Engineering Demonstrator (ED)	1 x 385 x 400	10	425	200 - 600	Water		50%
Commercial Demonstrator (CD)	2 x 385 x 450	17		600	Water		58 - 60%
CryoPower	PLANNED	7	PLANNED		LN2		>55%

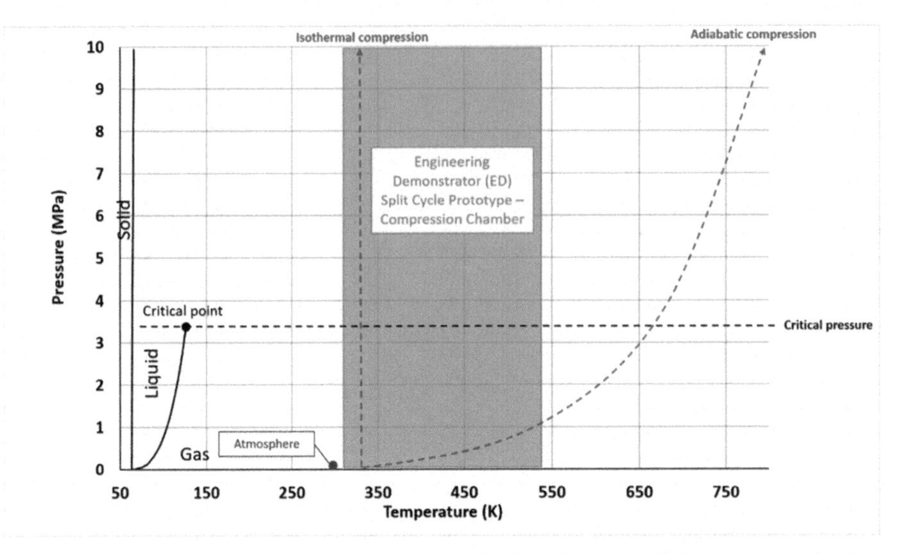

Figure 3. Phase diagram of nitrogen with chamber conditions in a typical isothermal compressor highlighting the critical point.

3.2 Current experiments at the University of Brighton

Due to the limited experiments available under the conditions relevant to the RSCE, an experimental effort has started at the UoB in the last couple of years in order to enhance existent data bases. Two series of experiments have been undertaken to study the behaviour of a liquid nitrogen spray. The first experiments were performed under steady state conditions through a plane orifice into ambient air. The second series of experiments used a gasoline injector modified to operate on liquid nitrogen. In both cases, the liquid nitrogen was supplied from a 120 litre tank that could be pressurised up to 18 bar. The nozzle and injector were submerged in a bath of liquid nitrogen to sub-cooled the assembly and prevent the cryogen boiling in the injector. With reference to Figure 4, at a macroscopic scale the cryogenic spray has the appearance of a gasoline fuel spray. However, at the micro scale no spherical 'blobs' and ligaments are observed. Although simpler structures are observed in other sprays, such as in the dense region of a diesel fuel spray, the fundamental origin for these structures in the cryogenic spray is as yet unexplained. The fundamental spray formation and in particular break up and evaporation processes must be understood to enable accurate models to be built to support the efficient design of the isothermal compressor.

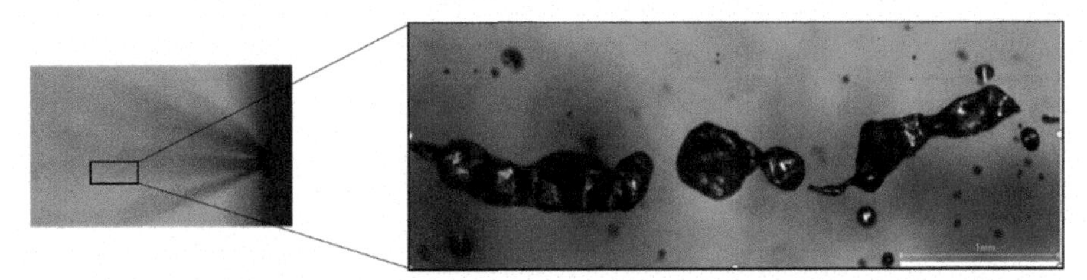

Figure 4. LN_2 spray into ambient air. The left image shows the spray in whole. The right image is the magnified image of the location marked by black rectangle in the whole spray.

4 CHALLENGES ARISING FROM THE THERMO-PHYSICAL PROPERTIES OF CRYOGENIC FLUIDS

As mentioned above, the primary goal of using either water or cryogenic fluids in the compression chamber of RCSE is to efficiently absorb the heat/energy arising due to the rise in pressure. The requirement is to absorb the heat in a way that will enable the compression chamber to remain at an ideally constant temperature during compression. How this absorption will occur depends on the thermo-physical properties of the fluids at the specific operating conditions, which are both sub-critical and super-critical depending on the fluid itself, the piston position and the injection timing. Water which was previously used to absorb the heat during compression has a much higher critical pressure of around 22 MPa. This high pressure was not encountered in the compression chamber, hence water remained sub-critical for the whole compression process. LN_2 on the other hand, has a critical pressure of 3.39 MPa. The pressure in the compression chamber of RSCE starts from an atmospheric pressure and goes up to the maximum pressure of above 10.0 MPa. Depending on the timing of injection of liquid nitrogen, it can either boil at sub-critical pressures or transform into super-critical fluid where the pseudo-boiling influences the heat absorption. This difference in the use of water or LN_2 as coolant is expected to influence the design parameters of injection for good heat absorption.

In an effort to explore further what is/are the optimal conditions to inject the LN_2, the two most important thermodynamic properties for the case of phase change, i.e. the latent heat of vaporisation (the energy absorbed by the cryogen in order to change phase under constant temperature) and the heat capacity C_P (the energy (ΔH) required to change the temperature (T) of the cryogen) (Eqn 1)) are examined closely. In order to better demonstrate this Figure 5 has been included. The figure demonstrates the isobaric specific heat capacities and specific enthalpies vs temperature for nitrogen at pressures from 0.1 MPa to 10 MPa. A reminder that the latent heat (Enthalpy of vaporisation ΔH_{vap}) is the enthalpy difference $(H_{vap} - H_{liq})$ between two phases of a substance at the same temperature (Eqn 2). However, the definition of the latent heat is more relevant for sub-critical cases, as will be demonstrated in the following paragraphs.

$$\Delta H = C_P \Delta T \qquad (Eqn\ 1)$$

$$\Delta H_{vap} = H_{vap} - H_{liq} \qquad (Eqn\ 2)$$

Starting with the specific heat capacity and looking at Figure 5, note that for sub-critical pressures (0.1, 1, 2, and 3 MPa), as the temperature approaches the boiling temperature at a given pressure, the specific heat capacity increases and at the boiling temperature there is a discontinuity. The discontinuity at the boiling point is represented by a red line. For super-critical pressures on the other hand (pressure higher than 3.39 MPa) there is a maximum heat capacity at some temperature (higher temperature for higher pressure), which is less than the heat capacity at critical pressure and critical temperature. The peak is continuous and shows no discontinuities signifying the inseparability of liquid and gas phases. The general trend of the profile for super-critical pressures is, with increase in pressure, the peaks tend to fall and flatten out. When the point that the heat capacity picks for super-critical pressures, the fluid can absorb a large amount of heat without significant rise in temperature but increases in volume without undergoing phase transition. This is what in the literature indicated as the pseudo-boiling point. For super-critical pressures, in addition to the specific heat maximum at pseudo-boiling point, other properties such as density, viscosity, thermal conductivity and enthalpy very rapidly. Although the large specific heat results in very little change in temperature with the absorption of heat, this small temperature change results in large gradients of thermos-physical properties as seen for 4.0 MPa (curve 2) in Figure 6.

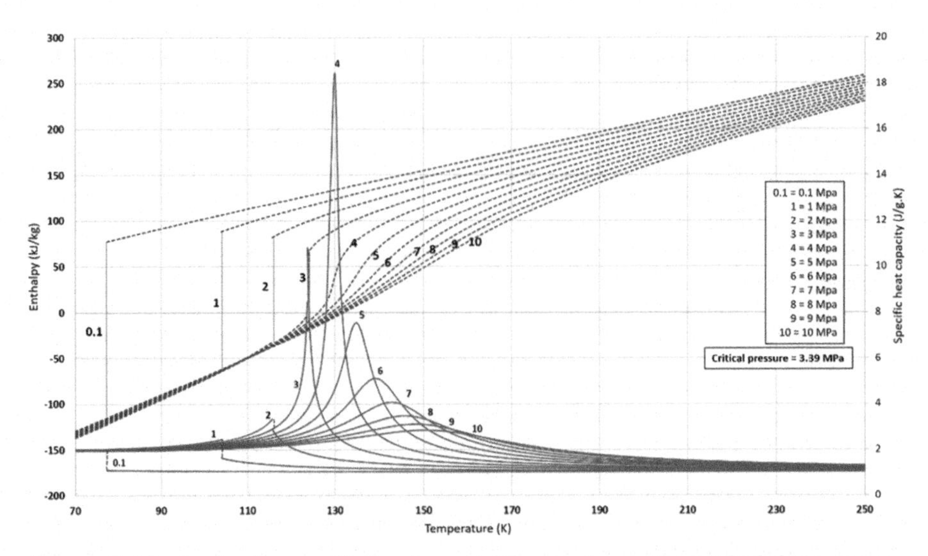

Figure 5. The isobaric specific heat capacities and specific enthalpies vs temperature for nitrogen at pressures from 0.1 MPa to 10 MPa. The dashed line in purple represents enthalpies and the continuous line in blue represents heat capacities. The discontinuities in specific heat capacity corresponding to the red line and discontinuities in enthalpy corresponding to green line locate the boiling point.

In addition to specific heat capacity, latent heat is also of interest. For sub-critical pressures as temperature increases enthalpy of vaporisation decreases and reaches zero once the fluid attains its gas phase. It is thus a very useful quantity for sub-critical liquids to quantify the heat absorbed due to phase change at constant temperature but not of much interest for super-critical fluids since it becomes zero. This of course does not mean that supercritical fluids do not change phase. As subcritical fluids, they start from a liquid-like phase at lower temperatures and as the temperature further increases, they eventually reach the gas phase. Banuti [12] describes that at low sub-critical pressures the latent heat is concentrated at the boiling point, whereas at super-critical pressures this latent heat is distributed over a temperature range around the pseudo-boiling point. This is the reason why in the graph enthalpy difference is used to visualise this, instead of latent heat of vaporisation. The discontinuities in the enthalpy at sub-critical pressures directly quantify the latent heat or enthalpy of vaporisation.

To understand the distributed latent heat and the heat absorption associated with it, an isobaric process (assume the cylinder pressure almost constant for a small change in volume dV) with heat generated is considered. At sub-critical pressures far below the critical pressure such as 0.1 MPa in the Figure 5, when a liquid is injected, the heat absorbed by the liquid raises its temperature without phase change occurring. The temperature increase is proportional to the heat capacity of the liquid. As the temperature reaches the boiling point the liquid transitions into gas, during which it absorbs a large amount of heat without any further raise in temperature, which is the latent heat concentrated at the boiling point. After the transition into gas, the fluid has a different heat capacity than that of the liquid state. Further absorption of heat results in a raise in temperature of the fluid proportional to its gas phase heat capacity. At higher sub-critical pressures, the trend is similar, although gradients in heat capacity start appearing as the pressure increases towards the critical pressure as seen in 2 and 3 MPa in the Figure 5. The corresponding difference in the specific enthalpies decreases. This signifies that the heat absorption due to rise in heat capacity increases while heat absorption due to vaporisation decreases.

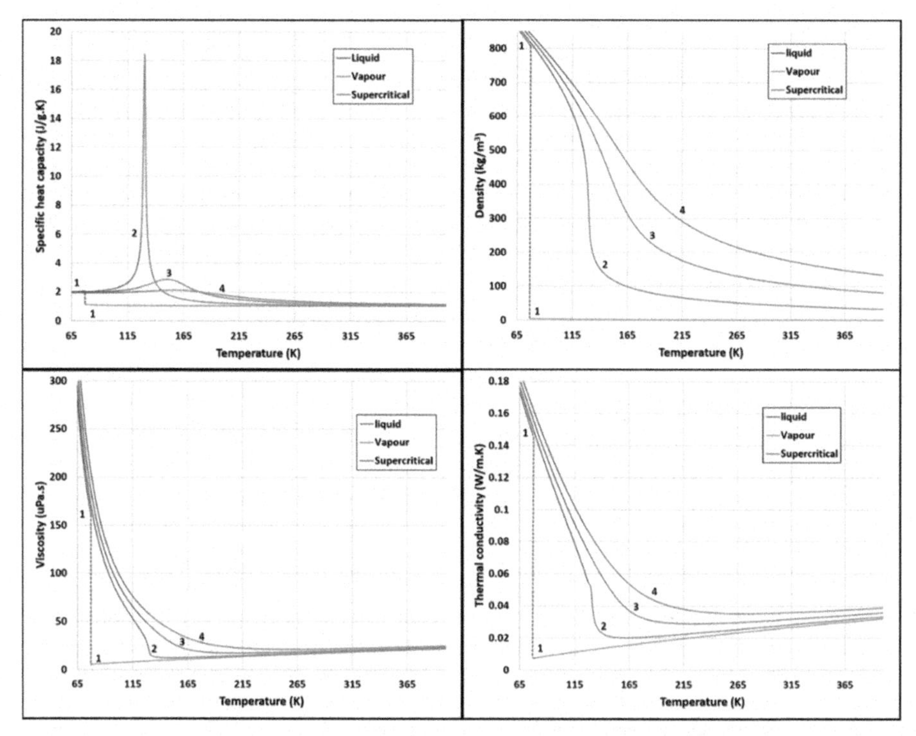

Figure 6. Thermo-physical properties of nitrogen at four pressures 1 - Atmospheric pressure (0.1 MPa), 2 - Just above critical pressure (4.0 MPa), 3 - Max pressure in the compression chamber of Engineering Demonstrator (ED) prototype of split cycle engine (10.0 MPa) and 4 - Max pressure in the compression chamber of planned Commercial Demonstrator (CD) prototype of split cycle engine (17.0 MPa). The red dashed line represents the discontinuity between liquid and vapour phase.

At super-critical pressures as the liquid is injected, the temperature of the liquid increases proportional to its heat capacity until it nears the pseudo-boiling temperature. Here the heat absorbed not only raises its temperature but also gradually changes the state from liquid to gas with several intermediate states over a relatively wide range of temperature. Here the heat is absorbed both due to the heat capacity increase and the intermediate phase transitions. This is because the latent heat is distributed over a range of temperature. This distributed latent heat or in other words, combination of heat absorption to raise the temperature and gradually transition the state from liquid to gas in super-critical pressures is attractive for application in RSCE compression. The combined rate of heat absorption is more than that of the rate of heat absorption due to the heat capacity of a fluid alone. Second, while vaporisation absorbs heat instantaneously, it is hard to distribute this absorption of heat throughout the compression process to achieve isothermal compression. Whereas this combined heat absorption spread over a temperature range, is advantageous to gradually but constantly cool the surrounding throughout the compression process.

In order to highlight the differences between the atmospheric conditions and the conditions taking place in the RSCE, some more thermo-physical properties (density, viscosity, thermal conductivity) of nitrogen corresponding to atmospheric pressure, near critical pressure and two prototypes of RSCE are represented in Figure 6. It should be reminded that the reason the properties at both the sub and super-critical region are of interest is because the pressure in the compression chamber starts from atmospheric pressure and goes up to the maximum pressure. It can be seen from the figure that at sub-critical pressures the thermo-physical properties suddenly change from that of the liquid to gas at the boiling point. Whereas, at super-critical pressure of 4 MPa which is just above the critical pressure, there is a large gradient in thermo-physical properties near the pseudo-boiling temperature. Here a small variation in temperature can result in large variation in these thermo-physical properties as described above. At higher supercritical pressures the gradients of thermo-physical properties become less steep.

Figure 7 compares the rise in specific heat capacity of cryogenic liquids and other naturally occurring liquids including water (that was previously used in the split cycle) at pseudo boiling point for various super-critical pressures. Evidently it can be seen that compared to fuels such as dodecane and iso-octane, the cryogenic fluids such as nitrogen, hydrogen and methane show a significantly higher rise in specific heat capacity at super-critical pressures.

In conclusion the challenges in designing the injection of cryogenic fluids into compression chamber is two-fold. First one is the distributed latent heat of cryogenic fluids around the pseudo-boiling point influencing the heat absorption. The second is the drastic rise in specific heat capacity and the associated gradients in thermo-physical properties influencing the flow evolution. The combination of heat absorption and flow evolution should be understood properly to achieve uniform constant temperature throughout the compression chamber.

5 NUMERICAL MODELLING

The variation of the thermo-physical properties of cryogenic fluids poses many challenges at the numerical modelling of these fluids. In the following paragraphs we will focus on the challenges associated with the calculations of these properties and the selection of an appropriate equation state. Until now in the literature it is not very clear which one is the optimum for cryogenic fluids under compression conditions. Other numerical challenges associated with for example the choice of the liquid break up model although are also very important are outside the scope of this work and will be investigated in future work.

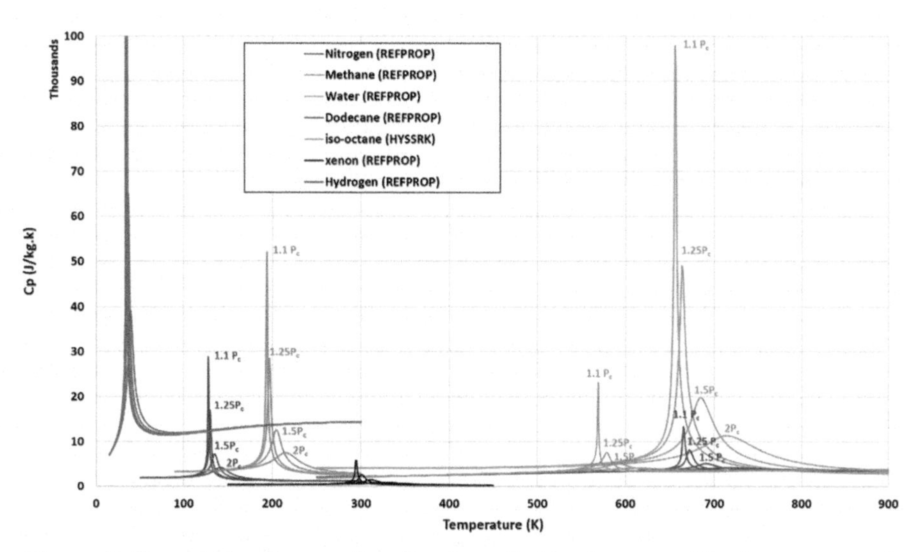

Figure 7. Specific heat capacity of various fluids at supercritical pressures of $1.1P_C$, $1.25P_C$, $1.5P_C$ and $2P_C$, where P_C is the critical pressure of the respective fluid.

5.1 Real fluid equation of state (EOS)

An Equation of State (EOS) which accurately describes the state of matter at both sub-critical and super-critical condition needs to be employed to model the thermodynamic anomalies which occur at near critical temperatures. It is a requirement that the EOS predicts the rapid changes in fluid's specific heat capacity density, viscosity, thermal conductivity and enthalpy near the critical point and the pseudo boiling point with sufficient accuracy in order to lead to accurate prediction of the heat absorbed by the compressed air.

Although Benedict-Webb-Rubin EOS has been used by few researchers due to its comparatively increased accuracy, Peng-Robinson (PR) and Soave-Redlich-Kwong (SRK) EOS are the preferred ones for numerical simulations by many researchers [13]–[15] for the case of cryogens due to reduced numerical complexity. Figure 8 shows Kim et al [13] and Muller et al [15] estimation of specific heat capacity and density of nitrogen using PR and SRK for 4 MPa. The NIST data are also plotted for comparison. It can be seen from the figures that both PR and SRK fail to capture the peak of specific heat capacity accurately. This might have significant influence on the flow simulation. For example, if the specific heat maximum is not captured accurately, then the simulation of injected cryogenic spray will be very different than the real one. This is due to the outer fluid quickly overcoming the pseudo-boiling point as temperature of the fluid can rise more rapidly than the real case. And the inner fluid will experience major heat transfer into it earlier than the real case which is further affected by the lower than actual specific heat maximum predicted by SRK or PR. The density predicted by PR and SRK are fairly accurate, though the density from NIST seems to lie in between the values of PR and SRK.

Additionally, the specific heat capacity and density of nitrogen at a high super-critical pressure of 17 MPa has been estimated using PR and SRK, and compared against REFPROP [16] from NIST (see Figure 9). REFPROP is NIST's Reference Fluid

Thermodynamic and Transport Properties Database. It can be seen that PR and SRK lie very close to the REFPROP plot of specific heat capacity. In the density plot SRK captures the density reasonably well whereas PR shows significant error at lower temperatures. Assessing from the low super-critical pressure predictions by Kim [13] and the high supercritical pressure predictions we calculated, the SRK would be the first choice to model the conditions in RSCE compression chamber.

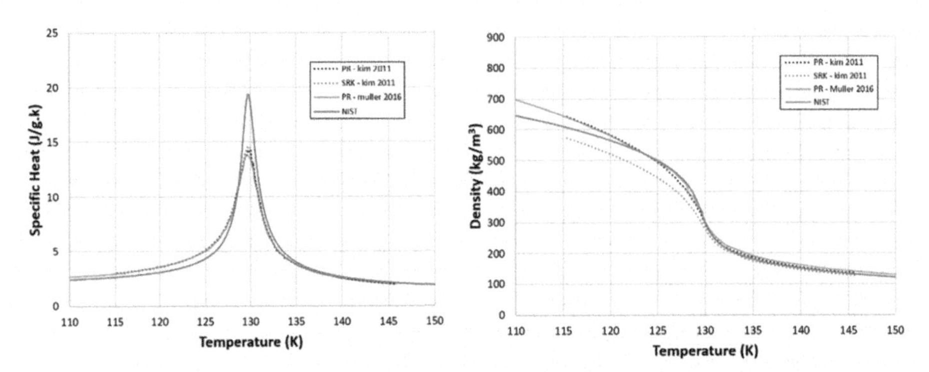

Figure 8. Estimation of specific heat capacity (left) and density (right) of nitrogen at 4 MPa using PR and SRK by Kim[13] and Muller[15].

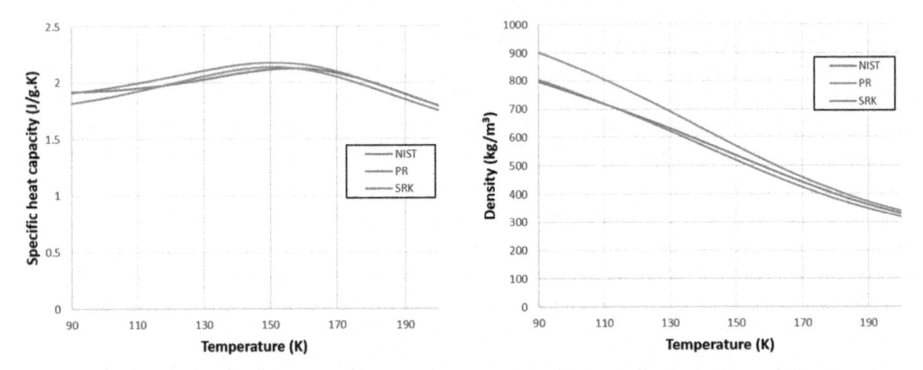

Figure 9. Estimation of specific heat capacity (left) and density (right) of nitrogen using PR and SRK compared against REFPROP (NIST) at 17 MPa corresponding to the maximum pressure in the planned Commercial Demonstrator (CD) split cycle engine.

6 CONCLUSION

In this paper the basic concepts of operations of a hybrid engine currently under development by Ricardo Innovations, Dolphin N2 and the University of Brighton, the recuperated split cycle engine, have been presented. The focus was the thermodynamics of the isothermal compression taking place in the engine with the injection of either water or cryogenic fluids. Injection of cryogenic fluids into the compression chamber to achieve isothermal compression could further enhance the efficiency of RSCE towards achieving their full potential. A RSCE with quasi isothermal compression can achieve up to 60% BTE. A critical review of existent literature showed that there are fundamental gaps in knowledge on cryogenic fluids at conditions relevant to the RSCE. We showed that the absorption of heat and evolution of spray is different than what has been examined in the literature since as the piston moves the fluid is found to be both at sun and super-critical state. We examined in detail the thermos-physical properties of both water and LN_2 at both sub and super-critical conditions and we highlighted the peculiarities relevant to the specific heat and the latent heat of evaporation. We also performed an analysis relevant to existent EOS and their accuracy both at low and high chamber pressures. In terms of future work further experiments on liquid nitrogen at conditions relevant to split cycle compression chamber needs to be undertaken. Moreover, the role of the breakup mechanism of the cryogenic liquids will be explored.

ACKNOWLEDGEMENTS

The work reported in this paper was funded by the EPSRC under grant award EP/S001824/1 Unveiling the injection dynamics of cryogenic energy carriers for zero-emission high-efficiency systems.

REFERENCES

[1] R. Morgan, N. Jackson, A. Atkins, G. Dong, M. R. Heikal, and C. lenartowicz, "The Recuperated Split Cycle - Experimental Combustion Data from a Single Cylinder Test Rig," *SAE Int. J. Engines*, vol. 10, 2017.

[2] G. Dong, R. E. Morgan, and M. R. Heikal, "Thermodynamic analysis and system design of a novel split cycle engine concept," *Energy*, vol. 102, pp. 576–585, 2016.

[3] G. Dong, R. Morgan, and M. Heikal, "A novel split cycle internal combustion engine with integral waste heat recovery," *Appl. Energy*, vol. 157, pp. 744–753, 2015.

[4] M. W. Coney, C. Linnemann, and H. S. Abdallah, "A thermodynamic analysis of a novel high efficiency reciprocating internal combustion engine—the isoengine," *Energy*, vol. 29, no. 12, pp. 2585–2600, 2004.

[5] N. Jackson, A. Atkins, J. Eatwell, and R. Morgan, "An alternative thermodynamic cycle for reciprocating piston engines," 2015.

[6] "SAE 2014 Heavy-Duty Diesel Emissions Control Symposium," *Johnson Matthey Technol. Rev.*, vol. 59, no. 2, 2015.

[7] W. Mayer, A. Schik, C. Schweitzer, and M. Schaeffler, "Injection and mixing processes in high pressure LOX/GH2 rocket combustors," in *32nd Joint Propulsion Conference and Exhibit*,.

[8] W. O. H. Mayer *et al.*, "Atomization and Breakup of Cryogenic Propellants Under High-Pressure Subcritical and Supercritical Conditions," *J. Propuls. Power*, vol. 14, no. 5, pp. 835–842, 1998.

[9] W. Mayer, J. Telaar, R. Branam, G. Schneider, and J. Hussong, "Raman Measurements of Cryogenic Injection at Supercritical Pressure," *Heat Mass Transf.*, vol. 39, no. 8, pp. 709–719, Sep. 2003.

[10] B. Chehroudi, D. Talley, and E. Coy, "Fractal geometry and growth rate changes of cryogenic jets near the critical point," in *35th Joint Propulsion Conference and Exhibit*, .

[11] M. Oschwald and A. Schik, "Supercritical nitrogen free jet investigated by spontaneous Raman scattering," *Exp. Fluids*, vol. 27, no. 6, pp. 497–506, Nov. 1999.

[12] D. Banuti, "The Latent Heat of Supercritical Fluids," *Period. Polytech. Chem. Eng.*, vol. 63, no. 2, pp. 270–275, 2019.

[13] T. Kim, Y. Kim, and S.-K. Kim, "Numerical study of cryogenic liquid nitrogen jets at supercritical pressures," *J. Supercrit. Fluids*, vol. 56, no. 2, pp. 152–163, 2011.
[14] T. S. Park, "LES and RANS simulations of cryogenic liquid nitrogen jets," *J. Supercrit. Fluids*, vol. 72, pp. 232–247, 2012.
[15] H. Müller, C. A. Niedermeier, J. Matheis, M. Pfitzner, and S. Hickel, "Large-eddy simulation of nitrogen injection at trans- and supercritical conditions," *Phys. Fluids*, vol. 28, no. 1, p. 15102, 2016.
[16] E. W. Lemmon, I. H. Bell, M. L. Huber, and M. O. McLinden, "NIST Standard Reference Database 23: Reference Fluid Thermodynamic and Transport Properties-REFPROP, Version 10.0, National Institute of Standards and Technology." 2018.

Internal Combustion Engines and Powertrain Systems for Future Transport 2019 – Institute of Mechanical Engineers, ISBN 978-0-367-90356-5

Forecasting the implications of Euro 7 on the powertrain supply chain

M. Southcott, G. Evans

IHS Markit, London, UK

ABSTRACT

Global emissions legislation is becoming increasingly stringent and challenging for both original equipment manufacturers (OEMs) and Tier 1 suppliers. The importance of tightened emissions standards is essential as the automotive industry moves towards zero emissions. In Europe, legislation has become complex in recent years with the introduction of Real Driving Emissions (RDE) and World Harmonized Light Vehicle Test Procedure (WLTP). The next round of emissions standards, likely to be called Euro 7, is expected to be implemented in 2024-25. In addition to Europe, Brazil, China, and India will all introduce new tighter emissions standards in the next few years.

Within this paper, IHS Markit explores global legislation and the key technologies OEMs will require to be tailpipe compliant to aide them with strategy and supply decisions. 48V architecture is increasing in usage which is enabling an increase in demand for some existing technologies, particularly e-cat technology in the exhaust aftertreatment.

On top of the emissions challenges, the automotive supply chain faces further uncertainties with ever-changing consumer demands and with the political outlook in Europe and the rest of the world.

1 INTRODUCTION

Governments around the world are beginning a substantial push towards a clean, zero-emissions society. While there are unquestionably challenges and obstacles to overcome in making this happen, the automotive industry is already showing its commitment to playing its part.

Global tailpipe legislation is on a clear path towards zero emissions, with major governmental bodies such as the European Commission also now beginning to look beyond the conventional tailpipe.

Original equipment manufacturers (OEMs) and Tier 1 suppliers are already showing their commitment to this; not only are existing clean air technology solutions being applied and developed, but new alternative zero-emissions solutions such as electric and fuel cell vehicles are coming to market.

As mentioned, despite the new zero-emissions solutions being presented, several obstacles must be overcome before they will become the dominant form of propulsion for the everyday car. As a result, the combustion engine and the underlying technology within this will continue to play a major role in the short to medium term.

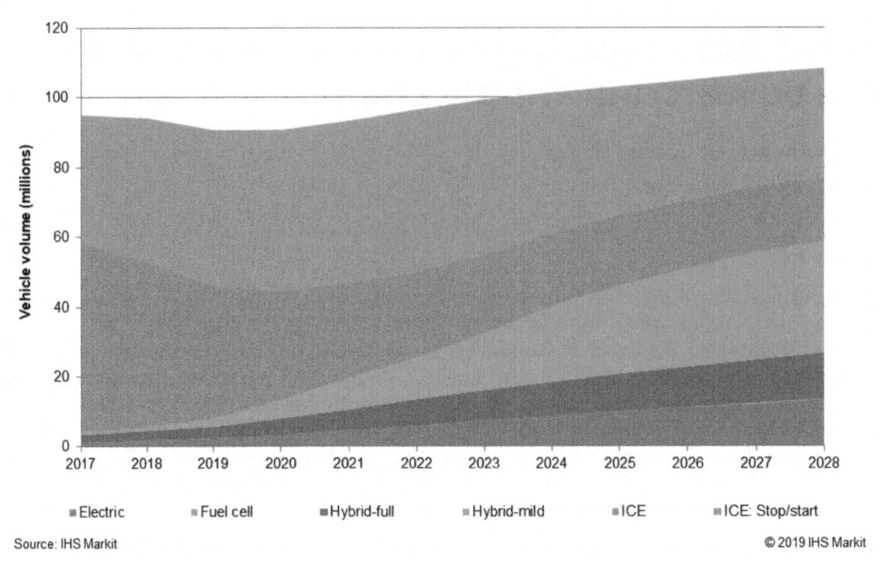

Figure 1. Global vehicle production split by propulsion system design (2017–28).

As shown in Figure 1, IHS Markit data reflects the importance of the internal combustion engine over the next decade. By 2028, approximately 88% of vehicles globally produced will require some form of combustion engine, albeit with the assistance of hybrid technology growing rapidly.

For the vehicles within that percentage, the aftertreatment and clean air technology fitted to them must be able to comply with the sales country they are intended for. With several regions such as Europe, China, and India set to introduce new emissions standards in the coming years, the exhaust aftertreatment market is going to be crucial for all OEMs.

2 GLOBAL EMISSIONS REGULATIONS

2.1 China

With the introduction of CN6a and CN6b so close together, it is increasingly likely that many OEMs will simply target CN6b emissions standards and bypass CN6a altogether.

The Chinese government has taken a large step toward zero NOx tailpipe emissions by setting the limit for CN6b at 35mg/km for Type I and for Type III it does vary from 35-50 mg/km in its proposed legislation in its legislation; at 35m/km, this is over half the level set for Euro 6d.

The decision to standardize gasoline and diesel emissions was widely discussed as a proposal for Euro 6, but China has now taken the lead with CN6b adopting a common emissions standard.

Despite this, the country still lags behind Europe in one key area: RDE testing. There are plans to introduce RDE testing in China; however, it is expected to initially conform to a factor of 2.1 for both NOx and PN based on EU6d RDE act 2 with an addition of N2O limits.

2.2 Europe

Although legislation has yet to be drawn up, Europe's path to enforcing Euro 7 emissions standards is becoming a lot clearer.

The final NOx target is likely to be influenced by CN6b, falling in the region of 30–35 mg/km. There is potential to further reduce this, but this seems unlikely due to the difficulty with reducing conformity factors. A 35mg/km NOx target will also be fuel neutral, representing a reduction of about 42% for gasoline and about 56% for diesel vehicles.

In addition to the NOx reduction, there are several other key features likely to feature in the Euro 7 standard. One of these—certainly from a gasoline perspective—will be a reduced particle measurement size from 23 to 10 nm. There is also potential for pollutants, such as brake dust, to be measured; the technology effect for this will be discussed in a later section.

IHS Markit anticipates other pollutants such as ammonia to be included in the legislation, while an increased emphasis on cold start and cold-running portions of the WLTP and RDE will influence technologies that OEMs use in their exhaust aftertreatment. The expectancy here is to extend legal limits for testing down to -7°C while also reducing the conformity factor for emissions such as CO in RDE tests.

An extension of durability requirements from 160,000 km to 200,000 km and more stringent in-service monitoring are also expected.

Figure 2 gives an overview of global emissions standards out to 2025.

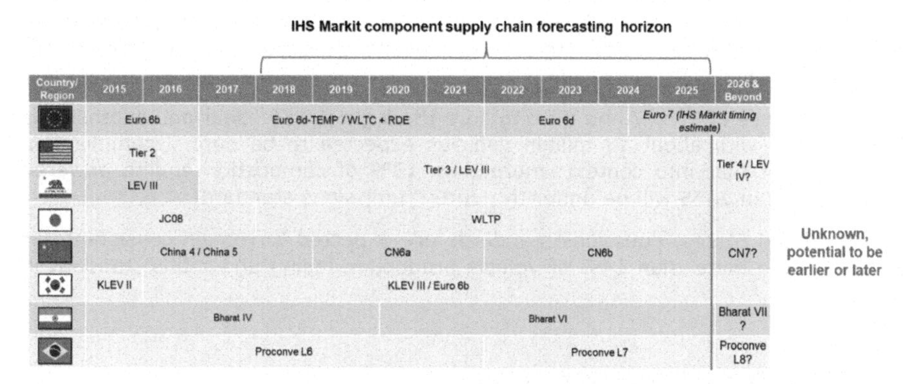

Figure 2. Global emissions legislation outlook to 2025.

Global emissions standards are being introduced in key automotive production regions at a rapid pace. Europe, Brazil, China, and India already have or will be introducing tighter emissions standards from 2022 onwards.

IHS Markit component forecasts currently look out to 2025 and therefore will capture the effect of these standards, reflecting the key suppliers in markets, as well as the technology OEMs and Tier 1 suppliers will adopt to remain compliant.

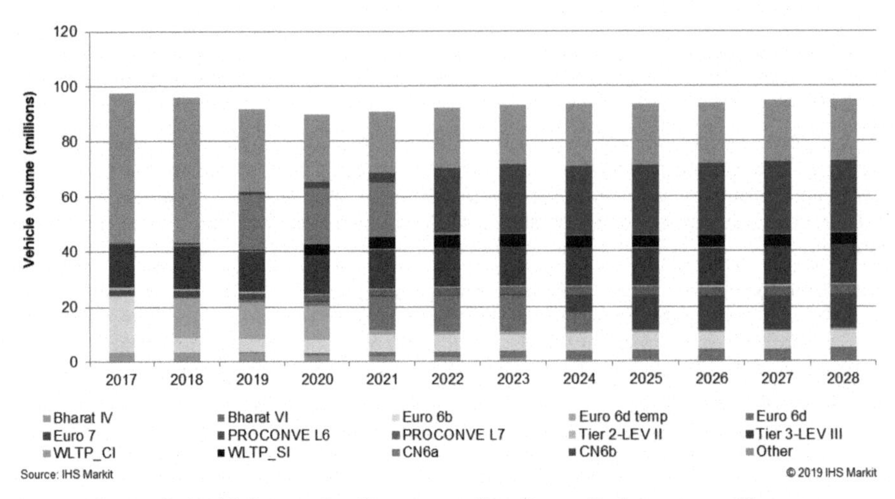

Figure 3. Vehicle production according to emissions compliancy.

Figures 2 and 3 above show the global picture by emissions standards and the corresponding vehicle production. For Figure 3, electric and fuel-cell powered vehicles have been removed so all vehicles included will require a combustion engine and therefore be required to meet the respective tailpipe emissions.

In 2018–19, the importance of the Chinese government implementing CN6a can be seen, with about 20 million vehicles forecast to be sold under the standard, although as discussed, many key OEMs in the country will make new launches of CN6b compliant from the outset.

The importance of Euro 7 and making sure the legislation is clear and extensive is shown here, with about 12 million vehicles expected to be Euro 7 compliant in 2025. To put that into context, more than 13% of combustion engine powered vehicles sold in 2025 will be under the Euro 7 emissions standard.

In the United States, emissions standards are expected to remain consistent out to 2025, with more than 14% of vehicle production falling under this standard in 2025.

3 EUROPEAN TECHNOLOGY TRENDS

3.1 Gasoline

3.1.1 *Overview*

When Euro 7 standards are introduced, gasoline aftertreatment systems will be affected by a few key details. The largest of these for OEMs is likely to be the decision to incorporate particles down to 10 nm in testing. At present, the market for gasoline particulate filters is predominantly a technology used on direct injection engines. With 10 nm particles combined with the potential to measure other emissions such as brake dust particles, GPFs will become mandatory across the gasoline industry on both MFI and DI engines.

With the gasoline market driving the uptake of 48V+ hybrid systems, the potential for gasoline electrically heated catalyst technology is high, albeit the size of the market is currently unclear. The extent to which cold-start and warm-up running is factored into the overall cycle numbers will ultimately determine this.

With the NOx limit being reduced to 35mg/km, it is likely that the number of vehicles and OEMs using more than one three-way catalyst will increase. As with all current aftertreatment systems, packaging the necessary equipment into a vehicle remains the biggest challenge. As such, the market for coated GPFs will grow when Euro 7 is introduced.

3.1.2 *Key component forecasts*

Please note, unless stated otherwise, the following data represent all vehicles globally with an emissions standard in line with either Euro 6a, b, d temp, d, or Euro 7. The data are based on an introduction of Euro 7 around 2024–25 and are based on the described legislation above.

Figures 4-6 below show some of the key aftertreatment technologies and their uptake out to 2028.

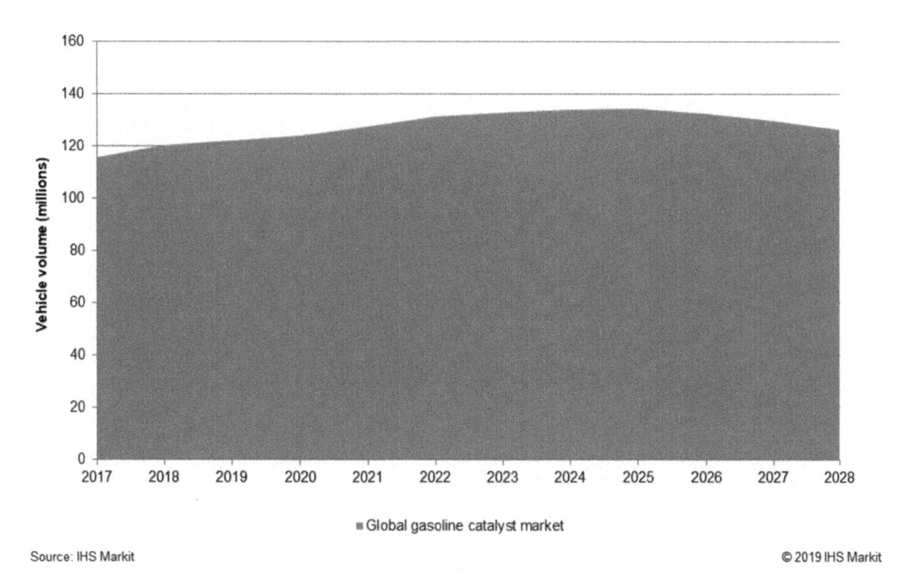

Figure 4. Global gasoline catalyst market (2017–28).

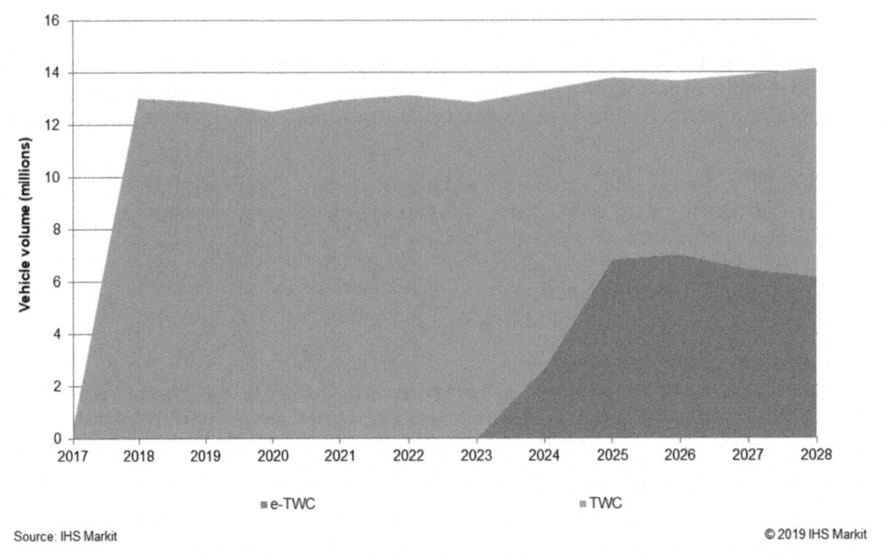

Figure 5. Gasoline catalyst market for Euro 6d, d temp, and Euro 7 vehicles.

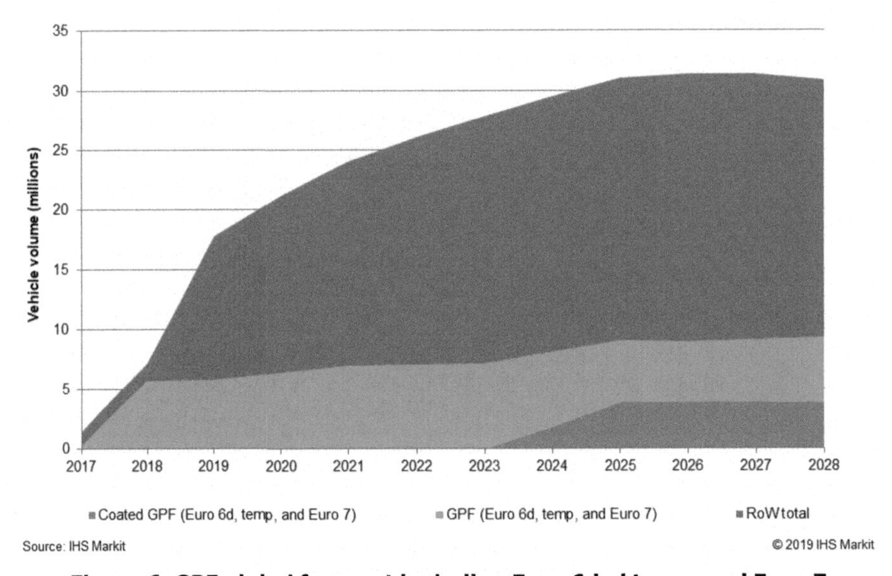

Figure 6. GPF global forecast including Euro 6d, d temp, and Euro 7 breakdown.

The outlook for the gasoline vehicle exhaust technology is very much a case of evolving what already exists by making better use of architecture and space in the vehicle.

A key challenge for any vehicle design is going to be packaging, with reductions in emissions target and an emphasis on cold running, a traditional single TWC will not be sufficient to reach Euro 7 emission standards.

Figures 4 and 5 show the global and more focused European markets for catalysts. TWC volume is expected to reach in excess of 130 million units worldwide.

In Figure 5, we can also see that the total market for TWC levels off with the introduction of Euro 7. One of the driving factors here is the introduction of coated GPFs. With packaging at a premium, introduction of 2xTWC+GPF is going to be a challenge for smaller passenger cars. As a result, expect many OEMs to adopt the TWC + cGPF approach to meet Euro 7 emission targets.

The uptake of electrically heated catalysts and the blanket use of GPFs are the two key technology challenges IHS Markit forecasts as a direct consequence of tightening emissions standards.

Figure 6 shows that the global GPF market will increase to more than 30 million units in 2025, with CN6b and Euro 7 two of the key contributors to the component volumes shown.

One further key technology, while not directly in the exhaust system, is the uptake of cooled EGR on gasoline engines.

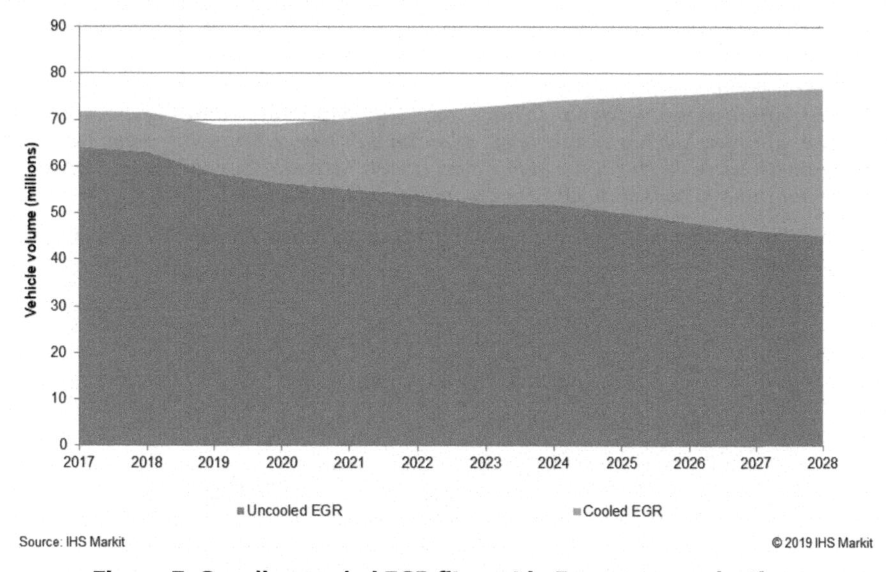

Figure 7. Gasoline cooled EGR fitment in European production.

With the requirement for lambda 1 across the entire engine map, Figure 7 shows how the uptake of cooled EGR on gasoline engines will reflect this. Despite not being the only technology available to aid an engine map of lambda 1, it is likely to be the technology of choice for many.

3.2 Diesel

3.2.1 *Overview*

The significant reductions upcoming in NOx, CO, and HC emissions will have a further cost impact on the already expensive diesel aftertreatment solutions.

To meet Euro 6d emissions standards, OEMs are already adopting policies of fitting the aftertreatment as close as they possibly can, and the expected legislation targeting cold ambient and warm-up emissions is likely to require either further close coupling, additional components, or sources of heating.

Several Tier 1 suppliers are also exploring the potential of reducing the desired light-off temperatures for catalysts, as well as the time taken to reach them.

As with gasoline vehicles, the introduction of 48V+ system architecture gives OEMs the potential to use an electrically-heated catalyst, enabling much faster light-off times. This diminishes the need for extensive close coupling of the aftertreatment. IHS Markit anticipates that diesel vehicles with 48V+ architecture will be fitted with e-cat technology.

Despite the benefits, 48V+ architecture with e-cat technology will add a large cost to what is an expensive exhaust layout. For the OEMs that do not adopt 48V+, one potential option will be for a much smaller SCR system, fitted with close coupling and with a reduced light-off temperature due to its size. This system would effectively see three SCR catalysts, including the SCR on filter.

As with all aftertreatment choices, cost versus packaging space is going to be a delicate balance that all OEMs must overcome, especially for smaller A- and B-segment vehicles.

3.2.2 *Key component forecasts*

Please note, unless stated otherwise, the following data represent all vehicles globally with an emissions standard in line with either Euro 6a, b, d temp, d, or Euro 7. The data are based on an introduction of Euro 7 around 2024–25 and are based on the described legislation above.

Figures 8 to 12 show some of the key aftertreatment technologies and their uptake out to 2028.

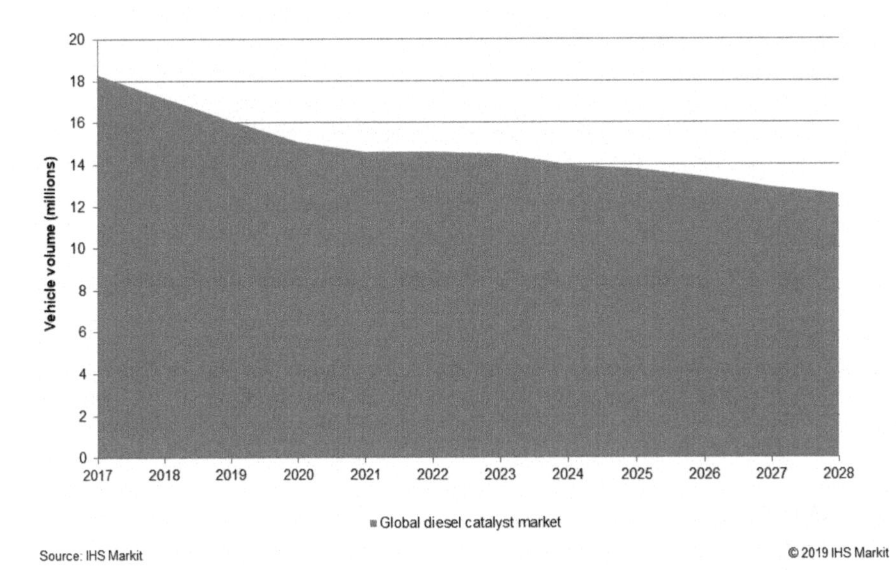

Figure 8. Forecast for global diesel catalyst demand (2017–28).

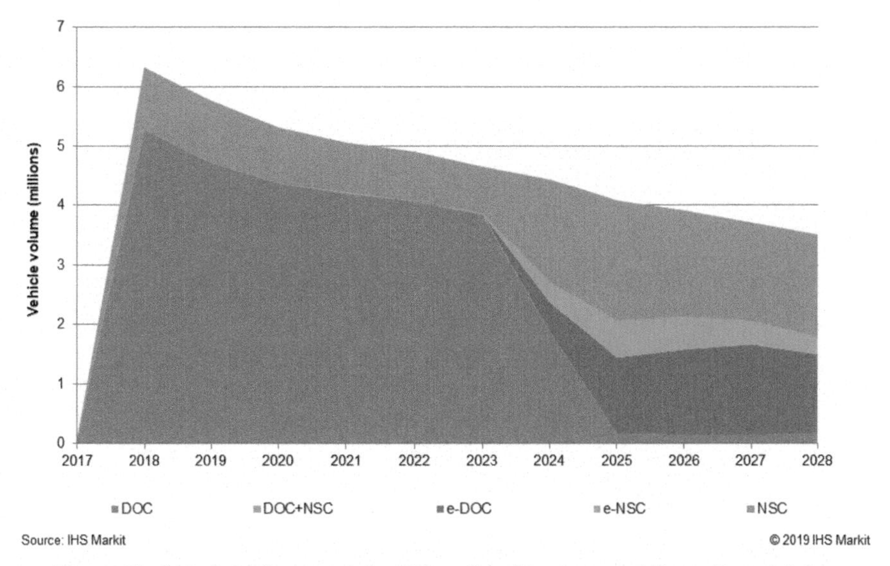

Figure 9. Catalyst forecast for Euro 6d, d temp, and Euro 7 vechicles (2017–28).

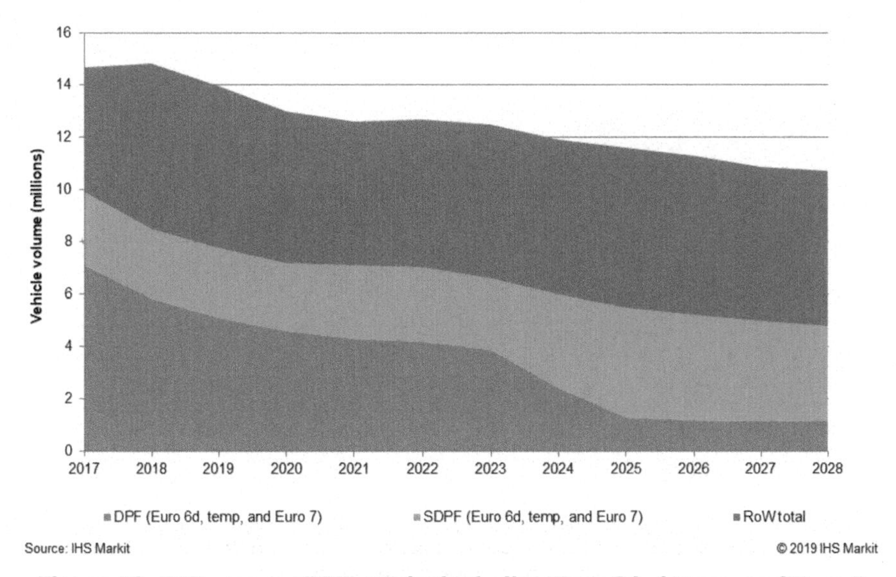

Figure 10. DPF versus sDPF uptake including Euro 6d, d temp, and Euro 7 breakdown.

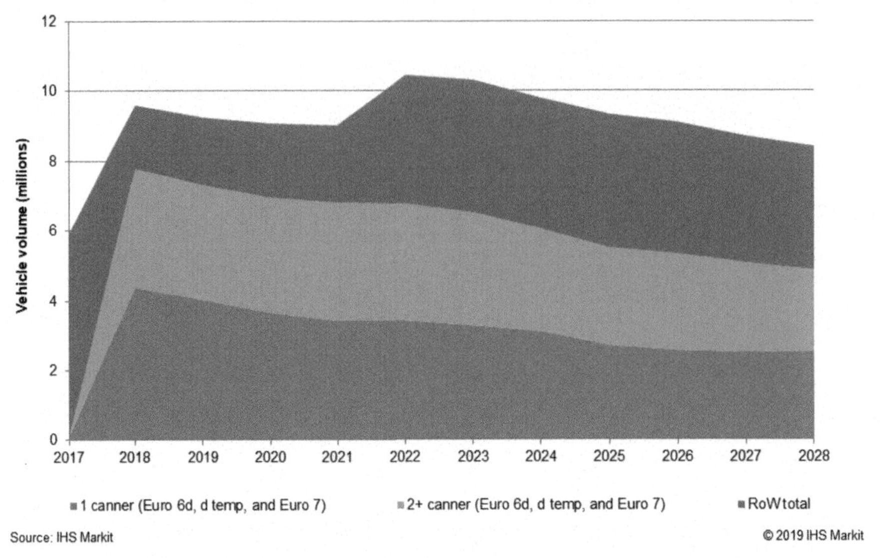

Figure 11. Global SCR volume including Euro 6d, d temp, and Euro 7 breakdown.

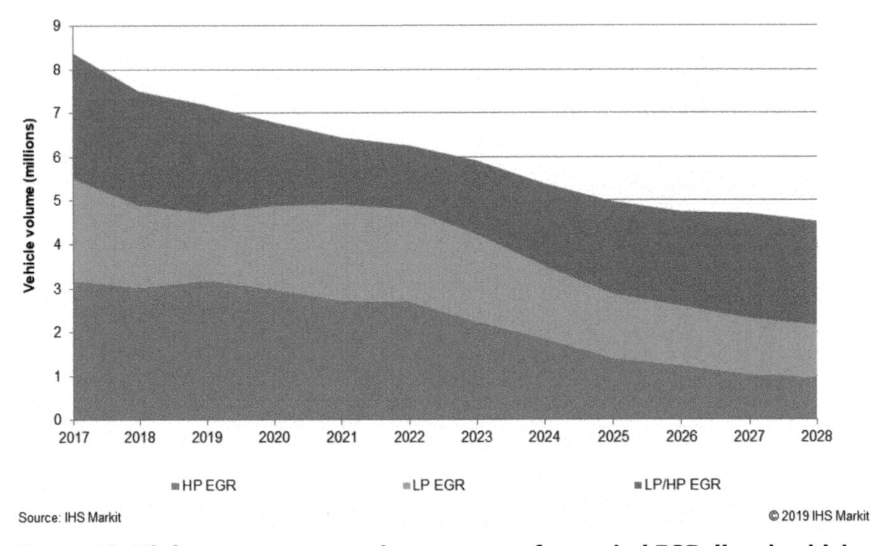

Figure 12. High pressure versus low pressure for cooled EGR diesel vehicles with European production.

As can be seen in Figures 8 to 12, the overall production of diesel vehicles globally and in Europe is forecast to significantly drop out to 2028. Despite this, the complexity and the cost of the aftertreatment fitted to diesel vehicles will continue to rise.

The catalyst market, shown in Figures 8 and 9, has traditionally been dominated by the diesel oxidation catalyst. For Euro 6, close coupling of the catalyst has been seen;

however, with the emissions reductions forecast for Euro 7, IHS Markit expects a rise in electrically-heated catalysts and NOx storage catalysts, with NSC production peaking above 2 million units in 2025 for Euro 6 and 7 compliant vehicles.

With the increased challenge of emissions reduction and space being a premium, Euro 7 introduction will also see the coated SCR on filter technology growing production to just more than 4 million units for Euro 6 and 7 compliant vehicles.

The market for SCR systems will remain. For vehicles without 48V+ architecture, potential remains for a further SCR system to be put into effect, meaning a smaller close-coupled SCR canner fitted alongside an underfloor SCR. The intention being the smaller, close-coupled device will have reduced light-off time and will cope with the light-load, low-temperature exhaust gases, while the larger underfloor system will be used for higher temperature exhaust gases such as on motorway driving.

Figure 12 shows that diesel vehicles will increasingly be fitted with a combination of both HP and LP EGR. Once again, this will provide suppliers and OEMs with the challenge of ensuring the correct size cooler for performance, but this will be traded off with the cost and packaging of the technology.

4 REST OF WORLD LEGISLATION

4.1 India

Bharat VI is due to be introduced in 2020. The Indian government has taken the decision to jump straight from Bharat IV to Bharat VI, aligning itself in the process more closely with Euro 6b standards.

Despite this, other incentives, initiatives, and public perception may well end up being the driving factor in the exhaust aftertreatment decisions in India.

Cost has always been a key driver, which is unlikely to change. Several OEMs are already making the choice to move away from small diesel engines due to the added cost of the aftertreatment to meet legislation.

Government incentives, such as low excise duties on selected vehicles less than 4 m in length, is one example that may also affect the small car diesel market. With the addition of a large particulate filter causing packaging issues, it may push some OEMs above the 4 m limit and cause them to rethink their powertrain strategy.

The market in India is forecast to be dominated by gasoline engine vehicles. With emissions standards from 2020 closely lining themselves with Euro 6b, IHS Markit expects TWC volumes in the country to continue rising.

4.2 United States

In the United States, Tier 3 with its BIN categories is now very much established and is forecast to remain for the foreseeable future into the next decade. There is common agreement within the industry that the link between real driving and tested results in the United States is far ahead of anywhere else.

Nevertheless, there is some discussion around changing particle emissions standards across the country, with 1mg/mile under discussion, in both California and potentially into federal emissions standards as well.

While no specific changes to legislation in North America are expected, IHS Markit believes that technology trends in the region will remain consistent, with TWC being the primary choice of gasoline aftertreatment.

If the US does introduce more stringent legislation on particulate number and particulate mass, a much greater uptake in GPF technology will occur.

4.3 China

The gasoline aftertreatment market is expected to evolve in the coming years with both CN6a and CN6b being introduced. As in Europe, TWC technology is required for CN6a, with a GPF not being necessary. However, this is trumped by most OEMs pre-empting that the close-following CN6b legislation will change this, and so most new powertrains in China will feature a GPF.

The diesel market in China continues to decline, with the country heavily pushing full electric vehicles. The combination of TWC+GPF will be the strategy of choice for OEMs out to 2028.

4.4 Discussion

While the underpinning technologies exist, the different approaches in combinations and sizing across the global markets are creating challenges for the industry. For suppliers and OEMs, meeting global demands is an obstacle from a technical and cost perspective, and it is quickly becoming clear that a global emissions standard would be beneficial; but the practicality of this is uncertain.

As legislation around the world becomes fuel neutral, and as powertrain technology shifts towards electrification, the big question that remains is whether we need technology-neutral legislation in future to rebalance the equation for combustion engines versus electric vehicles. While electrification is inevitable, whether it is the best solution in the near term must be considered because low-carbon fuels and the internal combustion engine still have a part to play in the coming decades.

5 SUPPLY CHAIN CHALLENGES AND OPPORTUNITIES

Suppliers, both at a Tier 1 level and below, must make changes to company structure as well as the technology they develop to address global changes in emissions technology.

One of the key challenges, certainly for larger Tier 1 suppliers, is the variations in global emissions regulations. From a research and development perspective, it means suppliers need to develop different solutions, adding to cost and complexity. Furthermore, with regulations changing at a rapid pace, suppliers are seeing large manufacturing expenses to cope with technology uptake. This is definitely the case in China, where a lot of OEMs are simply aiming for CN6b and bypassing CN6a.

Requirements for ever-increasing efficiency in filters and catalysts, as well as increasing complexity on key engine technology such as GDi and HCCI, are placing high costs on the supply chain. The challenge is how much of this cost can be absorbed by the end customer versus how much will need to come from OEMs and the supply chain.

Fitting all the required technology into the vehicle is becoming more difficult regarding exhaust aftertreatment. An increasing push from OEMs to have more close-coupled emissions devices has forced suppliers to get creative with their designs. One example of this is the ring catalyst turbocharger developed by Continental and presented at the 40th Vienna Motor Symposium.

Another challenge for the automotive industry is potential trade disruption. Ongoing disputes between China and the US, the US and the European Union, plus uncertainty around what Brexit will entail, is causing many Tier 1 suppliers to rethink manufacturing and sourcing locations. The knock-on effects of plant closures such as Bridgend for Ford and Swindon for Honda will be felt throughout the supply chain in the coming years.

Ultimately, the biggest threat to the industry is the shift away from diesel, which is posing a challenge in meeting fleet CO_2 emissions. A push towards electric vehicles is occurring at a large cost to suppliers. With the tipping point for public acceptance appearing in the distance, suppliers are having to come up with ways to maintain profit while also investing in the future.

However, there are opportunities for not only well-established suppliers but also smaller Tier 1 and 2 companies as electrification extends the life of existing, well-developed technologies. Components such as electrically-heated catalysts, e-compressors, and e-water pumps are examples of suppliers using the electrification drive to extend the life of existing technology and enter new product lines. From an e-cat perspective, Continental Emitec, Faurecia, and ElringKlinger have all revealed designs to enter this market.

We are already seeing the supply chain act to reduce costs and reassure investors. There seems to be three strategies for Tier 1 suppliers to adopt: invest, merge, or spin off divisions.

In the past few years, several Tier 1 suppliers have opened or expanded production facilities in emerging markets. Mergers such as Calsonic Kansei and Magneti Marelli to boost revenues and ensure competitiveness in key areas are becoming common place. Finally, several large Tier 1 companies such as Delphi (Aptiv and Delphi Technologies) and Continental (Vitesco Technologies) have begun to spin off their existing powertrain departments. This enables investment into new technologies and existing powertrain technology to be separated.

6 CONCLUSION

Emissions standards are changing at a rapid pace globally and creating challenges to OEMs and Tier 1 suppliers. Euro 7 legislation is expected to be introduced around 2024-25, following on from the introduction of new standards in Brazil, China, and India in the early part of the next decade.

While still under discussion, IHS Markit expects the final NOx target to be in the region of 35mg/km or under, with tighter conformity factors and the introduction of new emissions providing the biggest challenges to OEMs and Tier 1 suppliers.

Cost and packaging are likely to be two of the biggest challenges for the vehicle manufacturer when Euro 7 is implemented, with 48V architecture providing opportunities for existing technology to grow in usage, albeit at a cost. How these challenges are overcome will create an additional burden on the automotive supply chain.

On top of emissions legislation, there are several challenges for the industry to overcome; however, we are already seeing several Tier 1 suppliers begin to react and implement strategic decisions. Innovation within the automotive supply chain will be crucial in the coming years to ensure the automotive industry remains successful.

NOMENCLATURE

Table 1. Abbreviation definitions for IHS Markit data.

Abbreviation	Description
BEV	Battery Electric Vehicle
cGPF	Coated Gasoline Particulate Filter
DI	Direct Injection
DOC	Diesel Oxidation Catalyst
DPF	Diesel Particulate Filter
EGR	Exhaust Gas Recirculation
FCEV	Fuel Cell Electric Vehicle
GDI	Gasoline Direct Injection
GPF	Gasoline Particulate Filter
HC	Hydrocarbon
HCCI	Homogenous Charge Compression Ignition
MFI	Manifold Fuel Injection
NSC	NOx Storage Catalyst
RDE	Real Driving Emissions
SCR	Selective Catalytic Reduction
sDPF	SCR on Diesel Particulate Filter
TWC	Three Way Catalyst
WLTP	Worldwide Harmonised Light Vehicle Test Procedure

SESSION 4: FUELS AND FUEL INJECTION

Study on the behavior of liquid films with different viscosities after impact by single droplet

Lili Lu[1], Yiqiang Pei[1], Jing Qin[1,2], Zhijun Peng[1,3], Yuqian Wang[1], Qingyang Zhu[1], Zhenshan Peng[2]

[1]State Key Laboratory of Engines, Tianjin University, Tianjin, China
[2]Internal Combustion Engine Research Institute, Tianjin University, Tianjin, China
[3]Faculty of Creative Arts, Technologies and Science, University of Bedfordshire, UK

ABSTRACT

Research on single drop impact, especially in the past two decades, has been motivated by a need for better predictive capability in many industries. The objective of this work is to clarify the single droplet impingement behavior onto a liquid film with different physical properties. Fluorescent agent is added to the liquid film, and the experimental method of Laser induced fluorescence(LIF) is used to distinguish the liquid film from the incident liquid droplet. Within the experimental ranges tested, the liquid film can be divided into five types after the droplet impacts, and the instability mechanism responsible for the crown and splash formation was analyzed. New results on crown-splash thresholds are obtained on the basis of the Weber number of incident droplet (*We*) and the Ohnesorge number of the liquid film (*Oh**). Moreover, this article also explores the maximum height of the liquid film crown after the impact of the droplet and the crown diameter at the corresponding moment. They were analyzed by combining *We* with *Oh**, and experimental results allow to propose empirical correlations which can be used for prediction of crown parameters.

Keywords: Liquid film, GDI engines, splash, crown height, crown diameter

1 INTRODUCTION

In the field of industry and natural world, droplets impinging on a liquid film is a common phenomenon, such as raindrops falling, pesticide spraying [1], the spray in the internal combustion engine hits the combustion chamber wall [2]. Studying the phenomenon of droplet collision, exploring the critical boundary conditions of transitions of different impact results, giving a comprehensive and accurate physical description and scientific explanation, is of great value for understanding the multiphase fluid dynamics mechanism of droplet wall collision process.

Previous studies can be divided into different types according to the impact target: (a) dry solid surface, (b) thin liquid film and (c) deep liquid pool [3]. Moreover, the dynamics of colliding droplets are also different. Studies have shown that the droplet impact on the dry wall surface compared with the liquid wall surface, although some impact patterns are similar, but the mechanisms are fundamentally different [4]. Scholars have done a lot of experimental research on the impact of single droplets on dry and wet surfaces. Worthington and Cole [5,6] conducted their groundbreaking drop impact visualization experiment more than 100 years ago. They employed an ordinary quarter-plate camera and very short duration electric flash to illuminate and freeze the image. Since then, the phenomenon of droplet impingement has attracted many researchers' scientific attention. In 1954, Edgerton [7] used high-speed photography and photographed the famous milk drop crown splash photo.

Rioboo et al [8,9] found six different outcomes of a single droplet impacting a solid wall: Deposition, Prompt splash, Corona splash, Receding breakup, Partial rebound and Rebound. Stow and Stainer [10] measured the number and the size distribution of secondary drops produced by the impacts of single water drops onto thin water films. Rodriguez and Mesler [11] reported that the drop shape on impact had little effect on the criterion of the formation of central jet. Rioboo et al. [12], Wang and Chen [13] and Cossali et al. [14] conducted experiments on the drop impact on thin liquid film to reveal the limit conditions for the formation of secondary droplets from the edge of the liquid crown; A key dimensionless parameter is involved in these studies $K(= We\ Oh^{-0.4})$,which is critical to the outcome of the collision. In addition to classifying the collision outcomes, many researchers have also studied important geometrical parameters during crown evolution. The main geometrical parameters of crown formation are crown diameter, height, angle and thickness. Cossali et al. [14] qualitatively analyzed the effects of the Weber number of incident droplets and liquid film thickness on the crown parameters. The results show that the effect of Weber number on the crown height, diameter and splash droplet size is greater than the liquid film thickness, while the liquid film crown thickness only changes with time, independent of Weber number and liquid film thickness. Liang et al. [15] studied the dependence of crown size on drop Weber number and Reynolds number by changing the impact velocity and drop physical properties, respectively. They show that the crown diameter is independent with *We* and *Re*. The model by Roisman and Tropea [16] shows that non-dimensional the crown height increases appreciably with increasing liquid film thickness.

However, all of these previous analyses have focused on the drop impingement behavior on dry surface or on the interaction behavior between a liquid film and droplets of the same type of liquid. Few studies have collated a liquid film that has a large difference in physical properties from droplets, and little is known to distinguish the dynamics of droplets and liquid films after collision. However, such a situation is common in life. Considering the phenomena that occur in the cylinders of a gasoline direct injection engine, the impingement behavior of the fuel spray on the piston surface or combustion chamber head. We inevitably need to realize that this is the interaction between two kind of liquids, namely, the fuel spray and the oil film that forms on the cylinder wall. Studies [17,18,19,20] have shown that when the fuel spray hits the oil wall of the combustion chamber, the oil on the wall splashes into the combustion chamber, causing preignition. This can cause super knock of internal combustion engines.

In this study, we investigate the phenomenon that a single droplet impacts a liquid film with obvious difference in physical properties. This experiment distinguishes the incident droplet from the liquid film. The effects of different incident droplets Weber number, liquid film physical properties and thickness on the dynamics of the liquid film after collision were also investigated.

2 EXPERIMENTAL SETUP

In experiments to investigate droplet impingement on liquid film, it has been difficult to distinguish between the incident droplet and the liquid film by ordinary visualization of splash phenomena. In the present study, the method of laser-Induced fluorescence (LIF) was applied to discriminate the droplet and the liquid film , and the distribution of the liquid film after impacting was studied. Figure 1 shows the experimental setup. The impingement phenomena of a single droplet were observed by using a high-speed video camera (Photron, 5400fps, 1024×1024 pixels). An continuous laser (532 nm) was used for the light source. The laser is equipped with a Powell prism and a N2852-12 fiber, and a fan-shaped laser with a thickness of 15 mm can be produced directly above the liquid film. A syringe pump was used to produce droplets of uniform size, and they were released from a syringe toward the wall covered with a thin liquid film. By adjusting the lifting platform to change the free fall height of the fuel droplet to control the velocity of the incident droplet. In order to ensure the accuracy of the test, the liquid film was replaced after each shooting, and at least three times were taken under each working condition.

The composition of the incident droplets in this test is ethanol. It has a single composition and is similar in physical and chemical properties to gasoline. 30%, 60%,70%and 80% mass fraction (30 WT%, 60 WT%,70%, 80 WT%) of glycerol and water solution was used to simulate low, medium and high viscosity liquid films. The density and surface tension of these three liquid films are also different, but the difference in viscosity is most significant. Table 1 shows the physical properties of the different proportions of glycerin solution at room temperature and pressure (293K, 0.1MPa) and oil at 373K. Rhodamine-B, which is a fluorescent dye added to the liquid films. Since the dye density was lower than 0.1 WT% in this experiment, the dye had no appreciable effect on the physical properties of the liquid film. Experiments with and without dye were carried out and the results showed no significant differences between the two conditions. The center of the fluorescence wavelength was 580 nm Therefore, the crown and splash droplets produced by collision are irradiated by laser to produce orange fluorescence. After passing through the yellow longpass filter which cut the light at a wavelength below 590 nm, a fluorescent image that indicated the motion of the liquid film was captured by the highspeed camera.

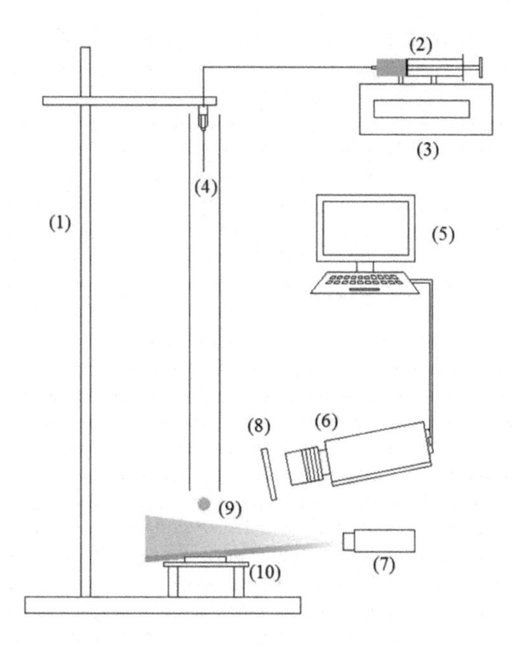

1-lifting platform 2- syringe 3- syringe pump 4-injection needle 5-computer 6-high speed camera 7-laser 8-filter 9-incident droplet 10-liquid film

Figure 1. Experimental setup.

Table 1. Physical properties of the liquid film and oil.

Liquid film (293K)	Density ρ (kg/m^3)	Viscosity μ (mPas)	Surface tension σ (mN/m)
30 WT% Water Glycerite	1070	2.2	66.5
60 WT% Water Glycerite	1150	8.8	64.7
70 WT% Water Glycerite	1188	21.2	64
80 WT% Water Glycerite	1210	45.9	63.4
Oil (373K)	900-950	5.1-24.8	40

Table 2 shows the experimental conditions of this experiment. The dimensionless liquid film thickness H^* is defined as the ratio of liquid film thickness h to incident droplet diameter d.

Table 2. Experimental conditions.

Single drople	Ethanol
Droplet diameter (d)	2.2mm
Droplet velocity (V)	4.4, 4.1, 3.9, 3.7, 3.4, 3.1, 2.8, 2.4 m/s
Liquid film	30WT%, 60WT%,70WT%, 80%WT%
Dimensionless liquid film thickness (H^*)	0.1,0.3,0.5,0.7,1

Compared with ethanol, the physical properties of isooctane and gasoline are more similar. We did a set of comparative experiments. The Weber number of the incident droplets and the dimensionless thickness of the liquid film are guaranteed to be the same, and the viscosity of the liquid film is guaranteed to be similar. Figure 2 is a comparison of the experimental phenomena of isooctane droplets and ethanol droplets impacting on the liquid film of glycerol solution (60WT%). Comparing the behavior of the two kind of droplets, we found that the crown and splash are very similar in the short term after the impact. However, when it was later developed, we saw that isooctane was incompatible with the glycerol solution liquid film, resulting in significant delamination. The ethanol droplets and the glycerol solution liquid film were mutually soluble. We consider that the gasoline and oil in the actual engine are mutually soluble, and the pre-isooctane droplets after the collision with the wall are similar to the collision of the ethanol droplets. Therefore, we chose ethanol as the droplet component in the test, simulated the fuel in the engine, and selected the glycerol solution as the liquid film to simulate the oil film on the cylinder wall in the engine. The glycerol solution with different mass fractions was used to simulate the engine oil at different working temperatures.

It will be noted that the droplet size was larger than the typical droplet size of the fuel spray in a DI gasoline engine in order to eliminate the influence of environmental conditions. The ratio of the droplet diameter to the liquid film thickness was on the same order as that of typical DI gasoline engines.

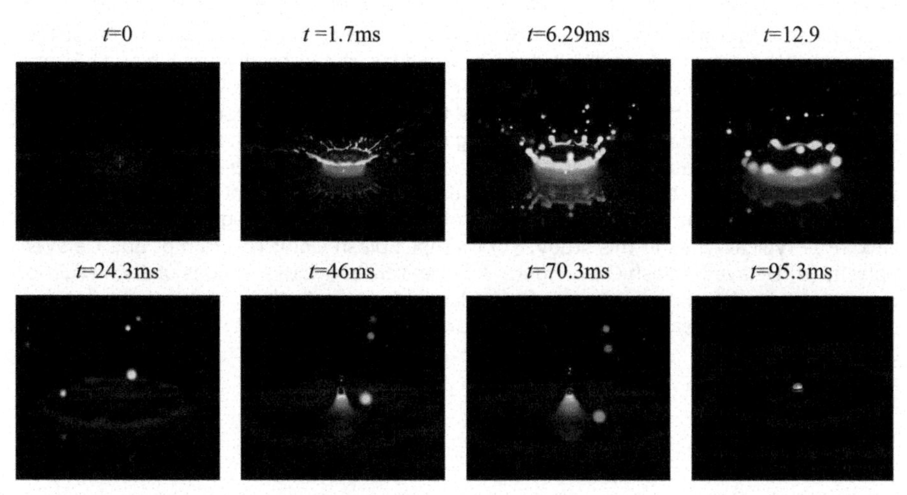

(a) Impingement of isooctane droplets hit on the liquid film of glycerol solution (60 WT%)

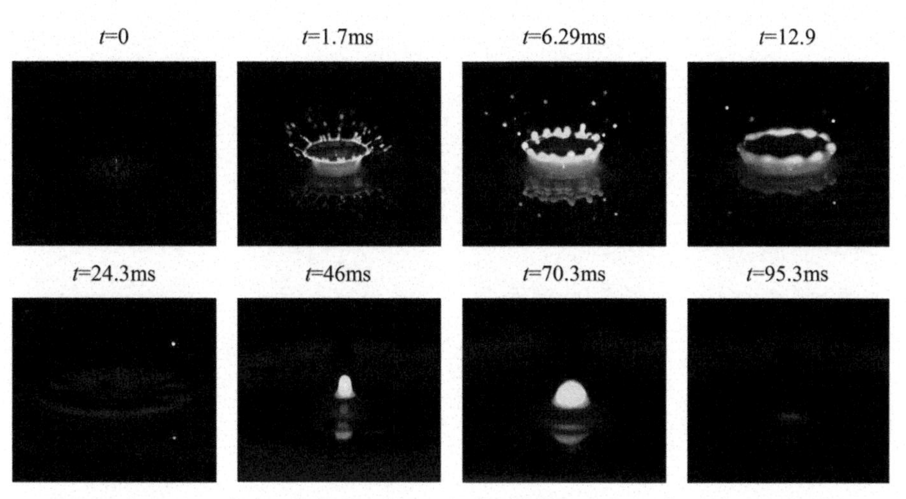

(b) Impingement of ethanol droplets hit on the liquid film of glycerol solution (60 WT%)

Figure 2. Different fuel droplets impingement on the same oil film (We=1075, H^*=0.7).

3 RESULTS AND DISCUSSION

3.1 Morphology classification of liquid film

The behavior of the liquid film after the fuel droplets hitting was observed by high-speed images. Observations were conducted under various conditions. Based on an analysis of the macroscopic observation results for the impingement behavior, After the droplets impingement, the behavior of the liquid film can be divided into two categories according to whether secondary droplets are generated: splash and no splash. When the liquid film does not splash, it will only have a crown-like structure after being impacted. According to whether the crown edge will produce bifurcations, it can be divided into: (a) Stable Crown; (b) Bifurcation Crown; When splash phenomena occurs, two types of splashes can be distinguished: the prompt splash and the delayed splash. For the prompt splash, liquid atomization already takes place during the jetting phase with the ejection of droplets from the crown edge while it is still advancing. The classic splash, called "delayed splash", corresponds to the formation of droplets at the end of the corolla growth [14,21,22]. Based on the two splash forms of liquid film, splash regimes were classified into three typical types in this study: (c) Prompt Splash Only; (d) Prompt and Delayed Splash; (e) Delayed Splash Only. Figure 3 shows representative images of each regime. Images obtained at different times were selected to indicate the distinctive characteristics of each regime clearly. The details of each type of impingement regime are described below.

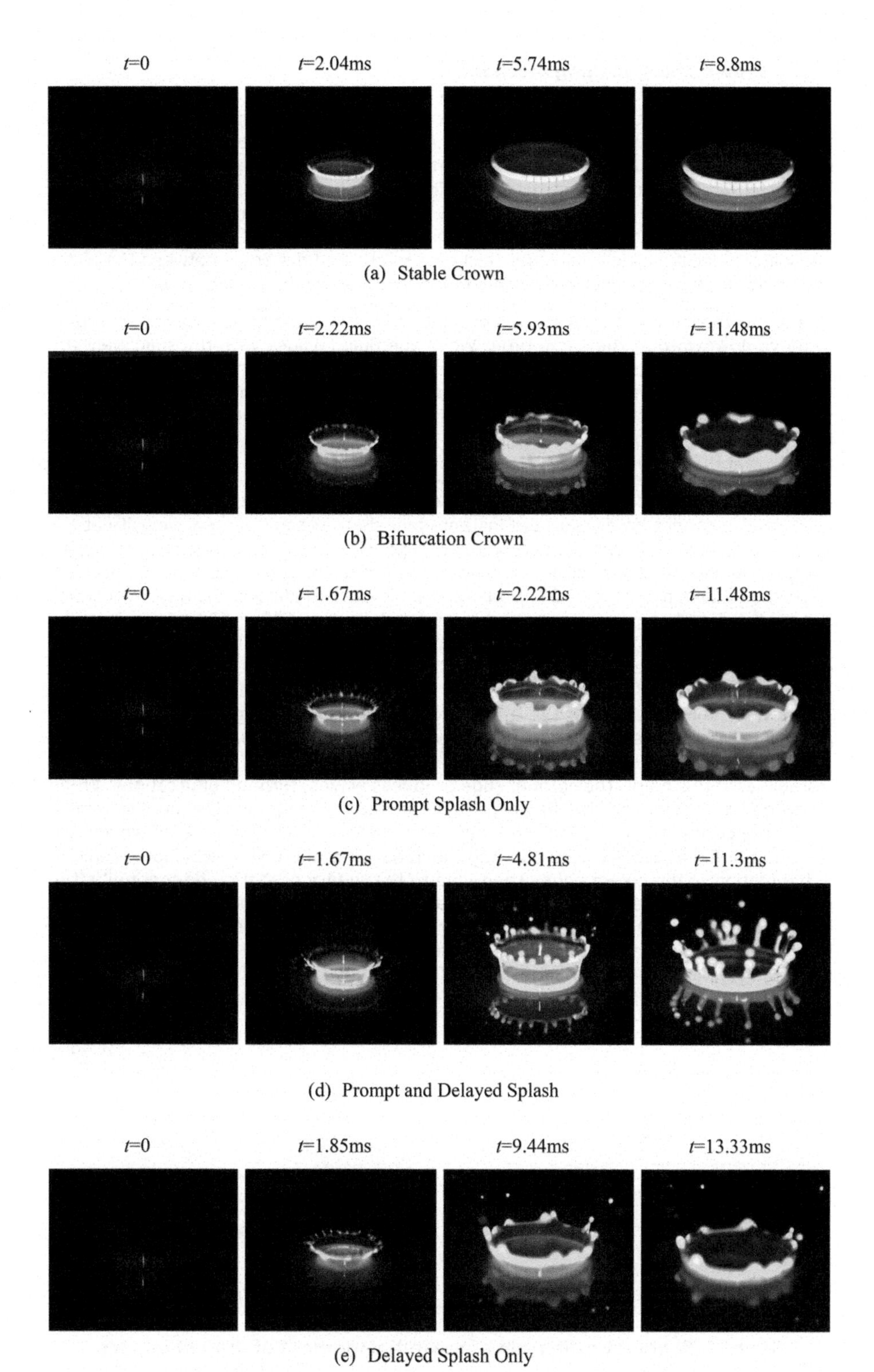

Figure 3. The comparison of different liquid film crown/splashing regimes.

3.2 Crown and splashing mechanism

It can be seen from the Figure 4 (a) and (b) that when the droplet impinges on the liquid film vertically at the velocity V, the liquid film produces an ejecta near the contact line between the droplet and the liquid film. An ejecta is a thin sheet of liquid, is thrown out radially and vertically outward very shortly after initial impact. This is also the basis for the formation of a crown structure. Yarin and Weiss [23] proposed theoretically in 1995 that the coronal splash was caused by the kinematic discontinuity of fluids in the liquid film. Roisman and Tropea [16] then extended this theory of motion discontinuity. The water velocity model, the water thickness model and the angle between the water and liquid film formed under the action of motion discontinuity are presented.

Due to the impact of the droplets, the droplets drive the liquid film in the impact region to move downward at the same time. When the fluid collides with the solid wall, the flow direction changes and gradually deflects horizontally, forming a radial flow in the liquid film. After the fluid flowing in the radial direction encounters the surrounding liquid film, it pushes the liquid film to expand outward. Due to the impact of the radial flowing fluid on the stationary liquid film, an upward fluid flow is formed inside the film, so that the crown structure is continuously expanded.

Assuming the impact time $t=0$, the ejecta originates from the liquid film is at rest for $t \leq 0$, and it cannot be accelerated instantaneously to reach non-zero velocity at t = 0, since this would contradict the fundamental physical law of energy conservation. As analyzed above, forces must be involved to bring it to the moving state, and acceleration should take place for $t \to 0+$. Under certain impact conditions,the time over which the acceleration happens is very short, certainly less than 100 μs [24]. Therefore, we need to pay attention to the instability mechanism to illustrate the occurrence of crown bifurcation. We can see from the test results that the ejecta exhibits some distinct properties at the early stages which in turn dictate the dynamics of the later ejecta evolution. First of all, the ejecta experiences a very large acceleration in an approximately impulsive manner at the early stage of an impact the heavier fluid (liquid film) is accelerated into the lighter one (air), thus is Richtmyer-Meshkov instability. At this point, the coronal edge produces waves, forming bifurcations. When the interface decelerates, the Rayleigh-Taylor instability magnifies the interface ripple due to the curvature effect, but it will not cause a change in the number of waves. We believe that RM instability occurs during the initial time of crown production, resulting in bifurcation of the crown edge. Then, due to the theory of motion discontinuity, the crown continues to develop and produces a splash. Under certain working conditions, such as the impact energy of the droplet is very low, and the viscosity of the liquid film is large, since the acceleration of the liquid film rushing into the air is small. There is no instability, and the smooth crown edge does not fluctuate.

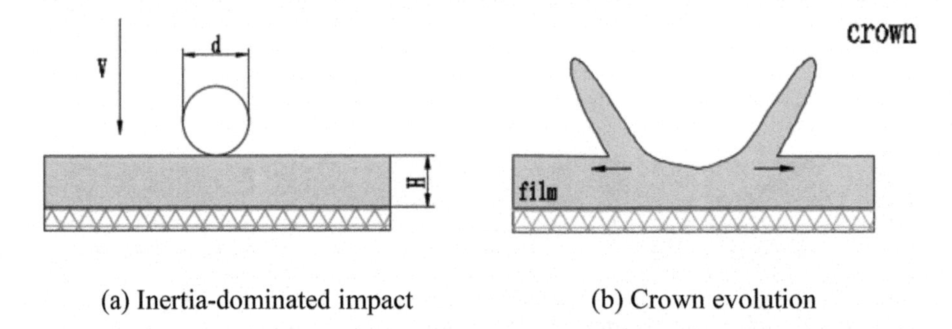

Figure 4. Schematic diagram of the initial moment of droplet impact.

3.3 The crown/splashing limit

Previous studies have shown that the physical parameters of the incident droplet, such as the surface tension, and the velocity have a great influence on weather or not splashing occurs after the incident. Here we define the few classical parameters related to this topic as usual. The Weber number is defined as the ratio of inertia force to surfaces:

$$We = \rho_0 V^2 d / \sigma_0 \tag{1}$$

Where ρ_0, V, d, σ_0 are the incident droplet density, the velocity, the diameter, and the surface tension, respectively.

The Reynolds number is defined as the ratio of inertia force to viscous force:

$$Re = \rho_0 dV / \mu_0 \tag{2}$$

Where μ_0 is the viscosity of the liquid.

The Ohnesorge number *Oh* is defined as:

$$Oh = We^{0.5} / Re \tag{3}$$

Many studies [12,13,25] have shown that the larger the *We* of the incident droplet, the greater the probability of splashing after the collision. Since both surface tension and viscosity have a large influence on the splashing threshold, a separate *We* or *Re* is not sufficient. Therefore, many researchers use the *K*-type relationship to combine surface tension with viscous forces and use the parameters *We* in combination with *Oh*. It was introduced by Mundo et al [26], Yarin and Weiss [23], and defined by:

$$K = We \cdot Oh^{-0.4} \tag{4}$$

Above a critical *K* value, splash is expected. Walzel et al. [27] experimented with water glycerite mixture and concluded that splashing at K=2500. Okawa et al. [28] experimented with water and concluded that splashing at K=2100. However, these conclusions are obtained under different liquid film conditions, such as the specific liquid film dimensionless thickness, H^*. Researchers [22,29,30] have also studied the effect of of H^* on splashing after impact.

In this study, the droplets and the liquid film are not the same kind of liquid, so the previous empirical formula splashing threshold cannot be directly adopted. According to lots of pictures in this experiment and previous research experience, we found that the *We* of the incident droplets, the thickness and the physical parameters of the liquid film all have an influence on the outcome of the liquid film after the collision. So we define the Ohnesorge number of the liquid film:

$$Oh^* = \mu / (\rho \sigma H)^{0.5} \tag{5}$$

Where ρ, H, σ, μ are the liquid film density, thickness, surface tension and viscosity respectively. The liquid film Oh^* value involved in this experiment ranges from 0.00556 to 0.3533.

Figure 5 shows the comparison of different crown or splashing thresholds under various H^*, in the form of *We* versus Oh^*. It can be summarized from the figure that the larger the incident Weber number of the droplet, the more the liquid film tends to splash after the impact. In this study, the maximum Weber number is *We* = 1521, at this *We*, the liquid film splashes in any condition (Within the range of this study), the difference is the type of splash. The Oh^* of the liquid film also has a significant effect on the threshold of liquid film after impact. The smaller Oh^*, the more the liquid film tends to splash. The smaller Oh^* indicates lower viscosity of the liquid film. When the viscosity of the liquid film is large, the internal shear force is enhanced, and the ability of the liquid film to maintain the crown shape stability is stronger. Comparing the graphs (a) to (e) of Figure 4, it is found that as the thickness of the liquid film increases, the We-Oh^* region where the liquid film splashes shrinks, and the splash region is concentrated with a large number of *We* and a small Oh^*. This means that in the range of 0.1 to 1 of H^*, the increase in the thickness of the liquid film suppresses the splash of the liquid film. When the film thickness is small, the type of splash in most splashes is Prompt and Delayed Splash. As the thickness of the liquid film increases, the proportion of Prompt Splash Only increases. Delayed Splash Only generally occurs under conditions where both Oh^* and H^* values are large.

We can conclude that the Weber number of the incident droplets, the viscosity of the liquid film, and the thickness of the liquid film all have an effect on the splash of the liquid film. From the point of view of mutual solubility, when the droplet diameter is constant, the droplet Weber number increases, which means that the incident velocity is increased, which accelerates the fusion speed of the droplet and the liquid film. Secondly, the smaller the thickness of the liquid film, the faster the droplet will reach the wall after it hits the liquid film, which also makes the droplet and the liquid film fuse with each other faster, because the total amount of liquid film is less, so it is diluted by the droplet. So that when the oil film thickness is small, the splash is more intense. When the thickness of the liquid film becomes large ($H^*>0.5$), the amount of the liquid film is large, and the degree of dilution by the droplet is small, so that the influence of the thickness is weakened. The smaller the viscosity of the liquid film, the easier the liquid film and the droplets fuse. Based on this analysis, we believe that the droplets that were splashed in the prompt are generated by the ejecta formed by the liquid film being hit by the droplets, so the composition of the prompt splash is the liquid film. The delayed splash is produced at the crown edge, which we consider to be a mixture of droplets and liquid film.

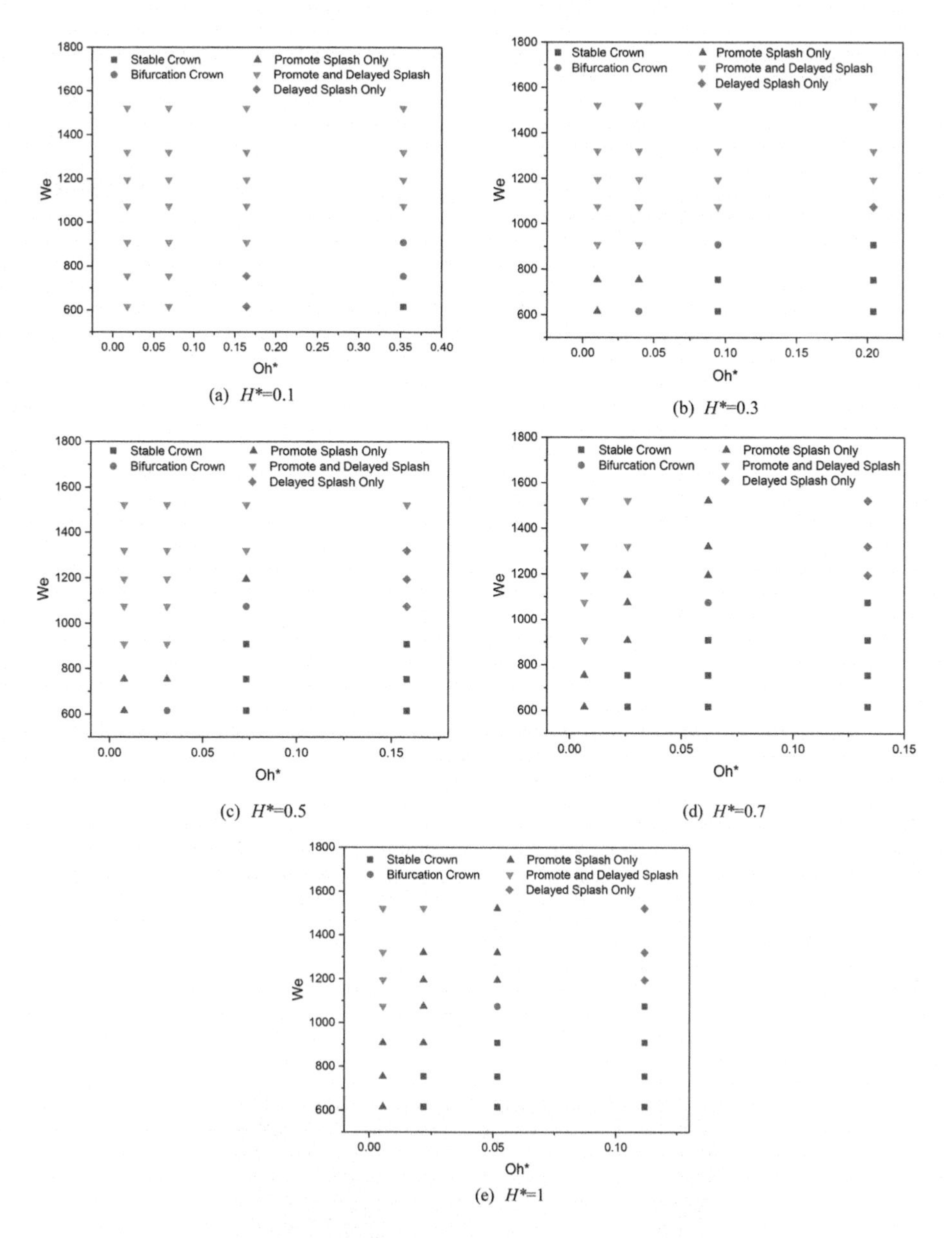

(a) H^*=0.1 (b) H^*=0.3 (c) H^*=0.5 (d) H^*=0.7 (e) H^*=1

Figure 5. Liquid film regime classification under various liquid film thicknesses.

Figure 6 shows all the cases studied combined on one graph. Based on the above conclusions, the experimental data were statistically analyzed and curve-fitting, and the relationship between the impact droplet critical Weber number and the liquid film Oh^* for liquid film regime transition from non-splash to splash in this experiment was obtained. Define We_{cr} as the critical Weber number of splashing , and define the curve of We_{cr} in the We, Oh^* map as

C-S limit. As shown in Figure 5, the C-S limit curve can be expressed as the following equation:

$$We_{cr} = 1182.92 - 637.22e^{-15.52Oh^*} \tag{6}$$

Applying the fitted formula, it is possible to predict whether the liquid film splashes when the incident liquid droplet impacts a different liquid film.

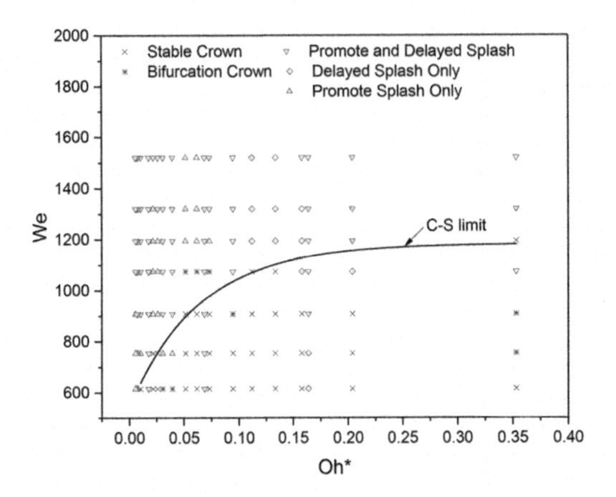

Figure 6. *We-Oh map of liquid film regimes under all experimental conditions.**

3.4 The crown evolution

As shown in Figure 7, we define the maximum height value at which the crown reaches its maximum: H_c, and the diameter of the coronal upper end at the corresponding moment is crown diameter D_C. The dimensionless crown height H_c* and the dimensionless crown diameter D_C* are defined as:

$$Hc^* = H_c/d \tag{7}$$

$$Dc^* = D_c/d \tag{8}$$

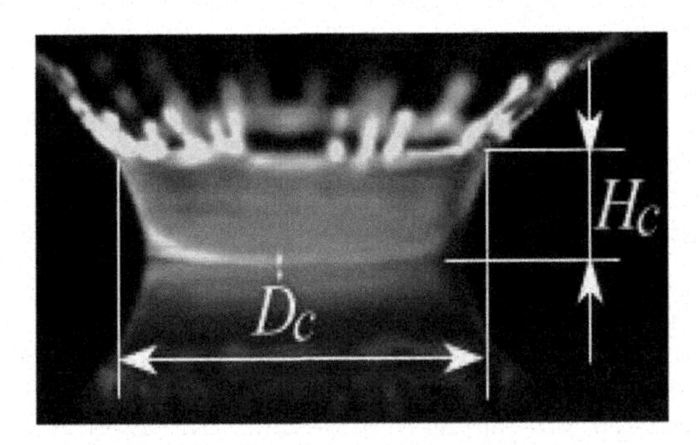

Figure 7. Parameters of the crown.

3.4.1 ***Crown height***

Figure 8 shows the relationship of H_c* with the droplet *We* in each of the liquid films *Oh** corresponding to $0.1 \leq H^* \leq 1$. From the figure, we find that H_c* increases linearly with the increase of *We* when the liquid film *Oh** is constant, which is consistent with the conclusions in previous studies [31]. This is because the larger the incident droplet *We*, the greater the inertial force and kinetic energy of the incident droplet, the more energy that is transmitted to the liquid film during impact, resulting in a higher crown. When the Weber number of the incident droplets and the dimensionless thickness of the liquid film are kept constant, H_c* decreases as the liquid film *Oh** increases. The ability of the liquid film to resist deformation increases with the increase of Oh*, so the increase of the liquid film *Oh** leads to a decrease in the H_c* value. Comparing Figures 6 (a) ~ (e), it is found that H_c* increases first and then decreases with increasing liquid film thickness. When the liquid film is thin, the impact energy transmitted by the droplets to the liquid film is small, and more energy is converted into crown radial expansion energy instead of the axial growth energy, so that a relatively low crown is produced. As the thickness of the liquid film increases, the ratio of the liquid film absorbing the droplet energy increases, the axial growth energy increases, and the crown height increases accordingly. When the thickness of the liquid film is increased to a certain value, the thickness of the formed crown and the quality of the liquid film contained therein increase. Limited by the conservation of energy, the crown rise requires more energy to increase the potential energy, so that the crown height begins to decline.

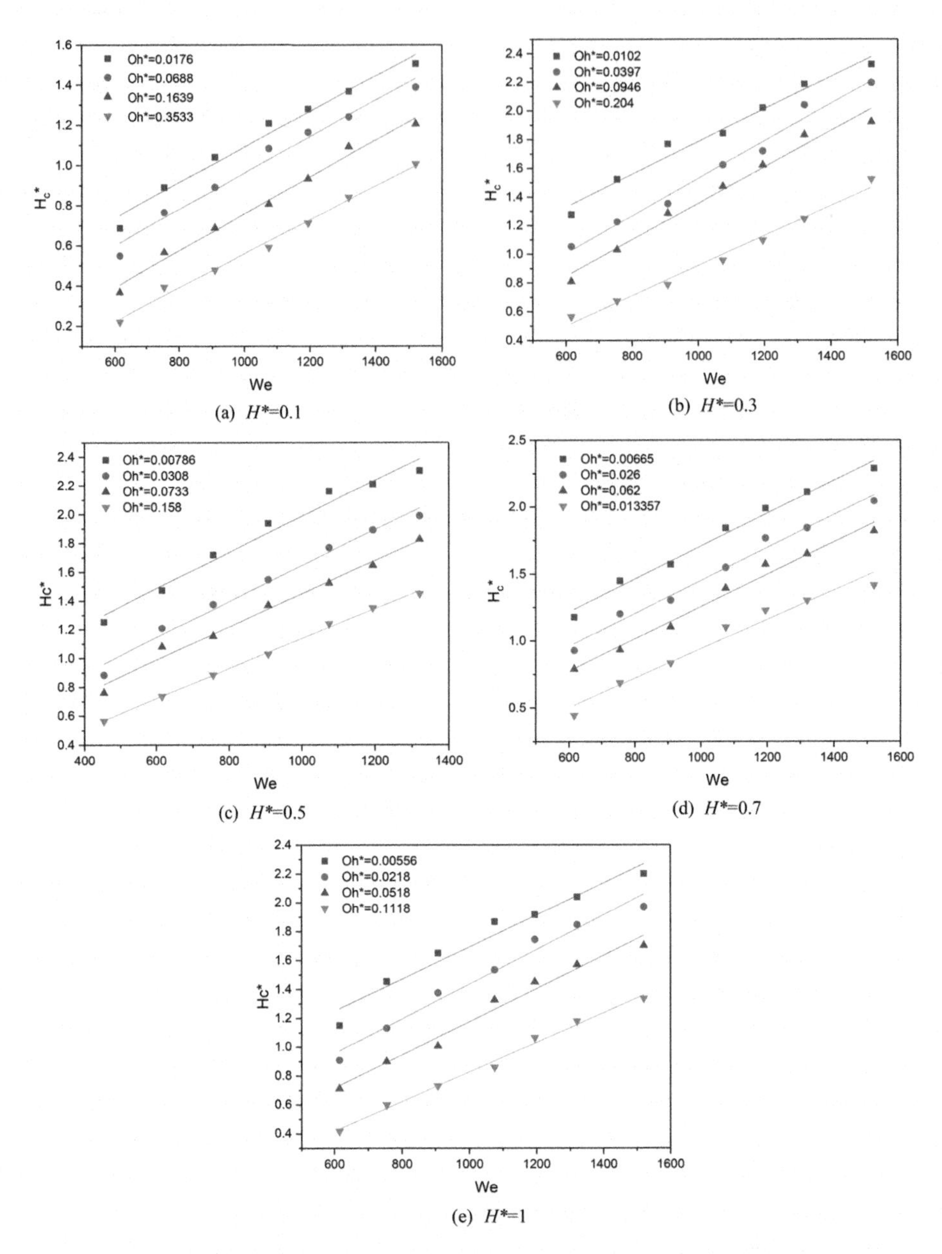

Figure 8. Dimensionless liquid film crown maximum height versus the droplet We under various liquid film thicknesses.

Assume that the relationship between H_c* and the *We* number of incident droplets is:

$$H_C{}^* = k_1 We + b_1 \tag{9}$$

According to the previous analysis, it was found that Hc* is related to the liquid film *Oh** in addition to the incident droplet *We*. Therefore, we explored the relationship between

k_1, b_1 in formula (9) and liquid film Oh^*. Figures 9 (a) and (b) are plots of k_1 and b_1 as a function of Oh^*, respectively. From image (a) we find that the value of k_1 is nearly constant with the change in the Oh^* value of the liquid film. By analyzing the data, we conclude that $k_1 = 0.00113$. The value of b1 in image (b) tends to decrease with increasing Oh^*. The relationship between b1 and liquid film Oh^* is obtained by fitting, as follows:

$$b_1 = -0.2\ln\left(Oh^* - 7.019 \times 10^{-4}\right) - 0.48 \tag{10}$$

Based on this result, H_c^* is expressed by the following equation:

$$H_C^* = 0.00113We - 0.2\ln\left(Oh^* - 7.019 \times 10^{-4}\right) - 0.48 \tag{11}$$

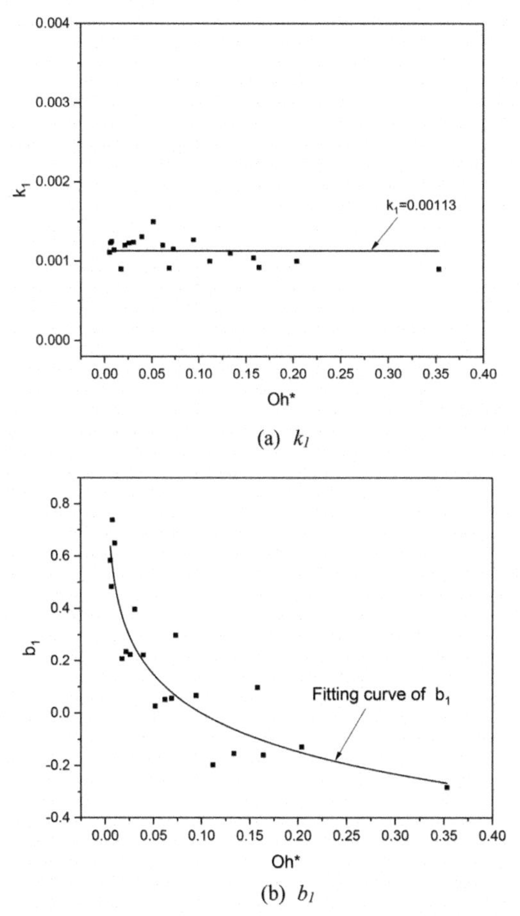

(b) b_1

Figure 9. Slope and Intercept in *Hc*-We* relation versus *Oh of liquid film.**

3.4.2 *Crown diameter*

Figure 10 (a)~(e) illustrates the relationship of D_c^* with the droplet *We* in each of the liquid films Oh^* corresponding to $0.1 \leq H^* \leq 1$. Similar to the height of the crown, D_c^*

also increases linearly with the increase of the incident droplet *We*. At the same liquid film thickness and the same incident droplet *We*, D_c* decreases as the *Oh** of the liquid film increases. By comparing the figures (a) ~ (e) we found that D_c* is not sensitive to the thickness of the liquid film.

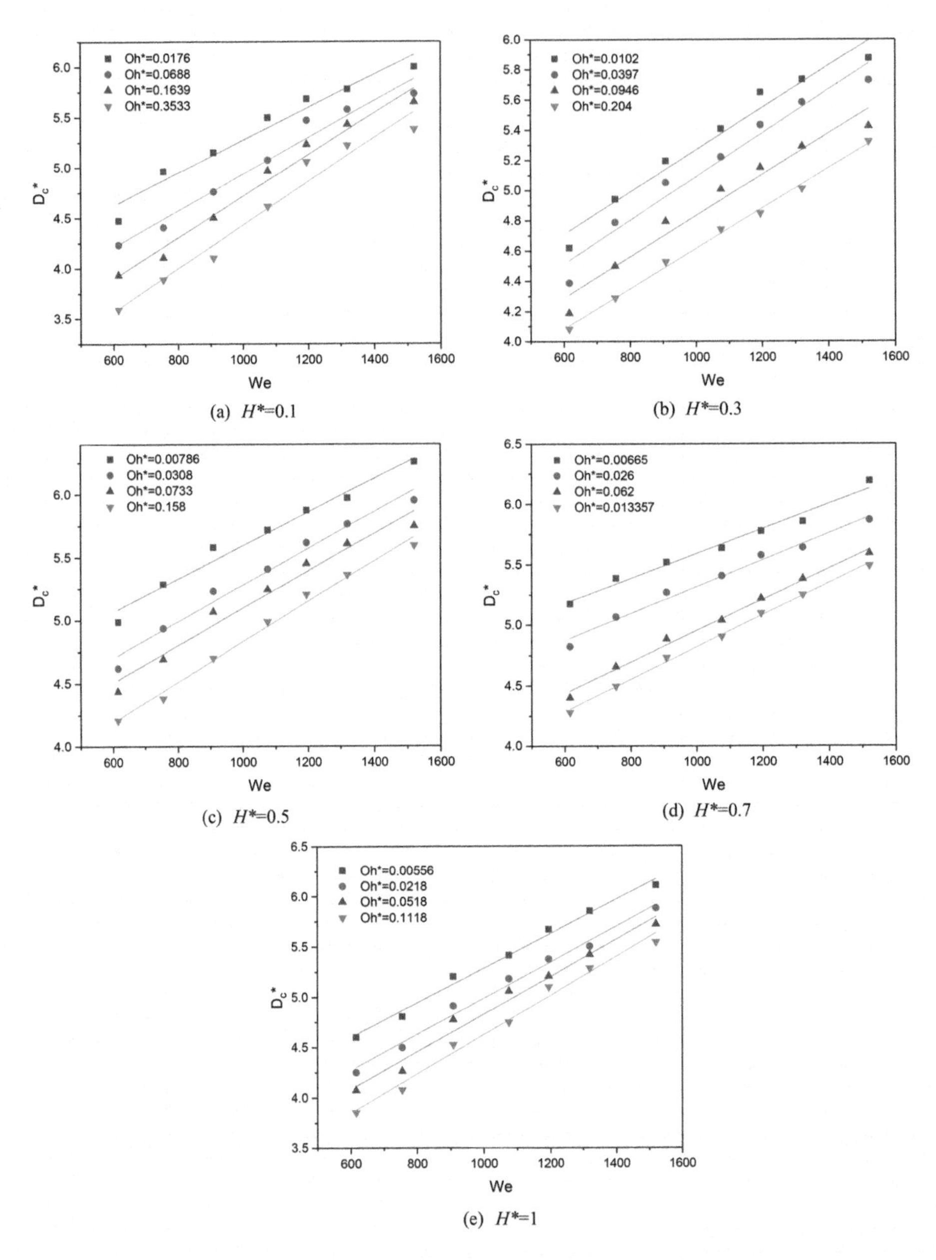

Figure 10. Dimensionless liquid film crown diameter versus the droplet *We* under various liquid film thicknesses.

We express the relationship between D_c* and the *We* number of incident droplets is:

$$D_C^* = k_2 We + b_2 \tag{12}$$

In order to explore the relationship between liquid film properties and k_2, b_2, we plot the relationship between k_2, b_2 and liquid film *Oh** in Figure 11 (a), (b), respectively. From figure (a) we can conclude that k_2 is not sensitive to *Oh** and remains almost constant at each *Oh** value. By fitting the data, as shown: k_2 =0.00155. Figure (b) shows that the value of b_2 decreases with increasing *Oh**. Through data analysis and fitting, the following approximation equations for b_2 can be obtained:

$$b_2 = -0.531 \ln(Oh^* + 0.215) + 2 \tag{13}$$

The following equation was then obtained by approximating the result:

$$D_C^* = 0.00155 We + -0.531 \ln(Oh^* + 0.215) + 2 \tag{14}$$

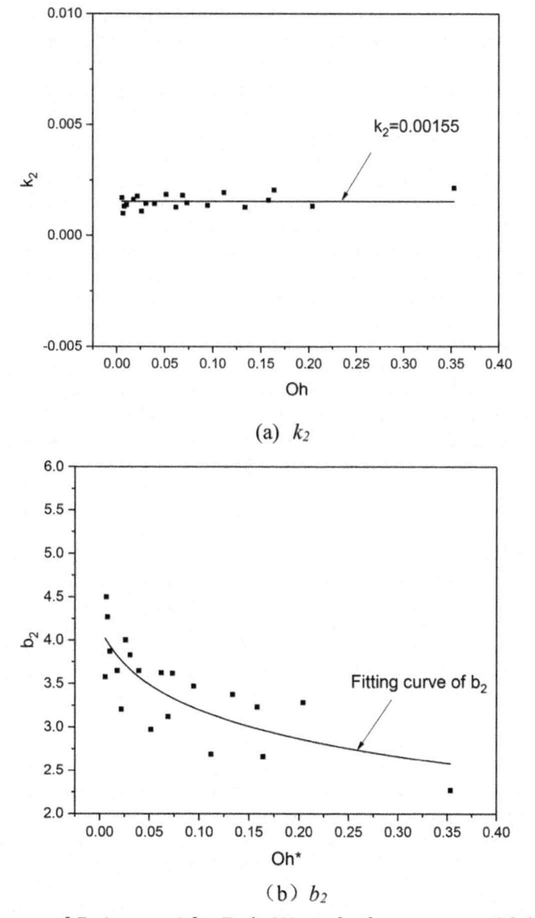

(a) k_2

(b) b_2

Figure 11. Slope and Intercept in D_c*-*We* relation versus *Oh of liquid film.**

4 CONCLUSION

The LIF method was used in this experiment to explore the development of liquid film after a single droplet impacted the liquid film with different physical properties. The effects of the droplet Weber number, the physical parameters of the liquid film such as viscosity, and the thickness of the liquid film on the dynamic of the liquid film after impact were systematically analyzed. The morphology of the liquid film after impact was classified, and the formation mechanism of various regimes was explained. According to the test results, the crown/splashing limit and the empirical formula of the dimensionless crown parameter value are obtained. It is possible to predict the dilution and the motion of the oil after the fuel in the engine hits the oil film on the cylinder wall. It provides experimental data and theoretical basis for optimizing the fuel spray impingement model. The analysis of the data revealed the following:

(1) The morphology of liquid film after being hit can be divided into: (a) Stable Crown; (b) Bifurcation Crown; (c) Prompt Splash Only; (d) Prompt and Delayed Splash; (e) Delayed Splash Only.
(2) RM instability occurs during the initial time of crown production, resulting in bifurcation of the crown edge. Due to the theory of motion discontinuity, the crown continues to develop and produces secondary droplets.
(3) The increase of the Weber number of incident droplets promotes dilution of the liquid film, while the increase in viscosity and thickness of the liquid film inhibits dilution of the liquid film. The component in the prompt splash is from liquid film, and the component in the delayed splash is a mixture of the liquid film and the droplet.
(4) A correlation for the crown/splashing limit (under the condition: $0.1 \leq H^* \leq 1$) was proposed in the form: $\mathrm{We_{cr}} = 1182.92 - 637.22e^{-15.52Oh^*}$
(5) The crown height and diameter of the liquid film are related to the Weber number of the droplets and the physical properties of the liquid film.
H_c* and D_c* can be predicted using the following empirical formula:

$$H_C{}^* = 0.00113\mathrm{We} - 0.2\ln(\mathrm{Oh}^* - 7.019 \times 10^{-4}) - 0.48\ ;$$

$$D_C{}^* = 0.00155We \pm 0.531\ln(Oh^* + 0.215) + 2$$

H_c* and D_c* increases linearly with the increase of We, while decrease as the liquid film Oh* increases, when H* is constant. when the liquid film Oh* is constant. H_c* increases first and then decreases with increasing liquid film thickness, while D_c* is not sensitive to the thickness of the liquid film.

ACKNOWLEDGMENT

This work was supported by the National Natural Science Foundation of China (51676136). The experimental facilities were supported by the State Key Laboratory of Engines, Tianjin University. The authors would like to express sincere thanks to Zhaoxu Chen, Xiang Li, Yi Liu for their assistance.

REFERENCES

[1] V. Bergeron, D. Bonn, J.Y. Martin, et al. Controlling droplet deposition with polymer additives[J]. Nature, 2000, 405:772–775.
[2] M.R.O. Panão, A.L.N. Moreira. Flow characteristics of spray impingement in PFI injection systems[J]. Exp. Fluids, 2005, 39:364–374.
[3] C. Josserand, S.T. Thoroddsen, Drop impact on solid surface, Annu. Rev. Fluid Mech. 48 (2016) 365–391.

[4] J.C. Bird, S.S.H. Tsai, H.A. Stone, Inclined to splash: triggering and inhibiting a splash with tangential velocity, New J. Phys. 11 (6) (2009) 063017.
[5] A.M. Worthington, A second paper on the forms assumed by drops of liquids falling vertically on a horizontal plate, Proc. R. Soc. Lond. 25 (171–178) (1876) 498–503.
[6] A.M. Worthington, R.S. Cole, Impact with a liquid surface studied by the aid of instantaneous photography. Paper II, Philos. Trans. R. Soc. A 194 (1900) 175–199.
[7] H.E. Edgerton, J.R. Killian, Flash!: Seeing the Unseen by Ultra High-speed Photography, C.T. Branford Co., Boston, 1954.
[8] Rioboo R, Tropea C, Marengo M. Outcome from a drop impact on solid surfaces[J]. Atomization and Sprays, 2001, 11:155–165.
[9] Rioboo R, Tropea C, Marengo M. Time evolution of liquid drop impact onto solid, dry surfaces[J]. Experiments in Fluids, 2002, 33:112–124.
[10] Stow C D, Stainer R D. The Physical Products of a Splashing Water Drop[J]. Journal of the Meteorological Society of Japan. Ser. II, 1977, 55(5):518-532.
[11] Rodriguez F, Mesler R. Some drops don't splash[J]. Journal of Colloid & Interface Science, 1985, 106(2):347-352.
[12] Rioboo R, Bauthier C, Conti J, et al. Experimental investigation of splash and crown formation during single drop impact on wetted surfaces[J]. Experiments in Fluids, 2003, 35(6):648-652.
[13] Wang A B, Chen C C. Splashing impact of a single drop onto very thin liquid films[J]. Physics of Fluids, 2000, 12(9):2155.
[14] Cossali G E, Coghe A, Marengo M. The impact of a single drop on a wetted solid surface[J]. Experiments in Fluids, 1997, 22(6):463-472.
[15] G. Liang, Y. Guo, S. Shen, Y. Yang, Crown behavior and bubble entrainment during a drop impact on a liquid film, Theor. Comp. Fluid Dyn. 28 (2) (2014) 159–170.
[16] I.V. Roisman, C. Tropea, Impact of a drop onto a wetted wall: description of crown formation and propagation, J. Fluid Mech. 472 (2002) 373–397.
[17] K. Fujimoto, M. Yamashita, et al. Engine Oil Development for Preventing Pre-Ignition in Turbocharged Gasoline Engine [C]. SAE 2014-01-2785.
[18] S. F. Dingle, A. Cairns, et al. Lubricant Induced Pre-Ignition in an Optical SI Engine [C]. SAE 2014-01-1222.
[19] S. Palaveev, M. Magar, et al. Premature Flame Initiation in a Turbocharged DISI Engine - Numerical and Experimental Investigations [C]. SAE 2013-01-0252.
[20] O. Welling, N. Collings, et al. Impact of Lubricant Composition on Low-speed Pre-Ignition [C]. SAE 2014-01-1213.
[21] Motzkus C, Géhin E, Gensdarmes F. Study of airborne particles produced by normal impact of millimetric droplets onto a liquid film[J]. Experiments in Fluids, 2008, 45(5):797-812.
[22] Motzkus C, Gensdarmes F, E. Géhin. Study of the coalescence/splash threshold of droplet impact on liquid films and its relevance in assessing airborne particle release[J]. Journal of Colloid & Interface Science, 2011, 362(2):540-552.
[23] Yarin, A., Weiss, D. Impact of drops on solid surfaces: Self-similar capillary waves, and splashing as a new type of kinematic discontinuity. Journal of Fluid Mechanics, 1995,283, 141-173.
[24] Krechetnikov R, Homsy G M. Crown-forming instability phenomena in the drop splash problem[J]. Journal of Colloid and Interface Science, 2009, 331 (2):555-559.
[25] Rodriguez F, Mesler R. Some drops don't splash[J]. Journal of Colloid & Interface Science, 1985, 106(2):347-352.
[26] Mundo C, Sommerfeld M, Tropea C. Droplet-wall collisions: Experimental studies of the deformation and breakup process[J]. International Journal of Multiphase Flow, 1995, 21(2):151-173.

[27] P. Walzel, Zerteilgrenze beim Tropfenprall, Chem. Ing. Tech. 52 (1980) 338–339.
[28] T. Okawa, T. Shiraishi, T. Mori, Production of secondary drops during the single water drop impact onto a plane water surface, Exp. Fluids 41 (6) (2006) 965–974.
[29] X. Gao, R. Li, Impact of a single drop on a flowing liquid film, Phys. Rev. E 92 (5) (2015) 053005.
[30] Q. Huang, H. Zhang, A study of different fluid droplets impacting on a liquid film, Petrol. Sci. 5 (1) (2008) 62–66.
[31] S. Asadi, M. Passandideh-Fard, A computational study of droplet impingement onto a thin liquid film, Arab. J. Sci. Eng. 34 (2B) (2009) 505–517.

SESSION 5: INCREASING EFFICIENCY AND REDUCING EMISSIONS

Evolution of RDE Plus – the road to chassis to Engine-in-the-Loop vehicle and engine development methodology

P. Roberts, A. Mason, S. Whelan

HORIBA MIRA Ltd, Global Development & Application Centre, UK

R. Mumby, L. Bates

HORIBA MIRA Ltd, Horizon Scanning Department, UK

ABSTRACT

A Road to Rig (R2R) whole vehicle development, calibration and verification approach known as Real Driving Emissions Plus (RDE+) is being developed at HORIBA MIRA using road, chassis dynamometer, Engine-in-the-Loop (EiL) and virtual toolsets. This will enable real world scenarios to be deployed further upstream of the vehicle and engine programme with the aim of reducing development timescales and costs that will inevitably increase due to RDE regulations.

Reported in this paper are the methodologies developed thus far to achieve Real Driving Emissions (RDE) replication (driving style, altitude and temperature) using chassis and engine dynamometers, EiL co-simulation and use of RDE vehicle test validation.

1 INTRODUCTION

In late 2017, the Worldwide Harmonised Light Vehicles Test Procedure (WLTP) and Cycle (WLTC) were implemented throughout Europe. The WLTP addresses many of the inadequacies of the previous laboratory cycles, but the laboratory-based environment in which the WLTP is undertaken still carries many of the limitations associated with vehicles being tested in a controlled environment.

In a real-world driving scenario, changes to ambient temperature and pressure, driving style and road characteristics (which are not reflected within the WLTP) can result in substantially different real-world fuel consumption and emissions. To address this issue, a supplementary on road test procedure called Real Driving Emissions (RDE) utilising a Portable Emissions Measuring System (PEMS) is mandatory. At this time, RDE testing is only used to ensure criteria emissions compliance (NOx and PN) for type approval; CO_2 emissions are not currently used for any type of approval system.

By implementing RDE, the operating window of the engine is significantly expanded compared with the New European Drive Cycle (NEDC) and WLTC (Figure 1). Not only does this ensure that the engine and aftertreatment technologies selected by vehicle manufacturers must be capable of meeting emissions legislation across a much wider proportion of the vehicle operational map, it naturally results in increased engine and vehicle calibration activities.

In addition to the requirement for increased calibration effort, the implementation of RDE will require advances in engine technology by virtue of the engine operating in areas of the engine speed and load map that currently requires some level of component protection (low speed, high torque and full power).

For a manufacturer to demonstrate compliance for all worst case RDE conditions, many more prototype vehicles, powertrains and tests are required compared to the pre-2019 situation. This is cost-prohibitive and leads to increased development time-scales; a knock-on effect of this is a narrowing of vehicle and engine configurations that are available to consumers. To avoid increased time and cost to achieve RDE compliance, a Road to Rig (R2R) methodology known as RDE Plus (RDE+) is being developed. This methodology will enable the frontloading of vehicle and powertrain development and calibration programmes using on-road, rig (chassis dynamometer) and combined Engine-in-the-Loop (EiL) technology.

The R2R programme discussed in the current document is similar in design to that outlined in [1,2]. Whilst the authors accept that front-loading of vehicle and engine development is not a technological breakthrough, the design and development of the toolsets and testing procedures (some of which are outlined in the current report), are novel. This will enable an Original Equipment Manufacturer (OEM) to rapidly and efficiently explore all permutations of the RDE extended boundary conditions using both physical and virtual testing. Due to pending patent applications, it is not possible to disclose all aspects of the RDE Plus programme within the paper, but the authors aim to give a true and accurate overview of the whole approach.

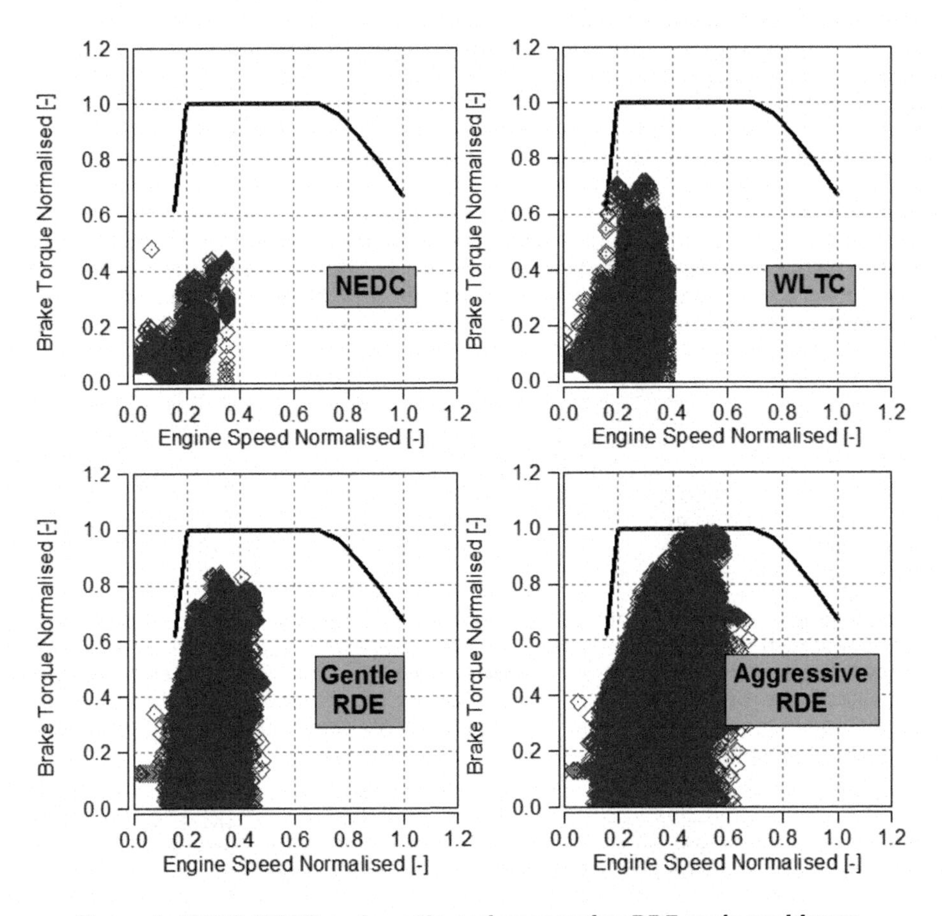

Figure 1. NEDC, WLTC and gentle and aggressive RDE cycle residence.

Discussed in this document is an introduction to the development vehicles, RDE test locations, instrumentation and the test processes developed for the road. Following this is a section dedicated to the chassis dynamometer testing that includes a description of the Advanced Emissions Test Centre (AETC) facility for chassis dynamometer testing, the replication of road load on a chassis dynamometer and tools development. Finally, an outline of the EiL configuration and initial EiL results are presented.

2 METHODOLOGY

2.1 Road testing methodology, vehicle, routes and procedure overview

The road-testing methodology and results for the RDE+ programme have been described in detail in [3]. For brevity, the road testing has been summarised in the bullet points and tables below with the initial results shown in the following section.

- Three vehicles (gasoline, diesel and diesel-hybrid) (Table 1), tested over five different ambient conditions over four locations throughout Europe (Table 2) covering the RDE extended boundary conditions of ambient temperature, altitude and driver "aggressivity".
- Vehicle and engine performance and emissions measured using bespoke instrumentation and HORIBA OBS-ONE PEMS units.
- Coastdown tests performed with each vehicle to calculate finite coastdown terms for use when each vehicle is tested on the chassis dynamometer.

Table 1. Vehicle and instrumentation summary.

Euro Vehicle Segment	**B (2016)**	**C (2015)**	**M (2017)**
Body Type	5 Door Hatchback		5 Door MPV
Engine (Power) and Transmission	Gasoline, 3 Cyl, 998cc, Turbo-charged (74kW) 5 spd manual	Diesel, 4 Cyl, 1499cc, Turbo-charged (88kW) 6 spd manual	Diesel, 4 Cyl, 1461cc, Turbo-charged, 48v Hybrid (81kW) 6 spd manual
EU Emission Std	EURO 6b		
Aftertreatment	TWC	EGR + DOC/ DPF/LNT	EGR + DOC/ DPF/LNT
Mass in Service [kg]	1130	1399	1583
PEMS	HORIBA OBS ONE GS12 Gas & Particle Number		
Amb. Temp. Rel. Humidity, Altitude	HORIBA OBS ONE Weather Station HORIBA OBS ONE GPS		
Base Instrumentation	National Instruments CompactDAQ System: NI 9185/9862/ 9214/9205		
Driveshaft Strain Gauges	Astech Electonics Rotary Telemetry System (RE3D)		

Table 2. RDE test matrix and route summary.

RDE Testing Phase	Location	Target Temperature [°C]	Average Altitude [m]	Total/Urban Cumulative Positive Elevation (CPE) [m/100km]	Distance Split [%] (Urban/ Rural/ MWay)
1	Innsbruck, Austria	-7-0	623.2	498.1/579	32.5/31.2/36.3
2	Nuneaton, UK	0-10	105.1	491.0/611.6	37/33.8/29.2
3		10-20			
4	Spain, Vera	30-35	103.9	837.1/950.4	39/30.4/30.6
5	Spain, Avila		1137.9	953.1/1022.4	33.3/30.3/36.4

2.1.1 *Emissions results overview*

Across all four different locations and five phases, a total of 124 RDE tests were conducted (including some repeats) with 9 out of 10 tests achieving all RDE regulation dynamic trip criteria in addition to successfully logging all emissions data.

To highlight the challenging conditions for complete RDE compliance across all combinations of ambient temperature, altitude and driving style, shown in Figure 2 (gasoline vehicle), Figure 3 (diesel vehicle) and Figure 4 (diesel-hybrid vehicle) are distance specific NOx and PN emissions for all RDE compliant tests at the locations listed in Table 2. Squares and circles denote gentle and aggressive drives respectively. All RDE emissions data from each vehicle is normalised to that of the corresponding WLTC emissions data (expressed mathematically as the ratio of RDE to WLTC emissions minus one) - represented by the diamond. In the case of multiple tests at the same condition, a mean value of all the results was taken. All emissions data shown includes a cold start as required in line with the RDE regulations.

In all cases, the distance specific CO_2 values give a good indication of work done, capturing the effects of both Cumulative Positive Elevation (CPE) and driving style; higher CO_2 is indicative of more "aggressive" driving compared with a "gentle" drive on an equivalent route. However, this trend does not always hold for NOx and PN where the results are sometimes contrary to expectations. In the case of the gasoline vehicle, NOx is typically higher for gentle drives compared to aggressive drives on the same route (the opposite trend is apparent for PN in most cases as is to be expected) yet this trend reverses for both diesel vehicles. One theory for the lower NOx emissions with more aggressive driving for the gasoline vehicle is that the TWC is operating at higher temperature and thus achieving maximum NOx reduction efficiency compared with gentle driving which may result in the TWC "lighting-out" and falling outside the NOx reduction temperature window. For the two diesel vehicles, gentle drives result in the engine operating with very high Air Fuel Ratios (AFR) which result in low NOx production whereas the aggressive drives result in close to stoichiometric operation where peak NOx formation occurs (approximately Lambda 1.1).

Since the vehicles on-test were not WLTP and RDE compliant, it is not within the scope of this study to evaluate the specific emissions trends of different powertrain technologies, but only to highlight the potential for their significant variance when RDE testing within the moderate and extended boundary conditions.

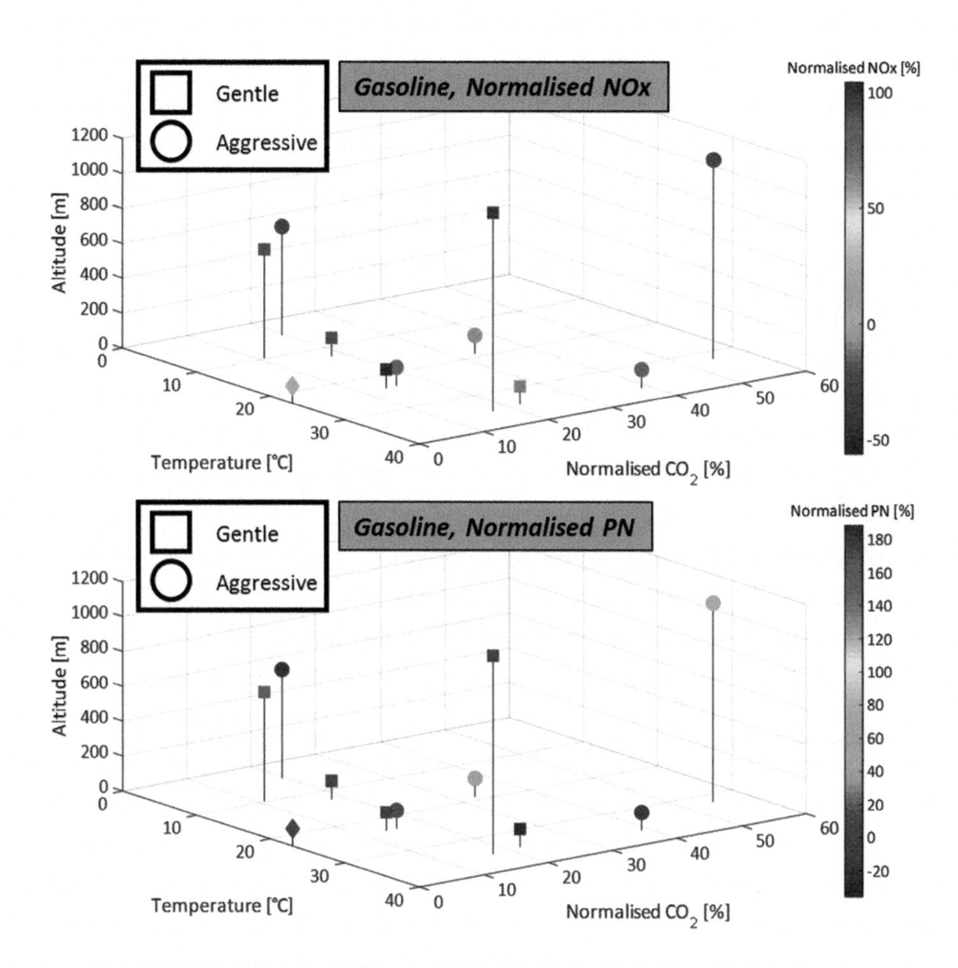

Figure 2. Gasoline vehicle normalised NOx and PN for all RDE results.

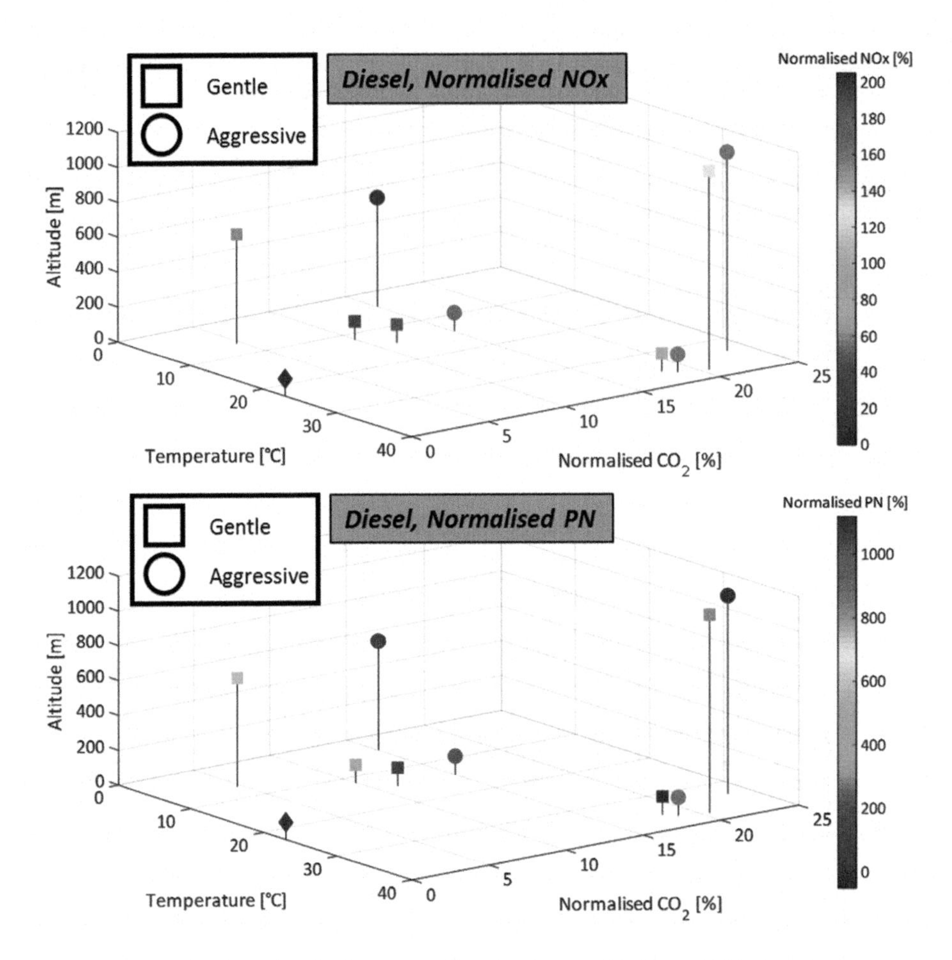

Figure 3. Diesel vehicle normalised NOx and PN for all RDE routes.

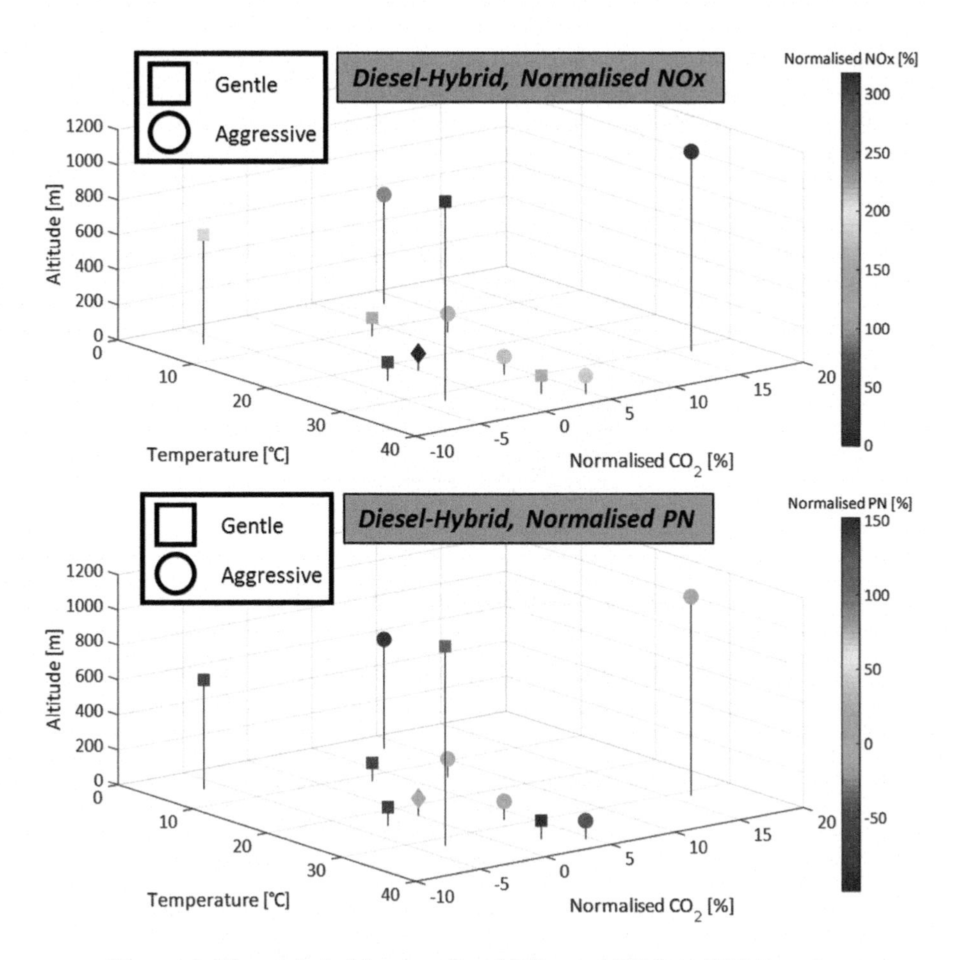

Figure 4. Diesel-hybrid normalised NOx and PN for all RDE routes.

2.2 Chassis dynamometer testing

2.2.1 *Chassis dynamometer specification and capability*

The end goal of performing RDE activities virtually and via an EiL system is likely to be extremely beneficial for OEMs developing new powertrain architectures to meet future emissions regulations. Moreover, the ability to perform RDE on chassis testbeds will also prove similarly useful. For example, understanding how development work on engine testbeds transfers to a full vehicle powertrain, aiding in benchmarking activities for third party suppliers (lubricants, fuels, aftermarket parts etc) and supporting regulatory bodies.

Chassis dynamometer testing for the RDE+ project has been conducted in the AETC located at HORIBA MIRA in Nuneaton, UK. This facility, commissioned in 2017, incorporates the latest emissions test equipment and was designed specifically for WLTP certification and RDE replication/development. As such, it embodies several features that enhance its capability in this regard.

Shown in Figure 5 is the general layout of the facility and listed in Table 3 are the major components and respective specifications. The dynamometer is a HORIBA Vulcan EVO all-wheel drive electric machine which includes a high capacity cooling fan. Fan speed is synchronized to the dynamometer to provide cooling for the vehicle. Test scheduling is regulated by the HORIBA STARS VETS test automation system with HORIBA SPARC being used to control the dynamometer's electric machinery.

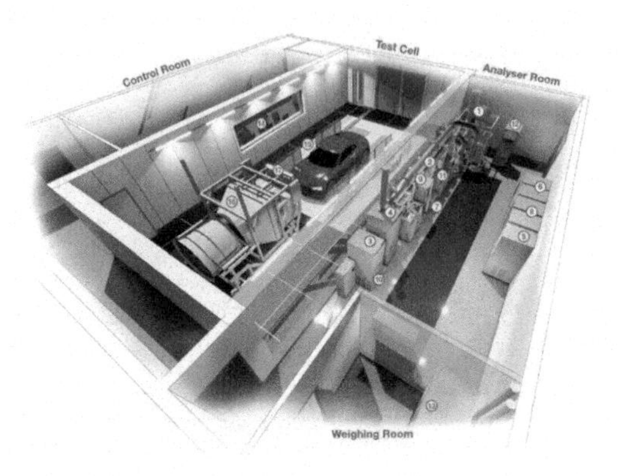

Figure 5. HORIBA MIRA AETC layout.

Table 3. HORIBA MIRA AETC equipment and specifications.

Analyser Room	HORIBA Supplied: DLT and DLS dilution tunnel and sampler CVS-ONE constant volume sampler MEXA ONE D2 EGR raw exhaust gas sampler MEXA ONE C1 SL OV dilute exhaust gas (bag) analyser MEXA 2000 SPCS solid particle counter MEXA ONE QL NX Quantum cascade laser system GMC ONE particulate measurement PFS ONE robotised particulate filter weighing system OBS ONE PEMS kit (gaseous and particle)
Test Cell	FWD, RWD, AWD -20°C to 35°C (plus 3 climatic soak chambers) 30-60% relative humidity Vehicle cooling fan
Altitude Simulation	HORIBA MEDAS 5012V + temperature and humidity 1200kg/hr maximum air flow 0 – 5000m altitude simulation -10°C to +40°C temperature simulation 0-100% relative humidity
Dynamometer Data	230kW per axle 300km/h maximum speed

2.2.2 *Initial methodology – human test driver*

To begin the R2R RDE+ development methodology, the chassis dynamometer testing was focused on the gasoline vehicle owing to its much simpler aftertreatment system (all data shown here-on-in is for this vehicle). The chassis dynamometer RDE cycle shown in this section was conducted using a human driver.

The approach to replicating road tests on the chassis dynamometer was to work backwards from known measured road data (strain gauges on the vehicle drive shafts). From this measured road load, the calculated road load using the measured coast down terms (f0, f1 and f2) and coast down inertia is subtracted. This leaves a residual load value. This residual load is equated to the road gradient or slope (see Equations 1-4) – and incorporates all the additional forces caused by aspects such as wind, lateral loading and changes in road surface.

Where:

$$RF_{Act} = \frac{(T_{RSG} + T_{LSG})}{R_{wheel}} \tag{1}$$

$$RF_{Res} = RF_{Act} - \left[f_0 + (f_1.V_{RDE}) + (f_2.V_{RDE}^2) + (I_{CD}.V_{RDE})\right] \tag{2}$$

$$\theta = \sin^{-1}\left(\frac{RF_{Res}}{m_{TM}.g}\right) \tag{3}$$

$$S_{Eff} = 100 \cdot \tan\theta \tag{4}$$

RF_{Act}	Actual road force	[N]
RF_{Res}	Residual road force	[N]
$T_{RSG/TLSG}$	Torque from right/left strain gauges	[Nm]
R_{wheel}	Wheel radius	[m]
f_0, f_1, f_2	Coastdown terms	[-]
V_{RDE}	Vehicle velocity on RDE	[km/hr]
I_{CD}	Coastdown inertia	[kg]
m_{TM}	RDE test mass	[kg]
θ	Effective road slope	[degrees]
S_{Eff}	Effective road slope	[%]

This road gradient value (S_{Eff}) is programmed into STARS VETS with the vehicle speed trace and gear selection. The additional roller load using these gradient values is then computed by the HORIBA SPARC system. In this way it is possible to playback the exact road load experienced by the vehicle on the chassis dynamometer.

A 90 minute RDE cycle was completed by a single driver following a speed trace and gear selection indicator on a STARS VETS driver aid system. Additionally, the driver aid displayed the effective road slope synchronized to the vehicle speed trace. This modification enabled the driver to pre-empt somewhat, the loading to be applied to the rollers and achieve a more accurate following of the speed trace.

Approaching the problem from this direction allowed rapid assessment of the capability of the test equipment and its ability to faithfully recreate real-world scenarios. Test results showed great promise with accurate replication of vehicle speed, the cumulative work done, and a cumulative CO_2 result within 2.1% of the equivalent full RDE road test. However, whilst there was a strong correlation for cumulative CO_2 across the full RDE cycle, CO, NOx and PN emissions did not show the same level of second by second correlation; although cumulative emissions totals were in good agreement as shown in Table 4. Note that figures relating to the data shown in Table 4 have been published in [3]; they are archived in the current paper for completeness alongside the fluid temperatures measured for the road and chassis dynamometer tests which were in excellent agreement.

Table 4. Summary cumulative emissions results for road and chassis drive with human driver, Nuneaton, UK.

	Work [MJ]	CO [g]	CO_2 [g]	NOx [g]	PN [#]
Road	54.3	43.1	14830	23.0	2.13E+14
Chassis Human	55.7	44.3	15139	21.3	1.90E+14
Delta [%]	2.6	2.8	2.1	-7.4	-10.8

Discrepancies in second by second emissions were caused partly by the pre-test vehicle preparation but more predominantly by the accelerator pedal tip-in/tip-out behavior of the driver. The latter problem is related to the chassis dynamometer test driver having to drive in a reactionary style. This led to an under or over-pressing of the accelerator pedal and subsequent corrections resulting in a significantly different pedal dynamic (accelerator pedal position and accelerator pedal position rate of change), where the on-road accelerator pedal tip-in and tip-out was far more aggressive than the chassis dynamometer tests (Figure 6). This had the knock-on effect of producing very different second-by-second emission values – an artifact that could be resolved by accurately replicating the accelerator pedal dynamics witnessed on the road tests.

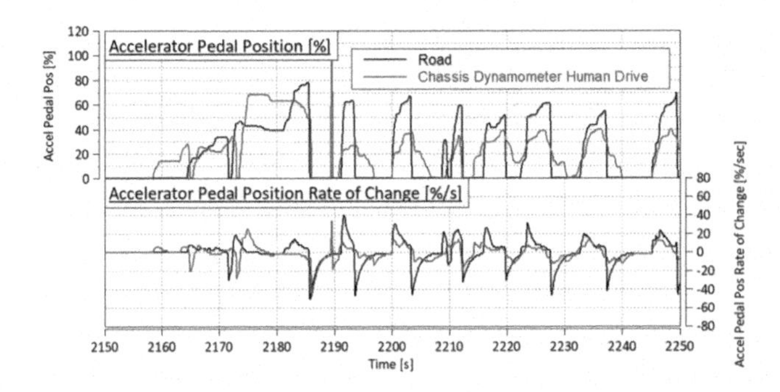

Figure 6. Road vs. chassis dynamometer drive comparison - accelerator pedal position and accelerator pedal position rate of change, Nuneaton, UK.

2.2.3 *Progression to robot driver*

A decision was taken to perform RDE replication drives with a robot driver. Reasons include test repeatability, removal of driver to driver variability (tip-in/tip-out behaviour for example) and the ability to playback actual pedal manipulations if desired. At the time of writing the robot driver has only performed drives as the human driver would; performing closed loop control on a target speed with the chassis dynamometer providing road load simulation. During the next project phase a change of chassis dynamometer control mode will enable playback of recorded pedal inputs via open loop control, with chassis dynamometer control performing the speed control.

The robot driver used throughout the project thus far was the Stähle SAP 2000. For replication drives the robot controlled all 3 pedals, the gear selection and the vehicle ignition/starter. A summary of the development processes required for the complete integration of the robot driver with the chassis dynamometer and vehicle is detailed below.

- RDE Importer Tool: A HORIBA software tool for converting OBS-ONE PEMS data into a chassis test schedule including vehicle speed, gear change points, road grade and ambient conditions.
- Drive Style Controls: Stähle control software allows for the creation of different driver styles. The styles alter PID values and control the amount of trace smoothing applied. It was necessary to utilise several drive styles during a test to achieve acceptable pedal applications.
- Clutch Operation: The robot controller allows for customisation of clutch speeds on a gear pair basis with separate speeds for upshift or downshift operations. These were populated with averaged clutch actuation measurements taken from RDE drive data.
- Throttle-by-Wire: For best control a direct electrical connection from the testbed to the accelerator pedal. A large improvement was seen by switching the accelerator pedal actuator to a direct electronic input (throttle by wire).
- External Gradient Control of Chassis Dynamometer: HORIBA SPARC is instructed by the robot controller with gradient setpoints to alleviate synchronisation issues between robot and chassis dyno test scheduler.

Presented in Figure 7 is a comparison of the vehicle speed measured on the road compared to that measured when the vehicle was driven on the chassis dynamometer by the human driver and robot driver – this is data for the Nuneaton, UK RDE route. Cumulative work done and CO_2 is presented in Figure 8. In all aspects, there is an excellent correlation for road and chassis dynamometers drives driven by the human and robot drivers.

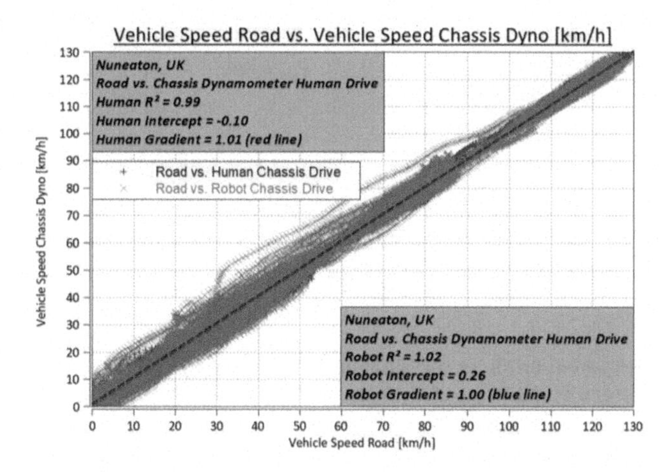

Figure 7. Road vs. chassis dynamometer drive comparison – human driver and robot driver correlations, Nuneaton, UK.

Presented in Figure 9 are further cumulative emissions results for the robot driven chassis dynamometer drive - a summary table of all cumulative emissions is shown in Table 5. There are noticeable differences in CO and PN compared with the road drive and the human driven chassis drive - although NOx appears to be equivalent for both the human and robot drives. All emissions for these cycles and throughout the road and chassis dynamometer sections were recorded with HORIBA OBS-ONE PEMS units. Unfortunately, the unit utilised for the road and human chassis dynamometer drives was not used during the robot driver testing; hence the indifferent CO and PN results are likely to be attributed to unit to unit variation[1]. However, whilst CO and PN results indicate that use of a robot driver delivers poorer replication, significantly improved repeatability and endurance are also important aspects to consider when developing the road to chassis RDE+ methodology. Hence the reason for utilising the robot driver during further testing outlined in the following section and the remainder of the RDE+ programme.

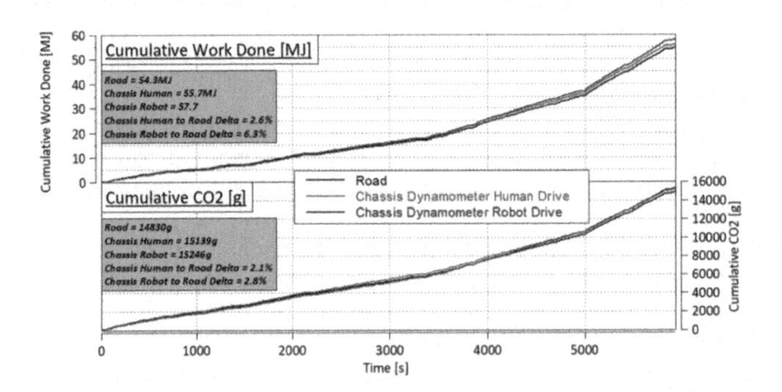

Figure 8. Road vs. chassis dynamometer drive comparison - human and robot drivers cumulative work done and cumulative CO_2 emissions, Nuneaton, UK.

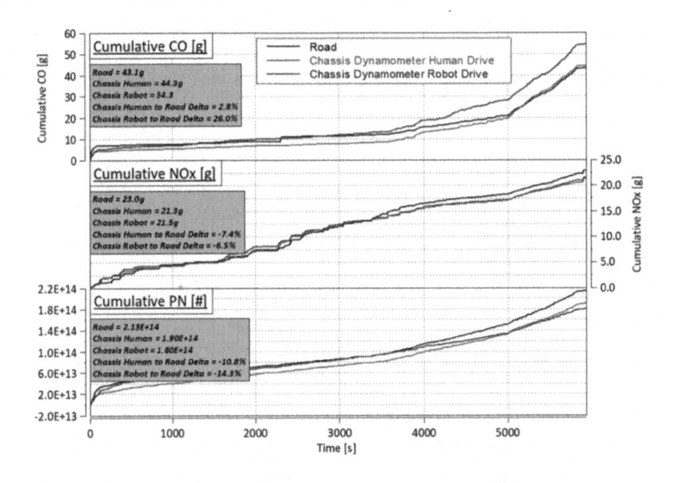

Figure 9. Road vs. chassis dynamometer drive comparison - human and robot drivers cumulative CO, NOx and PN emissions, Nuneaton, UK.

1. All PEMS kits were calibrated and adhered to the regulatory standards throughout all testing reported. Nevertheless, it is well understood that there is PEMS unit to unit variation, particularly with regards to PN.

Table 5. Summary cumulative emissions results for road, chassis drive with human driver and chassis drive with robot driver, Nuneaton, UK.

	Work [MJ]	CO [g]	CO_2 [g]	NOx [g]	PN [#]
Road	54.3	43.1	14830	23.0	2.13E+14
Chassis Human	55.7	44.3	15139	21.3	1.90E+14
Chassis Robot	57.7	54.3	15246	21.5	1.80E+14
Road to Human Delta [%]	2.6	2.8	2.1	-7.4	-10.8
Road to Robot Delta [%]	6.3	26.0	2.8	-6.5	-14.3

As discussed in a previous section, a shortfall of the human driver is the vastly different accelerator pedal dynamic produced by pre-emptive and reactionary driving. Unfortunately, it was not been possible to fully eliminate this behaviour using the robot driver as it has shown instances of more aggressive pedal application or even wide-open throttle inputs in reaction to falling behind the trace. Importantly though, these instances are reduced over the human driver and drive repeatability has been seen to be excellent as shown in Figure 10 (compared with equivalent data for the human driver presented in Figure 6).

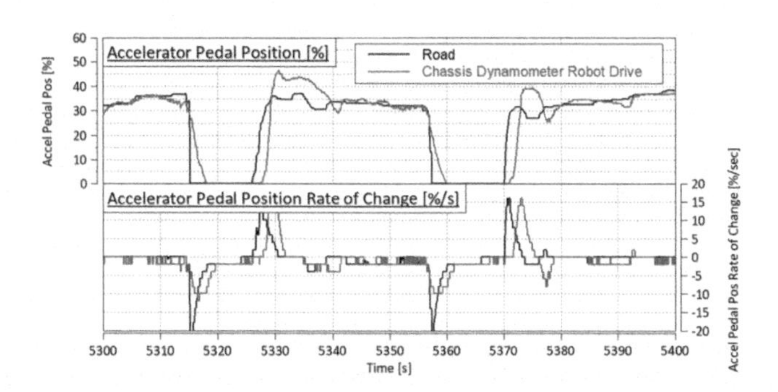

Figure 10. Improved accelerator pedal application by robot driver, Nuneaton, UK.

2.2.4 *Addition of altitude emulation*

Replication of a second RDE route taken from Innsbruck, Austria at elevations up to 800m required the use of altitude emulation via a Multi-Function Efficient Dynamic Altitude Simulator (MEDAS) system. It is not within the scope of the current paper to discuss the integration or application of MEDAS to the vehicle in question as both remain work in progress. However, replication from a drive and emissions perspective is detailed below. The robot driver was used to perform the driving and emissions were recorded with an OBS-ONE PEMS unit modified for use with the MEDAS system.

Shown in Figure 11 and Figure 12 are comparisons between the road drive and chassis drive vehicle speed and cumulative work done and cumulative CO_2

respectively; as shown, there is an excellent correlation for road and chassis dynamometer drives. As a result of the "gentle" nature of the drive (va_pos[95] as a percentage of the limit < 50%), it was easier to programme the robot driver to follow the route more accurately. This resulted in an excellent correlation in other emissions (with the exception of CO emissions) as presented in Figure 13 and summarised in Table 6. At the time of writing, further investigation of the poor correlation in CO is being undertaken.

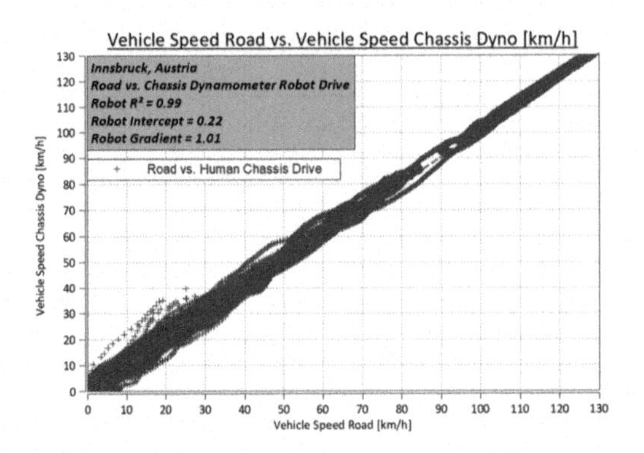

Figure 11. Road vs. chassis dynamometer drive comparison – vehicle speed match and cumulative positive work done (derived from CAN RPM and CAN torque values due to strain gauge failure), Innsbruck, Austria.

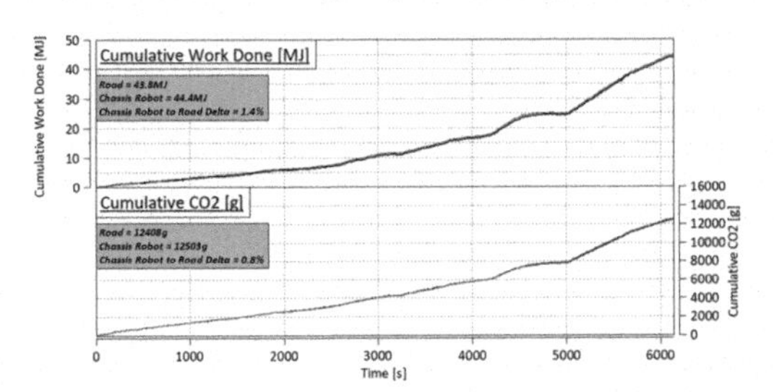

Figure 12. Road vs. chassis dynamometer drive comparison - cumulative work done and cumulative CO2 emissions, Innsbruck, Austria.

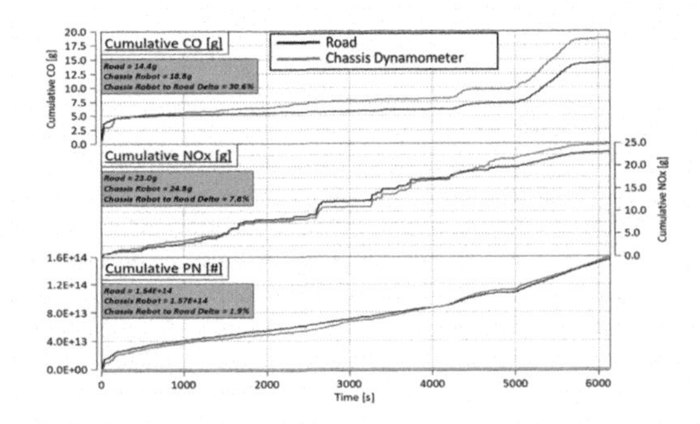

Figure 13. Road vs. chassis dynamometer drive comparison - cumulative CO, NOx and PN emissions, Innsbruck, Austria.

Table 6. Summary cumulative emissions results for road and chassis drive with robot driver, Innsbruck, Austria.

	Work [MJ]	CO [g]	CO_2 [g]	NOx [g]	PN [#]
Road	43.8	14.4	12408	23.0	1.54E+14
Chassis Human	44.4	18.8	12503	24.8	1.57E+14
Delta [%]	1.4	30.6	0.8	7.8	1.9

The environmental simulation using MEDAS is presented in Figure 14. Note that MEDAS has "looser" operating windows in the configuration used (coupled with humidity and temperature modules) because the control system is tasked with balancing the three inter-linked set points of pressure, temperature and humidity.

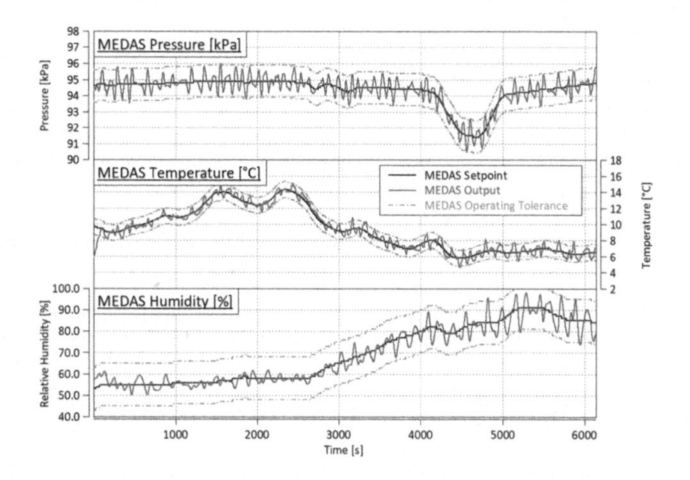

Figure 14. MEDAS performance for Innsbruck, Austria drive; pressure, temperature and humidity at engine intake.

2.3 Engine-in-the-loop configuration

2.3.1 *Overview*

Placing increased emphasis on virtual calibration will result in shorter developmental timescales (up to 1 year in duration) and fewer vehicle prototypes (20-25% fewer) whilst being more cost effective compared with traditional steady state and pseudo-transient engine and vehicle development programmes [4]. However, it has not been until the development of high-fidelity simulation models capable of replicating transmissions, road characteristics and real-time traffic situations (amongst others), that engine and vehicle developers could optimise engine configurations and calibrations in a virtual environment before validating these using fewer engine and vehicle derivatives.

To condense vehicle and engine development processes and move away from the cost prohibitive, labour intensive and time sensitive historic multi-engine/vehicle calibration approach, an EiL test methodology is being developed. This will allow frontloading of vehicle and engine hardware and calibration activities (particularly in relation to

RDE compliance) with a view to creating a test and development methodology that can be used for characterisation of the complete powertrain and vehicle. In addition, this approach will also be used for providing evidence of complete RDE compliance across all RDE boundary conditions reducing the need to send vehicles across the world to undergo typical climatic testing.

To develop the initial EiL methodology (without integration of the virtual environment) whilst allowing further chassis dynamometer work to be undertaken with the gasoline RDE+ vehicle (discussed in the previous section), a donor vehicle of the exact same specification was used. Prior to removing the engine and gearbox in readiness for installation into the engine testcell, this vehicle was tested across a WLTC on the chassis dynamometer in the AETC alongside the gasoline RDE+ vehicle. This was undertaken to ensure equivalent engine and emissions performance between the donor and RDE+ gasoline vehicles in the event the original RDE+ gasoline vehicle would be unavailable to complete the road to chassis to EiL correlation exercise outlined earlier in this paper.

2.3.2 *Initial results*

Following a successful correlation of engine performance and emissions between the donor and RDE+ gasoline vehicles, the engine, gearbox and cooling pack (radiator and charge air cooler) were removed from the donor vehicle and installed in the engine test-cell in the Propulsion Test and Development Centre (PTDC) at HORIBA MIRA. To re-connect the engine to the vehicle, the wiring harness was extended and the gearbox, accelerator pedal position and wheel speed sensors were coerced in order to fool the Engine Control Unit (ECU) that they remained as part of the vehicle or in the case of the wheel speed sensors, the vehicle was moving. By manipulating these signals, it was possible to operate the engine across the entire engine speed and torque range.

As highlighted earlier in this section, the EiL strategy at present is geared towards achieving a road, chassis dynamometer and EiL correlation with respect to engine performance and emissions across several RDE cycles. At the time of writing however, it was not possible to operate the EiL across a RDE cycle and hence report a correlation with the equivalent chassis dynamometer and road data presented earlier in this report. Nevertheless, shown in Figure 15 to Figure 19 is WLTC data recorded from the donor vehicle operated on the chassis dynamometer and the same engine operated over the same cycle on the engine testbed. To "drive" the WLTC on the engine testbed, a playback file of engine speed and accelerator pedal position (recorded from the vehicle when driven across a WLTC on the chassis dynamometer) was passed to the engine testbed controller at 10Hz. The feedback engine speed and accelerator pedal position trace for this cycle are shown in Figure 15 and Figure 16 respectively alongside the equivalent data recorded for the vehicle driven on the chassis dynamometer. As is shown, there is an excellent correlation for EiL and chassis dynamometer activities.

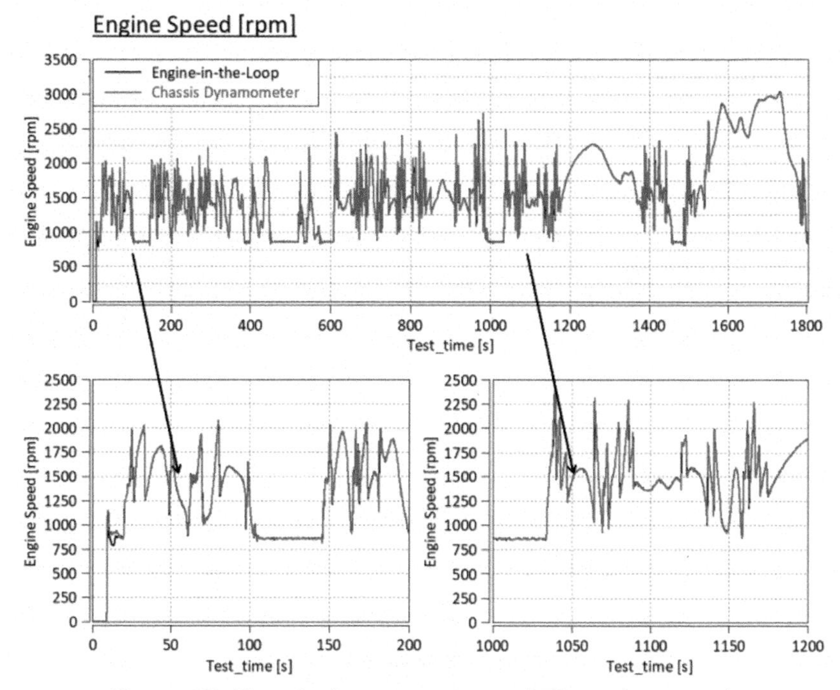

Figure 15. Chassis dynamometer and EiL engine speed.

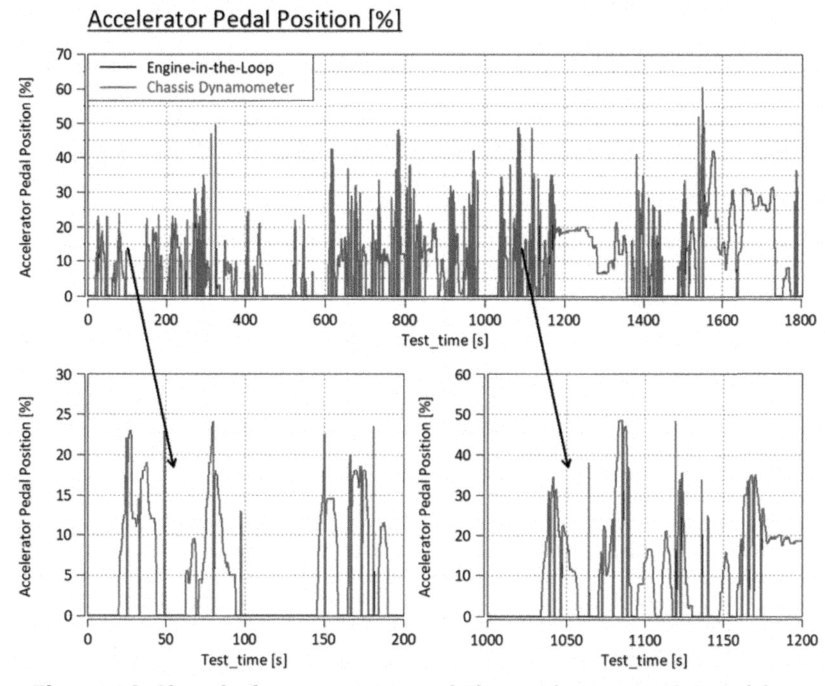

Figure 16. Chassis dynamometer and EiL accelerator pedal position.

Presented in Figure 17 are pre and post TWC temperature measurements alongside exhaust mass flow rate. A good correlation in pre-TWC temperature was achieved with the exception of the temperatures at approximately 400-600 seconds (40°C cooler for EiL). For post-TWC temperatures there was an offset of up to 100°C up to 1200 seconds after which the EiL and chassis dynamometers temperatures are in good agreement. The disparity in post-TWC temperature was likely to have been caused by inadequate replication the airflow across the engine as would normally have been encountered with the vehicle driven on the chassis dynamometer (this will be addressed at a later date using a cold-box solution). Unfortunately, oil and coolant temperatures were not logged during the chassis dynamometer drive. However, it is expected that the oil and coolant were at operating temperature (90°C) throughout the cycle due to performance of a pre-conditioning WLTC driven before the correlation WLTP.

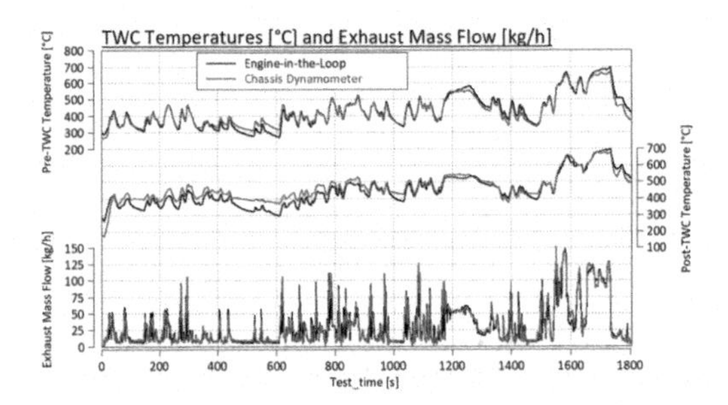

Figure 17. Chassis dynamometer and EiL pre and post TWC temperatures and exhaust mass flow.

Pre-TWC emissions mass flows for CO, CO_2, O_2, NOx are shown in Figure 18 (measured with a MEXA One gas bench). There is generally a good match in engine-out CO_2 and NOx emissions throughout the cycle yet there are excursions on the chassis dynamometer trace (at 1100 seconds for example) that do not correlate with the EiL data. Pre-TWC CO, O_2 and THC emissions are less well matched; the large spikes in these emissions during the EiL cycle indicating potential fuel rich misfire throughout a significant proportion of the cycle.

Fuel rich misfire could also be the reason for the significantly lower post-TWC NOx output during the EiL cycle as compared to the chassis dynamometer cycle (Figure 19). In this case, the excess fuel left over from misfire was utilised for complete NOx reduction over the TWC. Furthermore, the higher post-TWC O_2 for the EiL cycle further points to rich misfire conditions in-cylinder. Note however, that CO and THC emissions are broadly similar except for a few instances where EiL emissions are greater than the chassis dynamometer; this being mainly attributed to the high pre-TWC oxygen content (required for CO and THC oxidation) and the likelihood of the large space velocity operating window of the TWC.

Considering the emissions results discussed, the EiL methodology requires finessing to align EiL emissions with those recorded when operating the vehicle on the chassis dynamometer. One aspect thought to improve the correlation is to stream the engine speed and accelerator pedal position playback file to the engine dynamometer at a slower rate; thereby reducing the control transience thus potentially reducing the chance of misfire as a result of the micro-transient type engine-performance expected with fast rates of accelerator pedal position change. In addition, replication of engine and aftertreatment thermal conditioning will be improved by enclosing these systems using a thermal encapsulation cold box. This should result in improved replication of the under-bonnet engine and aftertreatment temperatures as experienced in the vehicle with a closer alignment of TWC temperatures expected.

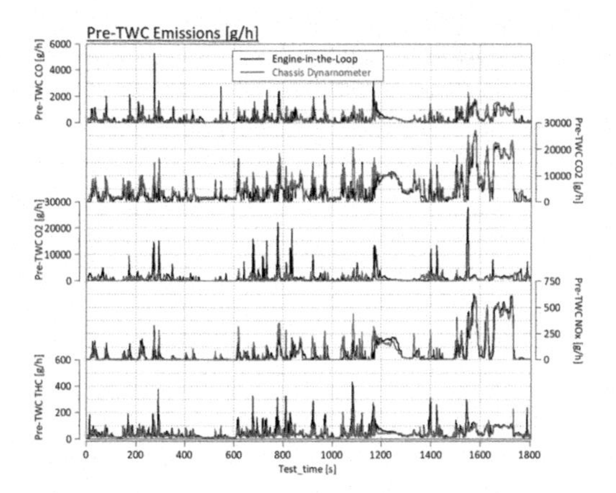

Figure 18. Chassis dynamometer and EiL pre-TWC emissions mass flow.

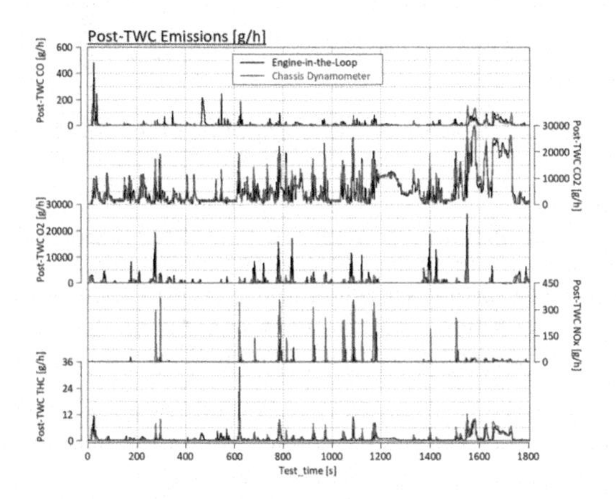

Figure 19. Chassis dynamometer and EiL post-TWC emissions mass flow.

3 FUTURE WORK

3.1 Cycle generators

An important tool needed to go beyond simple replication of RDE tests in the lab is the synthetic test generator. There are several approaches to such a tool, each suited to a different use case. The test generator can draw from a database of stored manoeuvres, re-assembling these short temporal sections into a drivable route that retains some part of the real world. An alternative approach is to generate an entirely synthetic route, choosing random speed and load targets using techniques such as Markov chains. This offers increased flexibility and allows a test to be generated for a vehicle that is not yet road ready.

A third approach utilises the K-means machine learning algorithm. Clustering is performed on a dataset that may include a single or multiple RDE tests. The Markov chain method (an example of which is shown in Figure 20) can be layered atop to generate a route by choosing between one of the many clusters identified by the algorithm; a manoeuvre that is representative of that cluster can then be generated. By setting the probability matrix to reflect cluster size, it would be possible to generate a test that is very similar to the mean of the training dataset. Modifying the probability matrix allows the test to be modified in a similar way to the fully synthetic method, while also retaining the real-world element.

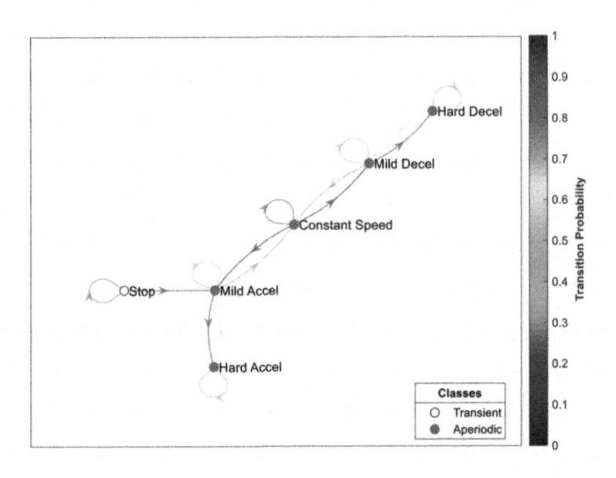

Figure 20. Simple Markov chain implementation.

3.2 Electrification

While an increasing number of vehicles are being developed with an electrified powertrain, many will still include an Internal Combustion Engine (ICE) component for the foreseeable future. Development of the RDE+ methodology includes provision for these electrified vehicles in a conventional engine test cell with minimal additional equipment required. Because electrical cells exhibit complex behaviours in response to variation in several parameters such as temperature and state of charge, it is not a simple task to substitute a large vehicle battery pack with a simple numerical model capable of running in real time. Instead a battery emulator is used, which uses a combination of numerical simulation and hardware to reproduce battery behaviour. The ICE component can then be developed without the rest of the physical powertrain in place while maintaining an integrated development process without the drawbacks of developing powertrain components in isolation. The test cell is controlled in the conventional way but has "some" virtual test drive tool working in the loop. The virtual environment handles the bulk vehicle properties and the simulated electric motor. Battery emulation in the loop is applied to maintain the accuracy of the electrified portion of the powertrain. The simulation generates the load demand for the physical ICE component, whether this is a propulsion component or a range extender.

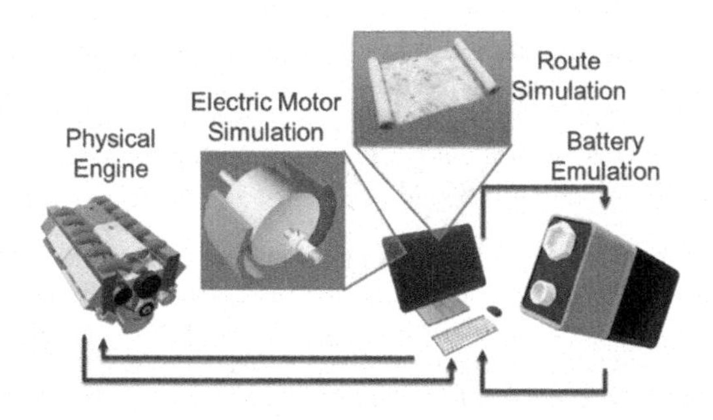

Figure 21. Powertrain emulations using virtual and physical testing.

3.3 Future markets

The RDE+ methodology is applicable to any road going vehicle and can be adapted to both large Heavy-Duty Vehicles (HGV) and smaller two and three wheeled applications. The limited weight and load space capacity on smaller vehicles means that a more minimalistic approach to data collection must be taken on road tests. With the use of a reduced number of sensors and smaller acquisition hardware, such as HORIBA Small Emissions Measurement System (SEMS) [5] the presented methodology allows a given road test to be repeated in a controlled environment with a good level of confidence. While this methodology is presented with a focus on engine and ICE-powered vehicles, the methodology is also applicable to the development of electric vehicles and their control strategies.

4 CONCLUSIONS

- The initial HORIBA MIRA RDE+ R2R (including Road-to-Chassis and Road-to-Engine) RDE replication test methodology has been described.
- Extensive road testing has been undertaken across Europe at locations that are within the RDE moderate and extended boundary conditions. Fully compliant RDE routes were developed and driven at each location utilising three heavily instrumented test vehicles.
- A high specification chassis dynamometer was used to replicate on-road performance and emissions. Dynamometer coast-down parameters were set using derivations from the actual vehicles on test.
- Initial testing with the gasoline vehicle showed that it was possible to recreate accurate road loads and achieve a good correlation of positive work done, cumulative CO_2 and fluid temperatures between the on-road and chassis dynamometer drives for one of the Nuneaton, UK RDE test routes. Cumulative regulated emissions of NOx and PN were also in good agreement.
- The application of a robot driver improved the instantaneous correlation of pedal control from chassis-dyno to the road test. However, whilst the robot driver achieved some level of correlation (albeit poorer correlation than the human driver) the repeatability and durability of the robot driver are superior.
- The robot driver will be developed further during subsequent stages of the RDE+ programme.
- A high-altitude RDE cycle (Innsbruck, Austria) was successfully replicated on the chassis-dyno using the HORIBA MEDAS altitude emulation system. Similar correlation to that observed over the Nuneaton, UK cycle was achieved.

- The MEDAS system emulated measured dynamic changes in altitude which included barometric pressure, temperature and humidity. This emulation was accurate enough to achieve a strong correlation of vehicle responses.
- The next phase of activity will connect vehicle simulation to the EiL rig and enable full simulation of RDE test drives with "real" feedback on CO_2 and regulated RDE emissions.
- EiL testing will form the initial stages of an engine and vehicle development programme. By integrating the virtual environment, high fidelity modelling and utilising highly dynamic test equipment, many aspects of engine and vehicle performance can be characterised before prototype vehicles are required.
- For the EiL test and the "road-to-engine" replication a production vehicle was used, and its engine installed on the engine testbed via an "umbilical" electrical harness. Significant and on-going issues were experienced with such an approach. Future activity will be supported by an OEM or Tier 1 supplier with access to control system architecture and calibration files.
- The initial EiL testing was reported. Good control correlation has been achieved from the engine testing to chassis-dyno over a WLTC.
- EiL correlation of RDE regulated emissions has been generally fair, but suspected misfire events have affected instantaneous emissions correlation. This situation is under investigation at the time of writing of this paper. It is suspected that this is another feature of running the engine unsupported by the OEM or Engine Management System (EMS) supplier.
- The EiL rig will be connected to a MEDAS system and enable the application and development of new transient Design of Experiments (DoE) development and calibration methods and tools. This approach will enable calibration over all RDE conditions in a controlled, repeatable test environment.
- It is likely that complete compliance with current and emerging emissions legislation in the short to medium term will almost always utilise real vehicles.
- By adopting a R2R RDE+ programme as per HORIBA MIRA's R2R RDE+ strategy, many of the unknown scenarios that arise through real testing can be mitigated much further upstream; thus reducing time, effort, money and pollution.

Definitions/Abbreviations

AETC	Advanced Emissions Test Centre
AFR	Air Fuel Ratio
CO	Carbon Monoxide
CO_2	Carbon Dioxide
CPE	Cumulative Positive Elevation
DOC	Diesel Oxidation Catalyst
DoE	Design of Experiments
DPF	Diesel Particulate Filter
ECU	Engine Control Unit
EGR	Exhaust Gas Recirculation
EiL	Engine-in-the-Loop
EMS	Engine Management System
GPS	Global Positioning System
HGV	Heavy Goods Vehicle
ICE	Internal Combustion Engine
LNT	Lean NOx Trap
MEDAS	Multi-Function Efficient Dynamic Altitude Simulator

NEDC	New European Drive Cycle
NOx	Oxides of Nitrogen
O_2	Oxygen
OEM	Original Equipment Manufacturer
PEMS	Portable Emissions Measurement System
PN	Particle Number
PTDC	Propulsion Test and Development Centre
R2R	Road to Rig
RDE	Real Driving Emissions
RDE+	Real Driving Emissions Plus
RPA	Relative Positive Acceleration
SEMS	Small Emissions Measurement System
TWC	Three Way Catalyst
va_pos[95]	Relative positive acceleration 95th percentile
WLTC	Worldwide Harmonised Light Vehicles Test Cycle
WLTP	Worldwide Harmonised Light Vehicles Test

Archive

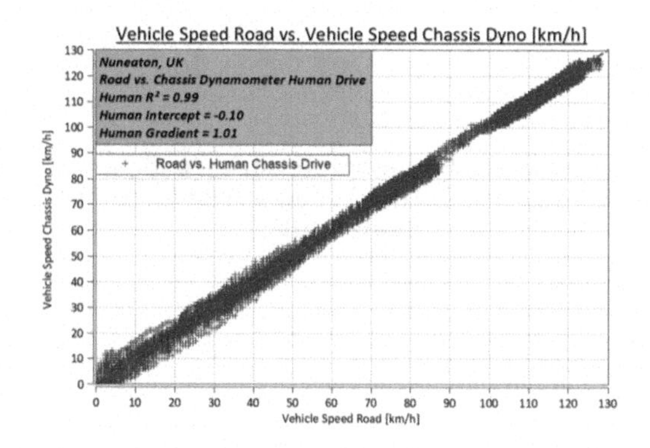

Figure A1. Chassis dynamometer drive vehicle speed vs. on-road drive vehicle speed, Nuneaton, UK.

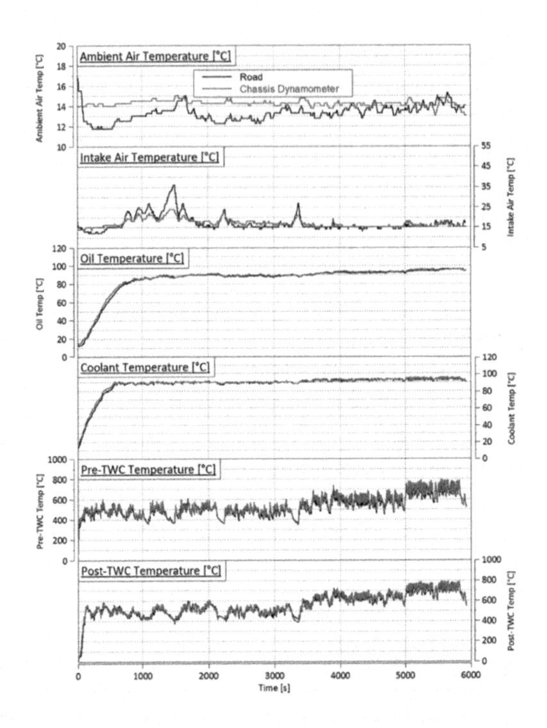

Figure A2. Road vs. chassis dynamometer drive comparison – temperature traces for ambient air, intake air, oil, coolant and pre and post TWC, Nuneaton, UK.

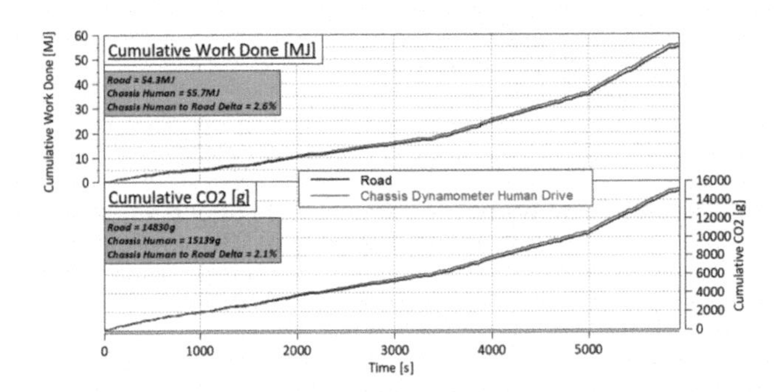

Figure A3. Road vs. chassis dynamometer drive comparison – cumulative positive work done and cumulative CO_2, Nuneaton, UK.

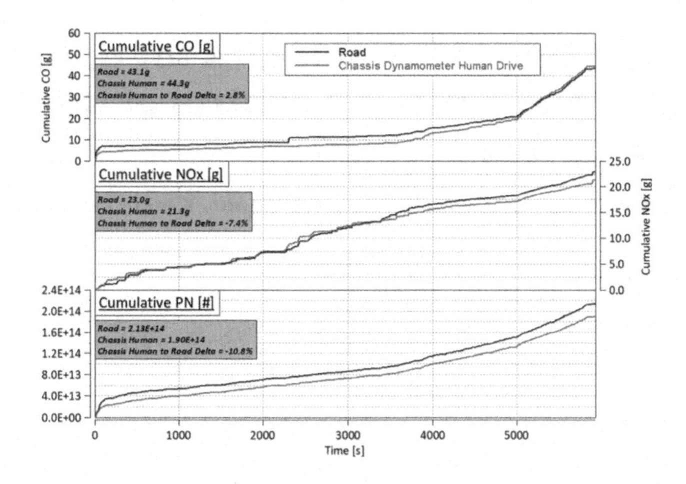

Figure A4. Road vs. chassis dynamometer drive comparison – cumulative CO, NOx and PN, Nuneaton, UK.

ACKNOWLEDGMENTS

The authors wish to thank the HORIBA MIRA Ltd team and our colleagues from the wider HORIBA Ltd group for contributions to the work published in this paper.

REFERENCES

[1] Jiang, S., Smith, M.H., Kitchen, J., and Ogawa, A., "Development of an Engine-in-the-Loop Vehicle Simulation System in Engine Dynamometer Test Cell," SAE Technical Paper 2009-01-1039, DOI10.4271/2009-01-1039.

[2] Lee, S.Y., Andert, J., Neumann, D., Querel, C., Scheel, T., Aktas, S., Ehrly, M., Schaub, J., and Kötter, M., "Hardware-in-the-Loop-Based Virtual Calibration Approach to Meet Real Driving Emissions Requirements," SAE Technical Paper 2018-01-0869, DOI:10.4271/2018-01-0869.

[3] Roberts, P,J., Mumby, R., Mason, A., Redford-Knight, L., and Kaur, P., "RDE Plus – The Development of a Road, Rig and Engine-in-the-Loop Methodology for Real Driving Emissions Compliance," SAE Technical Paper 2019-01-0756, 2019, DOI:10.4271/2019-01-0756.

[4] Trampert, S., Niijs, M., Huth, T., and Guse, D., "Simulation of real driving scenarios on the test bench," MTZ Extra 22 (S1),Pages 12-19, 2017, DOI: 10.1007/s41490-017-0008-5.

[5] Heepen, F. and Yu, W., "SEMS for Individual Trip Reports and Long-Time Measurement," SAE Technical Paper 2019-01-0752, 2019, DOI:10.4271/2019-01-0752.

Effective engine technologies for optimum efficiency and emission control of the heavy-duty diesel engine

Zhibo Ban[1], Wei Guan[1,2], Xinyan Wang[2], Hua Zhao[2], Tiejian Lin[1], Zunqing Zheng[2,3]

[1]Guangxi Yuchai Machinery Company, Yulin, China
[2]Brunel University London, London, UK
[3]Tianjin University, Tianjin, China

ABSTRACT

The new emissions legislation for the heavy-duty (HD) diesel engine will require cutting NOx emissions by 90% to 0.02 g/bhp.hr, which is heavily relied upon the effective operation of the SCR in the aftertreatment systems (ATS). However, the low exhaust gas temperature (EGT) at low-load operation usually impedes the efficient exhaust emissions reduction by these aftertreatment systems, which require a minimum EGT of approximately 200°C to initiate the emission control operations.

In this research, studies have been carried out on the effectiveness and trade-off of the advanced combustion control strategies, such as Miller cycle, internal (iEGR) and external exhaust gas recirculation (eEGR), on the engine efficiency, emissions, and EGT management at low-load operation. Experiments were performed on a single-cylinder HD diesel engine equipped with a high-pressure loop cooled eEGR and a variable valve actuation (VVA) system. The VVA system enables the Miller cycle operation with variable later intake valve closing (LIVC) and produces iEGR via a second intake valve opening (2IVO) event during the exhaust stroke.

The results show that common techniques such as the retarded injection timing, intake throttling, iEGR combined with high exhaust back pressure could increase the EGT but at the expense of high fuel consumption and deteriorating the combustion process. In comparison, the Miller cycle operation could increase EGT to more than 200°C with insignificant impact on the net indicated efficiency (NIE). However, the resulting lower effective compression ratio (ECR) decreased the combustion gas temperature, leading to higher hydrocarbon (HC) and carbon monoxide (CO) emissions. The combined Miller cycle with iEGR helped to reduce CO and HC emissions but this strategy demonstrated a limited NOx emissions reduction, particularly when the injection timing was optimised to achieve the maximum NIE. The introduction of 26%eEGR on the Miller cycle operating with iEGR decreased NOx emissions by 50%, on average, but presented insignificant impact on the NIE and EGT. When introducing a higher eEGR of 44%, the NOx emissions were substantially decreased while increasing the EGT to more than 200°C with a higher NIE. However, these were attained with an increase in soot emissions. The additional results demonstrated that the optimised Miller cycle operating with iEGR and eEGR of 44% achieved the highest EGT of 225°C and the lowest NOx emissions of 0.5 g/kWh but with a soot emissions of 0.026 g/kWh. Alternatively, Miller cycle operating with eEGR of 44% and with no iEGR achieved the highest NIE of 43.7% and the lowest total fluid (fuel and urea) consumption of 0.83 kg/h as well as increasing the EGT to 216°C. Meanwhile, the soot and NOx emissions were decreased to below 0.01 g/kWh and 0.79 g/kWh, respectively. Thus, the Miller cycle operating with iEGR and eEGR have been identified as the most effective means of achieving simultaneous higher engine efficiency, lower emissions, and desired EGT, substantially improving the effectiveness of ATS at the low-load operation.

Keywords: heavy-duty diesel engine, Miller cycle, EGR, aftertreatment, total fluid consumption, exhaust gas temperature

1 INTRODUCTION

Due to the superior traction properties with high torque output and low fuel consumption, diesel engines have been the dominant powerplant for heavy-duty (HD) vehicles and many other applications. The existence of locally fuel-rich and high combustion temperature zones resulted from the non-premixed diffusion-controlled combustion in the conventional diesel combustion, however, lead to the emission of particulate matter (PM) and nitrogen oxides (NOx) [1,2]. As more vehicles are produced and used with the increased economic development and prosperity in both developed and developing countries, it is necessary to reduce CO_2 emission and pollutant emissions as reflected by the introduction of more stringent emission regulations. In particular, the US EPA has recently finalized their Phase 2 HD greenhouse gas (GHG) regulation, which requires about 5% reductions in carbon dioxide (CO_2) compared to Phase 1 [3]. Additionally, the California Air Resource Board is continuing to move forward with cutting HD NOx emissions up to 90% to 0.02 g/bhp-hr from new trucks by about 2024, which is significantly lower than the current limit of 0.2 g/bhp-hr [4,5].

These legislative trends described above put forward higher requirements for diesel engine development and require further research work in engine and aftertreatment technologies in order to simultaneously decrease NOx emissions and improve net indicated efficiency (NIE). Moreover, aftertreatment systems such as NOx selective catalytic reduction (SCR) system, diesel particulate filter (DPF), and diesel oxidation catalyst (DOC) are strongly dependent on the exhaust gas temperature (EGT). A minimum EGT of approximately 200°C is required for catalyst light-off and to initiate the emissions control [6,7]. This is extremely challenging at low-load conditions as the EGT is too low to provide sufficient emissions reduction [8]. Advanced combustion technologies such as the multiple fuel injection strategy, higher fuel injection pressure, and higher boost pressure have been employed to improve upon NIE, however, these technologies are typically accompanied with a lower EGT [9].

Miller cycle has recently been adopted for in-cylinder NOx reduction and exhaust gas temperature management of diesel engines, though the Miller cycle was originally patented for the boosted spark ignition gasoline engine by Ralph Miller in 1957 [10]. Compared to the other more conventional techniques such as the retarded injection timing [11], post injection [12], and intake throttling [13], Miller cycle is an effective technology for NOx emissions and EGT control, particularly at the low engine load operation in the HD diesel engine [14–17]. This is due to the fact that Miller cycle implemented via either early or late intake valve closing (IVC) timings reduces the effective compression ratio and intake charge, resulting in a reduction in peak in-cylinder combustion temperature and air-fuel ratio. Previous studies have demonstrated the capability of Miller cycle to reduce the engine-out NOx emission, the level of NOx reduction achieved by Miller cycle alone is limited and less than EGR, particularly at low engine loads [18]. In addition, the resulting lower in-cylinder pressure and temperature at the start of combustion lead to poor combustion stability and higher CO and unburned HC emissions. Alternatively, the iEGR realised via a 2nd IVO during the exhaust stroke and/or exhaust valve re-opening (2EVO) during the intake stroke can retain hot residuals from the previous cycle. The combined use of iEGR and Miller cycle operation enabled exhaust thermal management with low levels of unburned HC and CO emissions [19–21], but with insufficient NOx emissions reduction, as reported in our previous works [22].

Cooled external EGR is a proven technology for the reduction in NOx formation in the cylinder, because of their thermal, chemical, and dilution effects [23,24]. Previous studies have investigated the combined use of Miller cycle and EGR and have demonstrated that this strategy is effective in curbing engine-out NOx emissions [25,26]. Kim et al. [18] showed that the combined use of Miller cycle with EGR could reduce the NOx emissions from 10 g/kWh to approximately 1 g/kWh at low load operation. Similar results were reported by Verschaeren et al. [27] on the use of Miller cycle and EGR in a HD diesel engine.

To address the challenges encountered by current HD diesel engines at low-load operation, research and development work is required to further optimise the combustion process to achieve simultaneous reduction in fuel and urea consumption. This paper will present the results obtained with different advanced combustion control strategies and analyse their effectiveness to improve upon exhaust gas temperatures and reduce emissions as well as to increase NIE and reduce total fluid consumption at low-load operation. The experimental study was carried out on a single cylinder HD diesel engine at an engine speed of 1150 rpm and engine load of 2.2 bar net indicated mean effective pressure (IMEP). The conventional strategies such as the retarded injection timing and intake throttling for EGT management were investigated and compared to the Miller cycle and iEGR. Moreover, the influence of Miller cycle operating with internal and external EGR on the engine combustion, performance, and emissions was investigated. Finally, an overall engine efficiency and exhaust emissions analysis of various combustion control strategies was analysed and the optimised engine combustion control strategies were identified.

2 EXPERIMENTAL SETUP

2.1 Engine specifications and experimental facilities

Figure 2 shows the schematic of the experimental setup in this work, which comprises of a single cylinder engine, the engine test bench, a boosting system, as well as measurement devices and a data acquisition system. In order to absorb the work inputted from the engine, a Froude-Hofmann AG150 Eddy current dynamometer with a rated power of 150 kW and maximum torque of 500 Nm was deployed.

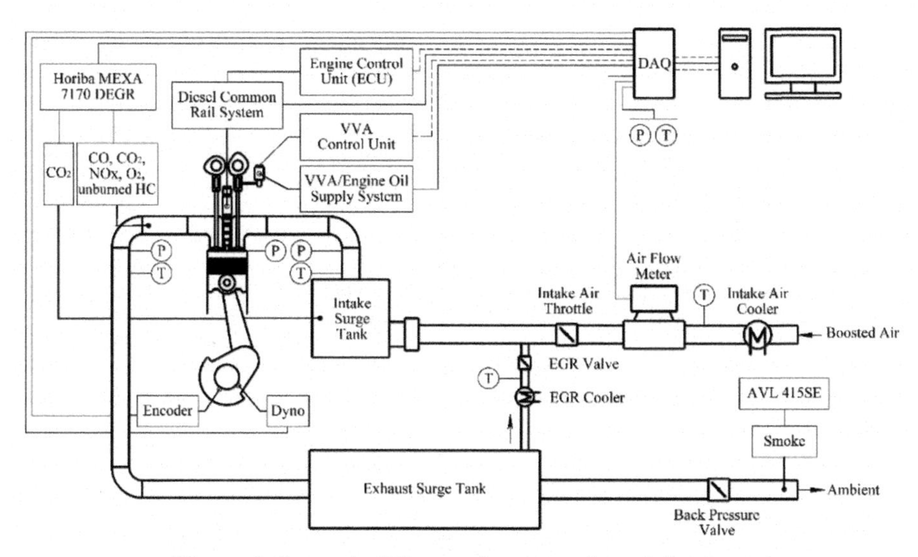

Figure 1. Layout of the engine experimental setup.

Table 1. Specifications of the test engine.

Displaced Volume	2026 cm^3
Stroke	155 mm
Bore	129 mm
Connecting Rod Length	256 mm
Geometric Compression Ratio	16.8
Number of Valves	4
Piston Type	Stepped-lip bowl
Diesel Injection System	Bosch common rail
Nozzle design	8 holes, 0.176 mm hole diameter, included spray angle of 150°
Maximum fuel injection pressure	2200 bar
Maximum in-cylinder pressure	180 bar

The testing engine is typically developed for a modern truck. The key specifications of the engine are listed in Table 1. Amid the composition of the engine, the design of the cylinder head with 4-valve and a stepped-lip piston bowl were based on the Yuchai YC6K six-cylinder engine, while the bottom end/short block was AVL-designed with two counter-rotating balance shafts.

The compressed air was supplied by an AVL 515 sliding vanes supercharger with closed loop control. Two surge tanks were installed to damp out the strong pressure fluctuations in intake and exhaust manifolds, respectively. The intake manifold pressure was finely controlled by a throttle valve located upstream of the intake surge tank. The intake mass flow rate was measured by a thermal mass flow meter. An electronically controlled butterfly valve located downstream of the exhaust surge tank was used to independently control the exhaust back pressure. High-pressure loop cooled external EGR was introduced to the engine intake manifold located between the intake surge tank and throttle by using a pulse width modulation-controlled EGR valve and the pressure differential between the intake and exhaust manifolds. Water cooled heat exchangers were used to control the temperatures of the boosted intake air and external EGR as well as engine coolant and lubrication oil. The coolant and oil temperatures were kept within 356 ± 2 K. The oil pressure was maintained within 4.0 ± 0.1 bar throughout the experiments.

During the experiments, the diesel fuel was injected into the engine by a high-pressure solenoid injector through a high pressure pump and a common rail with a maximum fuel pressure of 2200 bar. A dedicated electronic control unit (ECU) was used to control fuel injection parameters such as injection pressure, start of injection (SOI), and the number of injections per cycle. The fuel consumption was determined by measuring the total fuel supplied to and from the high pressure pump and diesel injector via two Coriolis flow meters. The specifications of the measurement equipment can be found in Appendix A.

2.2 Variable valve actuation system

The engine was equipped with a lost-motion VVA system, which incorporated a hydraulic collapsing tappet on the intake valve side of the rocker arm. This system allowed for the Miller cycle operation via LIVC. The intake valve opening (IVO) and IVC timings of the baseline case were set at 367 and -174 crank angle degrees (CAD) after top dead centre (ATDC), respectively. All valve events were considered at 1 mm valve lift and the maximum intake valve lift event was set at 14 mm.

In addition, this system enables a 2IVO event during the exhaust stroke for the purpose of trapping iEGR in order to increase the residual gas fraction. The earliest opening timing and the latest closing timing of the 2IVO strategy were set at 130 CAD ATDC and 230 CAD ATDC, respectively. The maximum valve lift of this configuration was 2 mm. Figure 2 shows the intake and exhaust valve profiles for the baseline engine operation as well as the LIVC and 2IVO strategies. The effective compression ratio, ECR, was calculated as

$$ECR = \frac{V_{ivc\_eff}}{V_{tdc}} \tag{1}$$

where V_{tdc} is the cylinder volume at top dead centre (TDC) position, and V_{ivc_eff} is the effective cylinder volume where the in-cylinder compressed air pressure is extrapolated to be identical to the intake manifold pressure [28].

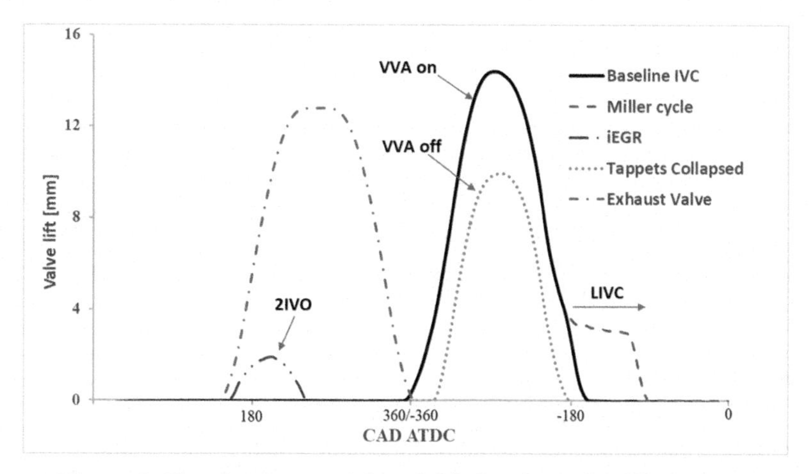

Figure 2. Fixed exhaust and variable intake valve lift profiles.

2.3 Exhaust emissions measurement

A Horiba MEXA-7170 DEGR emission analyser was used to measure the exhaust gases (NOx, HC, CO, and CO_2). In this analyser system, gases including CO and CO_2 were measured through a non-dispersive infrared absorption (NDIR) analyser, HC was measured by a flame ionization detector (FID), and NOx was measured by a chemiluminescence detector (CLD). To allow for high pressure sampling and avoid condensation, a high pressure sampling module and a heated line were used between the exhaust sampling point and the emission analyser. The smoke number was measured downstream of the exhaust back pressure valve using an AVL 415SE Smoke Meter. The measurement was taken in filter smoke number (FSN) basis and thereafter was converted to mg/m^3 [29]. All the exhaust gas components were converted to net indicated specific gas emissions (in g/kWh) according to [30]. In this study, the eEGR rate was defined as the ratio of the measured CO_2 concentration in the intake surge tank ($(CO_2\%)_{intake}$) to the CO_2 concentration in the exhaust manifold ($(CO_2\%)_{exhaust}$) as

$$eEGR\ rate = \frac{(CO_2\%)_{intake}}{(CO_2\%)_{exhaust}} * 100\% \tag{2}$$

2.4 Data acquisition and analysis

The instantaneous in-cylinder pressure was measured by a Kistler 6125C piezo-electric pressure transducer with a sampling revolution of 0.25 CAD. The captured data from the high speed and low speed National Instruments data acquisition (DAQ) cards as well as the resulting engine parameters were display in real-time by an in-house developed transient combustion analysis software.

The crank angle based in-cylinder pressure traces were recorded through an AVL FI Piezo charge amplifier and was averaged over 200 consecutive engine cycles and used to calculate the IMEP and apparent heat release rate (HRR). According to [1], the apparent HRR was calculated as

$$HRR = \frac{\gamma}{(\gamma - 1)} p \frac{dV}{d\theta} + \frac{1}{(\gamma - 1)} V \frac{dp}{d\theta} \tag{3}$$

where γ is defined as the ratio of specific heats and assumed constant at 1.33 throughout the engine cycle [31]; V and p are the in-cylinder volume and pressure, respectively; θ is the crank angle degree.

In this study, the CA10, CA50 (combustion phasing) and CA90 were defined as the crank angle when the fuel mass fraction burned (MFB) reached 10%, 50%, and 90%, respectively. Combustion duration was represented by the period of time between the crank angles of CA10 and CA90. Ignition delay was defined as the period of time between the SOI and the start of combustion (SOC), denoted as 0.3% MFB point of the average cycle. The in-cylinder combustion stability was monitored by the coefficient of variation of the IMEP (COV_IMEP) over the sampled cycles.

3 METHODOLOGY

3.1 Estimation of the total fluid consumption

An increase in engine-out NOx emissions can lead to a higher consumption of aqueous urea solution in the aftertreatment system of an SCR equipped HD diesel engine. This can adversely affect the total engine fluid consumption and thus the engine operational cost. Therefore, the total fluid consumption is estimated in this study in order to take into account both the measured diesel flow rate ($\dot{m}_{diesel}$) and the estimated urea consumption in the SCR system ($\dot{m}_{urea}$). As the relative prices between diesel fuel and urea are different in different countries and regions, the price and property of urea is simulated to be the same as diesel fuel in this study [32]. According to Charlton et al. [32] and Johnson [33], the required aqueous urea solution to meet the Euro VI NOx limit of 0.4 g/kWh can be estimated as 1% of the diesel equivalent fuel flow per g/kWh of NOx reduction, i.e.

$$\dot{m}_{urea} = 0.01 * \left(NOx_{Engine-out} - NOx_{EuroVI}\right) * \dot{m}_{diesel} \tag{4}$$

The total fluid consumption is then calculated by adding the measured diesel flow rate to the estimated urea flow rate as

$$\dot{m}_{total} = \dot{m}_{urea} + \dot{m}_{diesel} \tag{5}$$

3.2 Test conditions

In this study, the experimental work was carried out at the engine speed of 1150 rpm and light engine load of 2.2 bar IMEP, which represents a typical engine operating condition of a HD drive cycle characterised with low exhaust gas temperature below 200°C. Table 2 summarises the engine test conditions for the different engine combustion control strategies investigated.

The start of injection (SOI) of the baseline engine operation was swept between -11.5 and 2.5 CAD ATDC to achieve the maximum fuel efficiency and EGT of 200°C. In the intake throttling mode, the intake pressure was gradually decreased while maintaining the exhaust pressure constant. During the engine operation with iEGR via 2IVO, the exhaust back pressure was varied to increase the residual gas fraction while the intake pressure was held constant at 1.15 bar. Miller cycle operation was achieved via late intake valve closure (LIVC) and the IVC timing was swept to estimate the effectiveness of Miller cycle on the EGT management. For the Miller cycle operating with iEGR and eEGR, the SOI was optimised to achieve the maximum fuel efficiency. When the EGR rate of 44% was introduced, the Miller cycle was operated under naturally aspirated conditions with an intake pressure of 0.98 bar. This setting was necessary in order to isolate the influence on the turbocharger operation and to analyse the feasibility of these strategies in a production multi-cylinder engine. Stable engine operation was determined by controlling the COV_IMEP below 3%.

Table 2. Engine test conditions for various combustion control strategies.

Testing modes	Speed (rpm)	Load (bar IMEP)	Injection pressure (bar)	SOI (CAD ATDC)	Intake pressure (bar)	Exhaust pressure (bar)	IVC (CAD ATDC)	2IVO	eEGR (%)
Baseline	1150	2.2	510	Swept	1.15	1.20	-178	off	0
Intake throttling				-5.8	Swept	1.20	-178	off	0
iEGR				-5.8	1.15	Swept	-178	on	0
Miller cycle				Swept	1.15	1.20	Swept	off	0
Miller cycle +iEGR				Swept	1.15	1.20	-100	on	0
Miller cycle +iEGR+26% eEGR				Swept	1.15	1.20	-100	on	26
Miller cycle +iEGR+ 44%eEGR				Swept	0.98	1.08	-100	on	44
Miller cycle +44%eEGR				Swept	0.98	1.08	-100	off	44

4 RESULTS AND DISCUSSIONS

4.1 Estimation of the effectiveness of various combustion control strategies

In this section, conventional techniques such as the retarded injection timing, intake throttling as well as the VVA based strategies including Miller cycle and iEGR are analysed and compared, in terms of their effectiveness on exhaust gas temperature management and the impacts on the fuel conversion efficiency and exhaust emissions.

Figure 3 shows an overview of the increase in EGT versus the percentage variation in fuel consumption of the different combustion control strategies when compared to the baseline case operating with the maximum fuel efficiency. It demonstrated that by delaying the combustion phasing through retarded injection timing, the EGT was increased at the expense of fuel efficiency penalty, leading to the highest fuel consumption when the EGT was increased to reach 200°C. For the intake throttling and iEGR strategies, the EGT could be increased to more than 200°C, but with lower NIE. This was a result of the increased pumping loop areas attributed to the lower intake manifold pressure in the intake throttling strategy and higher exhaust back pressure in the iEGR strategy, as demonstrated in our previous work [22].

In comparison, Miller cycle via LIVC strategy operating with and without iEGR were more effective in increasing EGT with lower fuel efficiency penalty. However, the application of Miller cycle significantly increased the unburned HC and CO emissions, as shown in Figure 4. This is because of the lower effective compression ratio and a reduction in the combustion temperature. Although the introduction of iEGR to the Miller cycle operation could substantially reduce both CO and unburned HC emissions, the capability of curbing NOx emissions was insufficient. Therefore, a cooled external EGR was introduced in an attempt to achieve lower level of engine-out NOx emission while improving upon the fuel efficiency and EGT management in the next section of study.

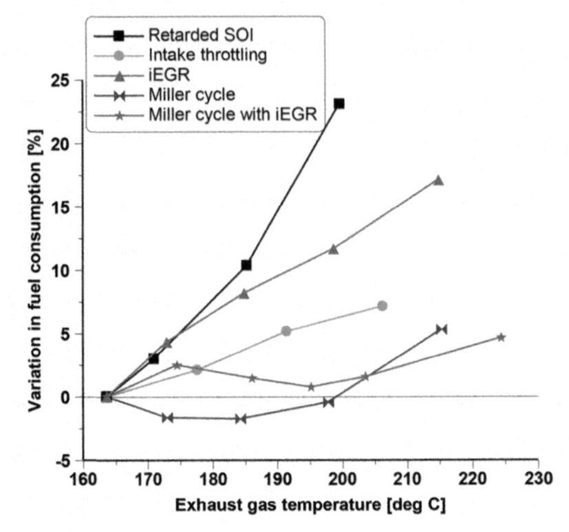

Figure 3. Comparison of the effectiveness of various strategies for EGT management.

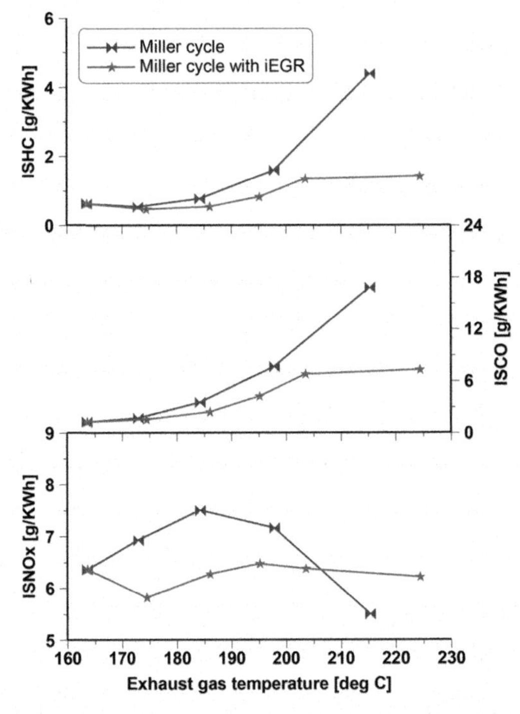

Figure 4. Effect of Miller cycle operating with and without iEGR on emissions.

4.2 Analysis of the in-cylinder pressure and heat release rate

According to the discussion and analysis in the above section, Miller cycle with iEGR strategy has been demonstrated as an enabling technology for efficient increase in the

EGT while maintaining reasonable fuel consumption penalty when compared to others. To further reduce the levels of engine-out NOx emissions, the eEGR was introduced to the Miller cycle operating with iEGR.

Figure 5 shows the in-cylinder pressure and HRR profiles for the baseline and the cases of Miller cycle operating with iEGR and eEGR. The comparison was performed with the maximum NIE achieved via optimising the injection timing (SOI). The optimised SOI of the Miller cycle combined with iEGR strategy was later than that of the baseline case. This strategy led to the most retarded heat release but the highest peak HRR attributed to the higher degree of pre-mixed combustion. The addition of eEGR advanced the optimised SOI and thus the start of combustion, especially when the eEGR of 44% was employed. It can be also seen that the iEGR advanced the SOC of the Miller cycle with eEGR of 44% at a constant SOI. This was due to the hot residual gas introduced via iEGR, which increased the gas temperature during the compression stroke. In comparison to the baseline operation, Miller cycle operating with iEGR and eEGR significantly decreased the in-cylinder pressure, particularly in the cases with the higher eEGR of 44%. The peak HRR was apparently higher than the baseline engine operation due to the increased percentage of pre-mixed combustion.

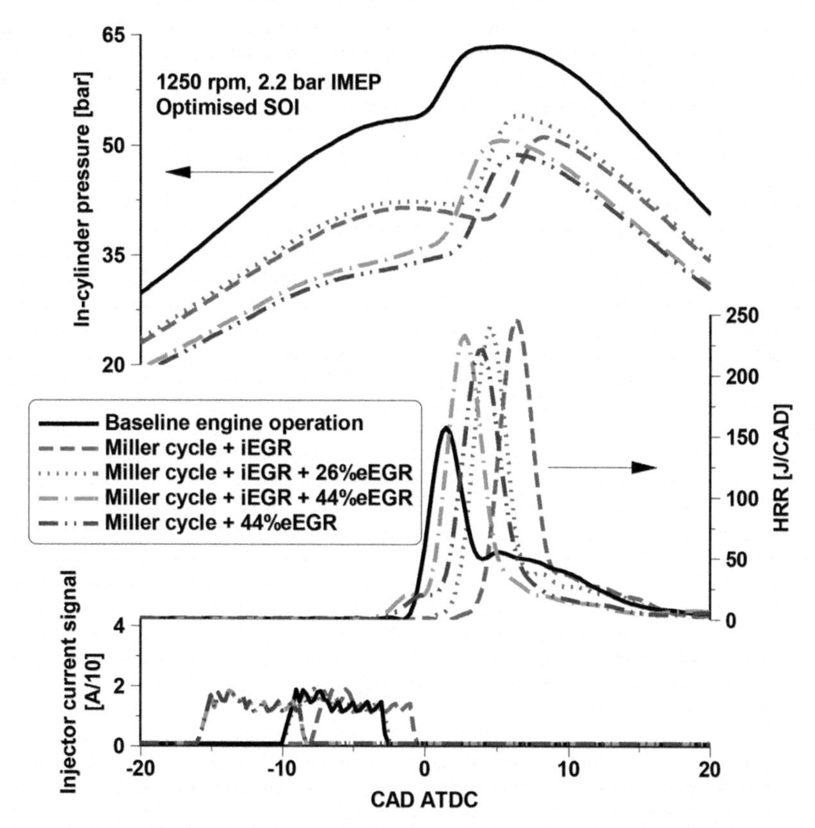

Figure 5. In-cylinder pressure, HRR, and diesel injection signal for the optimum baseline and Miller cycle operating with iEGR and eEGR cases.

4.3 Combustion characteristics

Figure 6 depicts the diesel injection timing and the resulting combustion characteristics as a function of the variation in EGT for the different combustion control strategies.

Compared to the optimum baseline case, the retarded SOI strategy achieved the minimum EGT requirement of 200℃ at a very late SOI. This led to a substantially later combustion phase and an unstable combustion process represented by an increase in the COV_IMEP to more than 3%. The combined use of Miller cycle and iEGR allowed for a relatively earlier SOI to achieve the minimum required EGT, but the combustion phasing and CA90 were delayed when compared to those of the optimum baseline case. The lower ECR resulted from the LIVC and the higher total heat capacity due to the use of iEGR lengthened the ignition delay as the EGT was increased. This increased the degree of pre-mixed combustion and thus led to a faster combustion rate, resulting in a shorter combustion duration.

The introduction of a moderate eEGR of 26% on the Miller cycle operation with iEGR produced insignificant impact on the combustion characteristics. When the eEGR was increased to 44%, the EGT was raised to more than 200℃ with the optimised SOI. In addition, the resulting CA50 and CA90 were held similar to those of the optimum baseline operation. These were because the use of a relatively higher eEGR of 44% produced higher impact on the combustion process attributed to the stronger dilution, heat capacity, and chemical effects. Consequently, the ignition delay was increased by 4 CAD, on average, compared to the baseline case with the optimised SOI. It can be also seen that the ignition delay was longer and the combustion duration was shorter when Miller cycle operating with eEGR of 44% and with no iEGR. Overall, all Miller cycle cases were performed with COV_IMEP of below 3%.

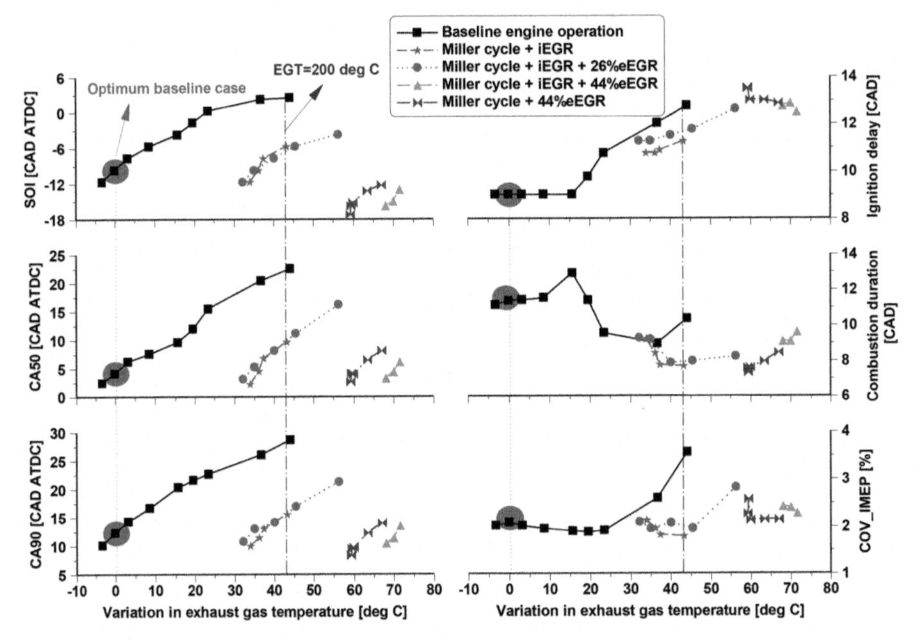

Figure 6. Injection timing and combustion characteristics of the baseline and Miller cycle operating with iEGR and eEGR.

4.4 Engine performance

Figure 7 shows the engine performance parameters for the baseline and Miller cycle operating with iEGR and eEGR. The intake pressure was held constant at 1.20 bar except for those cases performed with a higher eEGR of 44%, where the engine was operated under the naturally aspirated conditions. As a result, the intake air mass flow rate was dropped from 90 kg/h in the baseline operation to approximately 30 kg/h in those cases of Miller cycle operating with high eEGR. The variation in lambda presented a similar trend to the intake air mass flow rate, decreasing from 5.1 to 1.9, on average. The Miller cycle operating with high eEGR produced relatively higher pumping mean effective pressure (PMEP), as a higher pressure differential of 10 kPa between intake and exhaust manifolds was employed compared to the 4 kPa used in the other operating modes.

The retarded SOI strategy apparently decreased the combustion efficiency as the EGT was increased. The combustion efficiency was reduced from 99.6% in the optimum baseline case to 95.8% when the SOI was delayed to achieve the minimum required EGT of 200°C. This resulted in a lower NIE of 35.7% than the 42.1% in the optimum baseline case. The Miller cycle with iEGR strategy maintained the high combustion efficiency and achieved slightly higher NIE when achieving the minimum required EGT. This was attributed to the higher degree of pre-mixed combustion and shorter combustion duration, which improved the combustion process and minimised the heat loses. The introduction of a moderate eEGR of 26% produced insignificant impact on the combustion efficiency and NIE when the EGT was increased to 200°C. When a higher eEGR of 44% was introduced, however, the NIE was improved while achieving a higher EGT of more than 200°C, despite a higher PMEP and a slightly lower combustion efficiency than the optimum baseline case. This was also attributed to a higher degree of pre-mixed combustion and a shorter combustion duration. Figure 7 also shows that the Miller cycle operating with eEGR of 44% and with no iEGR achieved higher NIE than those cases operated with iEGR. This was because the trapped hot residual gas via 2IVO advanced the SOC and thus shortened the ignition delay, decreasing the degree of pre-mixed combustion and lengthening the combustion duration, as demonstrated in Figure 6.

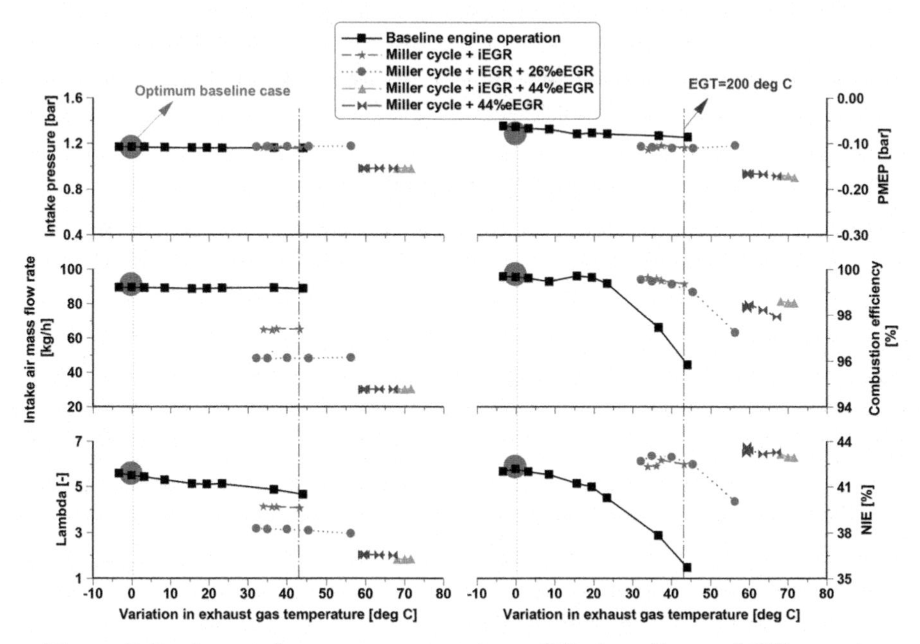

Figure 7. Engine performance parameters of the baseline and Miller cycle operating with iEGR and eEGR.

4.5 Engine-out emissions

Figure 8 depicts the engine-out emissions characteristics as a function of the variation in EGT for the different combustion control strategies. The retarded SOI decreased the NOx and soot emissions but produced substantially higher unburned HC and CO emissions due to the very late combustion process and much lower combustion temperature. The Miller cycle operating with iEGR achieved lower levels of soot, unburned HC, and CO emissions, but with a limited capability of NOx emissions reduction. The introduction of 26% eEGR decreased the NOx emissions by 50% from 6 g/kWh in the Miller cycle operation with iEGR to 3 g/kWh while maintaining the low levels of soot, unburned HC, and CO emissions. These were attained when the EGT was increased to 200°C. Nevertheless, the combination with the use of a higher eEGR of 44% significantly minimised the NOx emissions to about 0.5 g/kWh, which is very close to the Euro VI NOx limit of 0.4 g/kWh. However, this was at the expense of an increase in soot emissions. Alternatively, the Miller cycle operating with eEGR of 44% and with no iEGR decreased the soot emissions to below the Euro VI soot limit of 0.01 g/kWh at the expense of a slightly higher unburned HC and CO emissions.

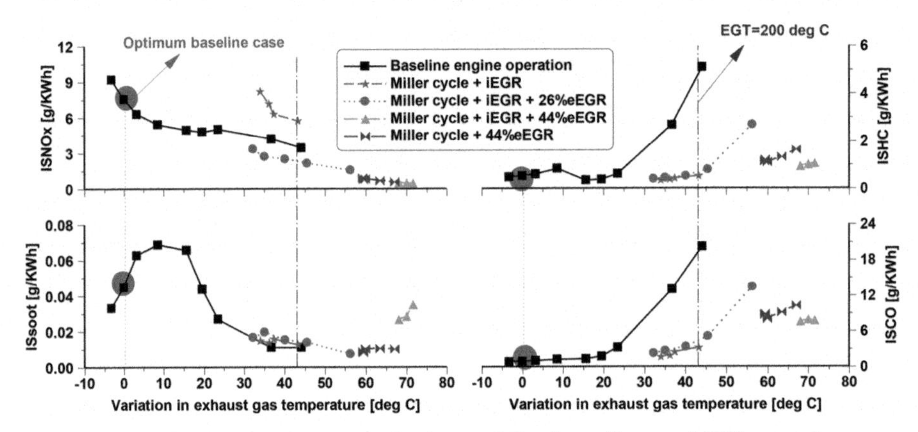

Figure 8. Engine-out exhaust emissions of the baseline and Miller cycle operating with iEGR and eEGR.

4.6 Overall engine efficiency and emissions analysis

In this section, the overall engine efficiency and engine-out emissions of different engine combustion control strategies were analysed by taking into account the consumption of aqueous urea solution in the SCR system. The effectiveness of Miller cycle operating with iEGR and eEGR was estimated and compared to the optimum baseline case.

Figure 9 provides an overall assessment of the potential of the Miller cycle operating with iEGR and eEGR in terms of exhaust emissions, engine performance, and total fluid consumption of a diesel engine operating at low engine load. The results of different combustion control strategies were compared when the SOI was optimised to achieve the maximum NIE. The optimum baseline case was characterised with low EGT of 157°C and NIE of 42.1% as well as high levels of NOx and soot emissions, and thus the highest level in total fluid consumption of 0.87 kg/h. It can be seen that the required urea consumption in the SCR systems closely linked to the level of the engine-out NOx emissions. The reduction in the engine-out NOx emissions by using advanced Miller cycle-based combustion control strategies substantially minimised the requirement on the urea consumption.

Compared to the baseline engine operation, the optimised Miller cycle combined with iEGR slightly increased the NIE and decreased the NOx emissions. Meanwhile, the soot emissions were substantially decreased. However, the EGT was lower than the minimum requirement of 200°C, although an increase in the EGT. The introduction of a moderate eEGR of 26% achieved higher NIE and lower NOx emissions, contributing to a noticeable reduction in the total fluid consumption due to lower fuel and urea consumptions. However, this strategy with optimised SOI produced insignificant impact on the EGT, which was still lower than the minimum requirement. Preferably, the addition of a higher eEGR of 44% achieved higher EGT of 225°C and NIE of 43.1% while significantly reducing the NOx emissions to 0.5 g/kWh. However, these were accompanied with an increase in soot, unburned HC, and CO emissions. Alternatively, the Miller cycle with eEGR of 44% and with no iEGR achieved the highest NIE of 43.7% and the lowest total fluid consumption of 0.83 kg/h while increasing the EGT to 216°C. In the meantime, the soot emissions were decreased to below 0.01g/kWh, but with an increase in unburned HC and CO emissions. The slightly higher CO and unburned HC concentrations could contribute to further increase in the EGT when they are oxidized in the DOC, which operates between 200 and 450°C [8,34].

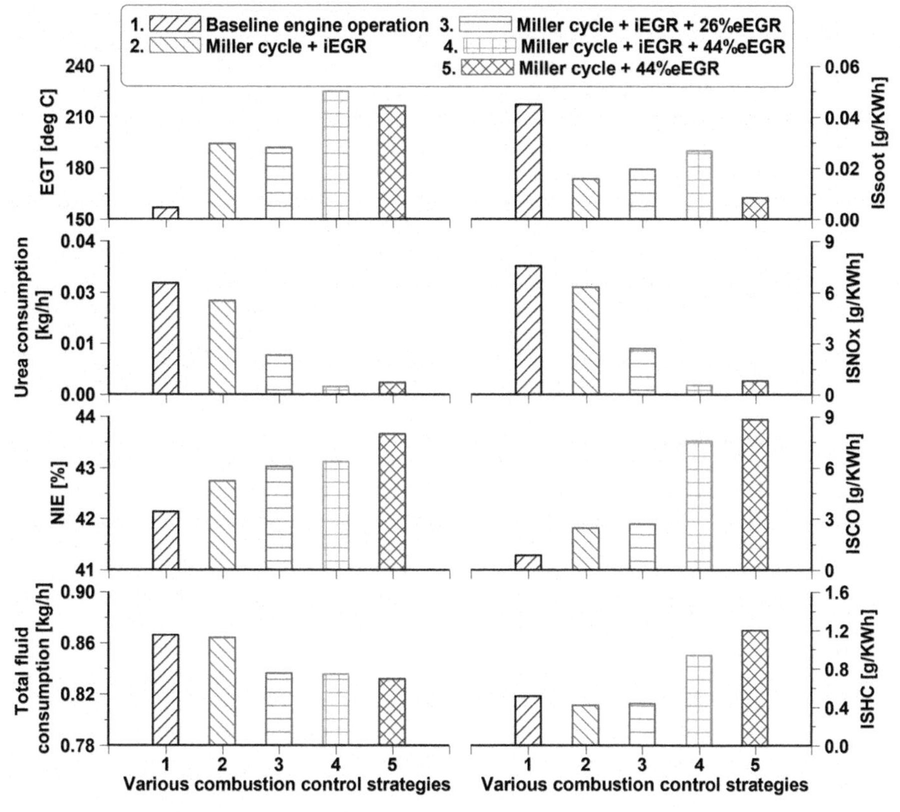

Figure 9. Comparison of the overall engine efficiency and engine-out emissions for various strategies under the condition of achieving maximum NIE.

5 CONCLUSIONS

In this study, the effect of Miller cycle operating with iEGR and eEGR on engine combustion process, performance, and exhaust emissions was investigated. Experiments were performed on a HD diesel engine operating at a typical light engine load of 2.2 bar IMEP with low exhaust gas temperature of below 200°C. The aim of the research was to explore alternative combustion control strategies as means to overcome the challenges encountered by current HD diesel engines operating at low engine loads, such as insufficient high EGT for efficient exhaust conversion, low fuel conversion efficiency, and high engine-out emissions. Both Miller cycle and iEGR operations were realised by means of a VVA system. Cooled external EGR and diesel fuel injection strategy were achieved via a high pressure loop EGR and a common rail fuel injection system, respectively. The primary findings can be summarised as follows:

1. Strategies such as retarded SOI, intake throttling, and iEGR were able to increase the EGT to reach the minimum requirement of 200°C, but these strategies decreased the net indicated efficiency (NIE) due to the significantly late combustion process and higher pumping losses accordingly.
2. The Miller cycle operation enabled a higher EGT with little impact on the NIE, but the lower in-cylinder combustion gas temperature resulted in higher levels of unburned HC and CO emissions. The Miller cycle with iEGR was able to improve the combustion process and thus lower HC and CO emissions, but its capability to minimise NOx emissions was very limited, especially when the SOI was optimised to achieve the maximum NIE.
3. The introduction of a moderate eEGR of 26% on the Miller cycle with iEGR operation produced insignificant impact on the engine combustion, performance, and exhaust emissions (except for NOx emissions). When increased the eEGR to 44%, however, the EGT was substantially increased with a higher NIE and significantly lower NOx emissions. These were attained with an increase in soot, unburned HC, and CO emissions.
4. When comparing at the optimised SOI of various combustion control strategies, the Miller cycle operating with iEGR and eEGR achieved higher NIE and lower engine-out NOx emissions, which contributed to a reduction in total fluid consumption (fuel and urea). Meanwhile, the EGT was noticeably increased and soot emissions were significantly reduced when compared to the baseline engine operation.
5. Among these strategies investigated, Miller cycle operating with eEGR of 44% and iEGR achieved the highest EGT of 225°C and the lowest NOx emissions of 0.5 g/kWh. Alternatively, Miller cycle operating with eEGR of 44% and with no iEGR achieved the highest NIE of 43.7% and the lowest total fluid consumption of 0.83 kg/h while increasing the EGT to 216°C and reducing soot emissions to below 0.01 g/kWh. However, these strategies produced higher levels of unburned HC and CO emissions than other engine operation modes.
6. Overall, strategies such as Miller cycle operating with eEGR and iEGR were identified as effective means for EGT management and emissions control as well as efficiency improvement at low-load operation, substantially minimise the total fluid consumption and potentially complying with the future fuel efficiency and ultra-low NOx emissions regulations.

ACKNOWLEDGMENTS

The authors would like to acknowledge the Guangxi Yuchai Machinery Company for supporting and funding this project carried out at Brunel University London.

DECLARATION OF CONFLICTING INTERESTS

The author(s) declared no potential conflicts of interest with respect to the research, authorship, and/or publication of this article.

FUNDING

The author(s) disclosed receipt of the following financial support for the research, authorship, and/or publication of this article: Funding for this project was provided by Guangxi Yuchai Machinery Company.

DEFINITIONS/ABBREVIATIONS

ATS	Aftertreatment System.
ATDC	After Firing Top Dead Center.
CA90	Crank Angle of 90% Cumulative Heat Release.
CA50	Crank Angle of 50% Cumulative Heat Release.
CA10	Crank Angle of 10% Cumulative Heat Release.
CAD	Crank Angle Degree.
CLD	Chemiluminescence Detector
CO	Carbon Monoxide.
CO_2	Carbon Dioxide.
COV_IMEP	Coefficient of Variation of IMEP.
$(CO_2\%)_{intake}$	CO_2 concentration in the intake manifold.
$(CO_2\%)_{exhaust}$	CO_2 concentration in the exhaust manifold.
DAQ	Data Acquisition
DOC	Diesel Oxidation Catalyst.
ECR	Effective Compression Ratio.
ECU	Electronic Control Unit.
EGR	Exhaust Gas Recirculation.
eEGR	External Exhaust Gas Recirculation.
EGT	Exhaust Gas Temperature.
EVO	Exhaust Valve Opening.
FID	Flame Ionization Detector
FSN	Filter Smoke Number.
FS	Full Scale
GHG	Greenhouse Gas.
HRR	Heat Release Rate.
HC	Hydrocarbons.
HD	Heavy Duty.
iEGR	Internal Exhaust Gas Recirculation.
IMEP	Indicated Mean Effective Pressure.
IVO	Intake Valve Opening.
IVC	Intake Valve Closing
ISsoot	Net Indicated Specific Emissions of Soot.
ISNOx	Net Indicated Specific Emissions of NOx.
ISCO	Net Indicated Specific Emissions of CO.
ISHC	Net Indicated Specific Emissions of Unburned HC.
LIVC	Late Intake Valve Closing.

(*Continued*)

(*Continued*)

MFB	Mass Fraction Burnt
NDIR	Non-Dispersive Infrared Absorption
NOx	Nitrogen Oxides.
NIE	Net Indicated Efficiency
SCR	Selective Catalytic Reduction.
SOI	Start of Injection.
SOC	Start of Combustion.
TDC	Firing Top Dead Centre.
VVA	Variable Valve Actuation.
WHSC	World Harmonized Stationary Cycle.

REFERENCES

[1] Heywood J.B, "Internal Combustion Engine Fundamentals," ISBN 007028637X, 1988.

[2] Kimura, S., Kimura, S., Aoki, O., Aoki, O., Ogawa, H., Ogawa, H., Muranaka, S., Muranaka, S., Enomoto, Y., and Enomoto, Y., "New combustion concept for ultra-clean and high-efficiency small DI diesel engines," *SAE Pap. 1999-01-3681* 3pp(724):01-3681, 1999, doi:10.4271/1999-01-3681.

[3] Environmental Protection Agency and National Highway Traffic Safety Administration, "Greenhouse Gas Emissions and Fuel Efficiency Standards for Medium and Heavy-Duty Engines and Vehicles - Phase 2," *Fed. Regist.* 81(206):73478–74274, 2016, doi:http://www.nhtsa.gov/Laws+&+Regulations/CAFE+-+Fuel+Economy/Fuel+economy+and+environment+label.

[4] Johnson, T. and Joshi, A., "Review of Vehicle Engine Efficiency and Emissions," *SAE Tech. Pap. Ser.* 1, 2018, doi:10.4271/2018-01-0329.

[5] California Air Resources Board, "Heavy-Duty Low-NOx and Phase 2 GHG Plans," https://www.arb.ca.gov/msprog/hdlownox/hdlownox.htm..

[6] Buckendale, L.R., Stanton, D.W., and Stanton, D.W., "Systematic Development of Highly Efficient and Clean Engines to Meet Future Commercial Vehicle Greenhouse Gas Regulations," *SAE Int.* 2013-01-2421, 2013, doi:10.4271/2013-01-2421.

[7] Stadlbauer, S., Waschl, H., Schilling, A., and Re, L. del, "Temperature Control for Low Temperature Operating Ranges with Post and Main Injection Actuation," 2013, doi:10.4271/2013-01-1580.

[8] Gehrke, S., Kovács, D., Eilts, P., Rempel, A., and Eckert, P., "Investigation of VVA-Based Exhaust Management Strategies by Means of a HD Single Cylinder Research Engine and Rapid Prototyping Systems," *SAE Tech. Pap.* 01(0587):47–61, 2013, doi:10.4271/2013-01-0587.

[9] Johnson, T., "Vehicular Emissions in Review," *SAE Int. J. Engines* 9 (2):2016-01–0919, 2016, doi:10.4271/2016-01-0919.

[10] Miller, R., "Supercharged Engine," *United States Patents* (US 2 817 322), 1957.

[11] Parks, J., Huff, S., Kass, M., and Storey, J., "Characterization of in-cylinder techniques for thermal management of diesel aftertreatment," *SAE Pap.* (724):01–3997, 2007, doi:10.4271/2007-01-3997.

[12] Honardar, S., Busch, H., Schnorbus, T., Severin, C., Kolbeck, A., and Korfer, T., "Exhaust Temperature Management for Diesel Engines Assessment of Engine Concepts and Calibration Strategies with Regard to Fuel Penalty," *SAE Tech. Pap.*, 2011, doi:10.4271/2011-24-0176.

[13] Mayer, A., Lutz, T., Lämmle, C., Wyser, M., and Legerer, F., "Engine Intake Throttling for Active Regeneration of Diesel particle," *SAE Tech. Pap.* (724), 2003, doi:10.4271/2003-01-0381.
[14] Imperato, M., Antila, E., Sarjovaara, T., Kaario, O., Larmi, M., Kallio, I., and Isaksson, S., "NOx Reduction in a Medium-Speed Single-Cylinder Diesel Engine using Miller Cycle with Very Advanced Valve Timing," *SAE Tech. Pap.* 4970, 2009, doi:10.4271/2009-24-0112.
[15] Benajes, J., Molina, S., Martín, J., and Novella, R., "Effect of advancing the closing angle of the intake valves on diffusion-controlled combustion in a HD diesel engine," *Appl. Therm. Eng.* 29(10):1947–1954, 2009, doi:10.1016/j.applthermaleng.2008.09.014.
[16] Guan, W., Pedrozo, V., Zhao, H., Ban, Z., and Lin, T., "Investigation of EGR and Miller Cycle for NOx Emissions and Exhaust Temperature Control of a Heavy-Duty Diesel Engine," *SAE Tech. Pap.*, 2017, doi:10.4271/2017-01-2227.
[17] Ding, C., Roberts, L., Fain, D.J., Ramesh, A.K., Shaver, G.M., McCarthy, J., Ruth, M., Koeberlein, E., Holloway, E.A., and Nielsen, D., "Fuel efficient exhaust thermal management for compression ignition engines during idle via cylinder deactivation and flexible valve actuation," *Int. J. Engine Res.* 17(6):619–630, 2016, doi:10.1177/1468087415597413.
[18] Kim, J. and Bae, C., "An investigation on the effects of late intake valve closing and exhaust gas recirculation in a single-cylinder research diesel engine in the low-load condition," *Proc. Inst. Mech. Eng. Part D J. Automob. Eng.* 230(6):771–787, 2016, doi:10.1177/0954407015595149.
[19] Pedrozo, V.B., May, I., Lanzanova, T.D.M., and Zhao, H., "Potential of internal EGR and throttled operation for low load extension of ethanol–diesel dual-fuel reactivity controlled compression ignition combustion on a heavy-duty engine," *Fuel* 179:391–405, 2016, doi:10.1016/j.fuel.2016.03.090.
[20] Fessler, H. and Genova, M., "An Electro-Hydraulic 'Lost Motion' VVA System for a 3.0 Liter Diesel Engine," 2004(724), 2004, doi:10.4271/2004-01-3018.
[21] Korfer, T., Busch, H., Kolbeck, A., Severin, C., Schnorbus, T., and Honardar, S., "Advanced Thermal Management for Modern Diesel Engines - Optimized Synergy between Engine Hardware and Software Intelligence," *Proc. Asme Intern. Combust. Engine Div. Spring Tech. Conf. 2012* 415–430, 2012, doi:10.1115/ICES2012-81003.
[22] Guan, W., Zhao, H., Ban, Z., and Lin, T., "Exploring alternative combustion control strategies for low-load exhaust gas temperature management of a heavy-duty diesel engine," *Int. J. Engine Res.* 146808741875558, 2018, doi:10.1177/1468087418755586.
[23] Ladommatos, N., Abdelhalim, S.M., Zhao, H., and Hu, Z., "The Dilution, Chemical, and Thermal Effects of Exhaust Gas Recirculation on Diesel Engine Emissions - Part 4 : Effects of Carbon Dioxide and Water Vapour," (412), 1997.
[24] Asad, U. and Zheng, M., "Exhaust gas recirculation for advanced diesel combustion cycles," *Appl. Energy* 123:242–252, 2014, doi:10.1016/j.apenergy.2014.02.073.
[25] Zhao, C., Yu, G., Yang, J., Bai, M., and Shang, F., "Achievement of Diesel Low Temperature Combustion through Higher Boost and EGR Control Coupled with Miller Cycle," (x), 2015, doi:10.4271/2015-01-0383.
[26] Sjöblom, J., "Combined Effects of Late IVC and EGR on Low-load Diesel Combustion," *SAE Int. J. Engines* 8(1):2014-01-2878, 2014, doi:10.4271/2014-01-2878.
[27] Verschaeren, R., Schaepdryver, W., Serruys, T., Bastiaen, M., Vervaeke, L., and Verhelst, S., "Experimental study of NOx reduction on a medium speed heavy duty diesel engine by the application of EGR (exhaust gas recirculation) and Miller timing," *Energy* 76(x):614–621, 2014, doi:10.1016/j.energy.2014.08.059.

[28] Stricker, K., Kocher, L., Koeberlein, E., Alstine, D. Van, and Shaver, G.M., "Estimation of effective compression ratio for engines utilizing flexible intake valve actuation," *Proc. Inst. Mech. Eng. Part D J. Automob. Eng.* 226(8):1001–1015, 2012, doi:10.1177/0954407012438024.
[29] AVL., "AVL 415SE Smoke Meter," *Prod. Guid. Graz, Austria;* 1–4, 2013.
[30] Regulation No 49 – uniform provisions concerning the measures to be taken against the emission of gaseous and particulate pollutants from compression-ignition engines and positive ignition engines for use in vehicles. Off J Eur Union, 2013.
[31] Zhao, H., "HCCI and CAI engines for the automotive industry," ISBN 9781855737426, 2007.
[32] Charlton, S., Dollmeyer, T., and Grana, T., "Meeting the US Heavy-Duty EPA 2010 Standards and Providing Increased Value for the Customer," *SAE Int. J. Commer. Veh.* 3(1):101–110, 2010, doi:10.4271/2010-01-1934.
[33] Johnson, T. V, "Diesel Emissions in Review," *SAE Int. J. Engines* 4(1):143–157, 2011, doi:10.4271/2011-01-0304.
[34] Chatterjee, S., Naseri, M., and Li, J., "Heavy Duty Diesel Engine Emission Control to Meet BS VI Regulations," *SAE Tech. Pap.* (x), 2017, doi:10.4271/2017-26-0125.

Appendix A. Test cell measurement devices

Variable	Device	Manufacturer	Measurement range	Linearity/ Accuracy
Speed	AG 150 Dynamometer	Froude Hofmann	0-8000 rpm	± 1 rpm
Torque	AG 150 Dynamometer	Froude Hofmann	0-500 Nm	± 0.25% of full scale (FS)
Diesel flow rate (supply)	Proline promass 83A DN01	Endress +Hauser	0-20 kg/h	± 0.10% of reading
Diesel flow rate (return)	Proline promass 83A DN02	Endress +Hauser	0-100 kg/h	± 0.10% of reading
Intake air mass flow rate	Proline t-mass 65F	Endress +Hauser	0-910 kg/h	± 1.5% of reading
In-cylinder pressure	Piezoelectric pressure sensor Type 6125C	Kistler	0-300 bar	≤ ± 0.4% of FS
Intake and exhaust pressures	Piezoresistive pressure sensor Type 4049A	Kistler	0-10 bar	≤ ± 0.5% of FS
Oil pressure	Pressure trans-ducer UNIK 5000	GE	0-10 bar	< ± 0.2% FS
Temperature	Thermocouple K Type	RS	233-1473K	≤ ± 2.5 K
Intake valve lift	S-DVRT-24 Dis-placement Sensor	LORD MicroStrain	0-24 mm	± 1.0% of reading using straight line
Smoke number	415SE	AVL	0-10 FSN	-
Fuel injector current signal	Current Probe PR30	LEM	0-20A	± 2 mA

Internal Combustion Engines and Powertrain Systems for Future Transport 2019 – Institute of Mechanical Engineers, ISBN 978-0-367-90356-5

Improvements in drive-cycle fuel consumption by co-optimisation of engine and fuel

Heather Hamje[1], John Williams[2], Andreas Kolbeck[3], Walter Mirabella[4], Jeff Farenback-Brateman[5], Cyrille Callu[6], Dolores Cardenas[7], Elena Rebesco[8]

[1]Concawe
[2]BP
[3]Shell
[4]Lyondell Basel
[5]ExxonMobil Research & Engineering
[6]Total
[7]Repsol
[8]ENI

ABSTRACT

Concawe has previously undertaken and published the results of two studies aimed at understanding the relationship between octane and the performance and efficiency of mainstream Euro 4 to Euro 6 vehicles. Whilst the performance and efficiency of these vehicles showed some small relationship to octane, it was important to note that most of these vehicles were not calibrated to take full advantage of fuels with a Research Octane Number (RON) in excess of 95.

To assess the full potential for higher octane fuels to lower vehicle CO2 output and fuel consumption when measured over current legislative drive-cycles, a test-bed and vehicle study was carried out using a highly downsized (30 bar BMEP), high compression ratio (12.3:1) engine with a series of 4 fuels with RON numbers ranging from 95 to 102.

Prior to measurement, the engine was calibrated specifically on each fuel over the full engine map, this ensured that the engine would experience the maximum benefit from changes in fuel properties. Based on these test-bed data, a GT-Drive model predicted the CO2 emission and fuel consumption over the New European Drive Cycle (NEDC), the Worldwide harmonised Light-duty Test Cycle (WLTC) and multiple Real Driving Emissions (RDE) cycles of differing severity.

The engine was subsequently fitted to a D-segment vehicle and NEDC, WLTC and RDE cycles were performed in order to validate modelled efficiency improvements and giving up-to 3.9% relative to the baseline 95 RON fuel.

1 INTRODUCTION

Gasoline combustion has traditionally been measured using Research Octane Number (RON) and Motor Octane Number (MON) which describe antiknock performance under different conditions. All European gasoline cars must be capable of running on the 95 RON petrol grade, however some vehicles are calibrated to be able to take advantage of higher octane fuels available in the market, typically by advancing spark timing or increasing boost pressure which allows more power and perhaps also better fuel consumption. In the future vehicles may be made available which have increased or variable compression ratio which can fully take advantage of higher octane but these are not commercially available at present.

Historically, increasing both RON and MON have been considered beneficial, however a large body of more recent literature suggests that while increasing RON still gives benefits in modern production cars, MON is less important and in fact lowering MON at the same RON level could improve vehicle performance.

It is now generally recognised that minimising energy consumption and CO_2 emissions in transportation needs consideration of both fuel production and vehicle efficiency, combining these factors into a 'well-to-wheels' approach. For the future, higher octane fuels could be used by engine designers to improve fuel efficiency using higher compression ratios, boost pressures, and other techniques. This needs to be balanced against the additional energy needed in the refinery to produce higher octane. For this reason, the optimum octane number for future fuels will come under discussion and the correct balance between RON and MON is clearly part of this process. However, such consideration of future vehicle possibilities cannot be addressed by testing vehicles in the market. Concawe carried out a study the first phase of which was reported during 2016 and the subject of several papers [1], [2] and a Concawe report published in 2016 [3]. The first phase of this study was to investigate the effect of RON and MON on the power and acceleration performance of two Euro 4 gasoline vehicles under full throttle acceleration conditions. Fifteen fuels covering RON levels 95 to 103 and sensitivities (RON minus MON) up to 15 were blended and tested. Both pure hydrocarbon and blends containing ethanol or ETBE were included so that any specific effects of oxygenates could be identified. Three additional fuels, covering RON as low as 86, were blended using primary reference fuels. The results confirm the findings of previous studies on older vehicles that MON is not a good predictor of vehicle acceleration performance and in fact high MON levels increase acceleration time under full throttle conditions. Both vehicles were tolerant of fuels in the 95-98 RON range, but reductions in performance were seen on lower octane fuels. It was found that fuel octane had no effect on the efficiency of the vehicle on the NEDC cycle, suggesting that either knock does not occur under these lighter load conditions or that adaptations to knock are not severe enough to impact on engine efficiency. Under more extreme full throttle acceleration conditions efficiency deteriorated on the lowest octane fuels tested as expected as the engine adapts to knock. It was also observed that efficiency increased up to higher octane levels than were expected for both vehicles.

A follow-on study to this screened a wider range of more modern Euro 5 and Euro 6 vehicles [4], [5]. Two vehicles were selected for further evaluation on the full fuel set of 22 fuels, again measuring acceleration performance at full load on a modified version of the test cycle used for the previous study. Both vehicles showed a strong appetite for octane in the range 86 < RON <95, with one vehicle also showing some further benefit beyond 95. This vehicle was tested for efficiency and regulated emissions on the WLTP and the US06 legislative test cycles. This vehicle was tested over the two legislative drive cycles on three fuels, to understand how the benefits, attributed to octane, at full-load would translate to vehicle efficiency over a representative drive cycle. Directionally improvements were seen beyond the octane to which the octane would be expected to respond particularly for the WLTP test cycle which appeared to show benefits up to beyond 99 octane. For the US06 some benefit was also seen up to 98 octane.

A criticism of these previous studies was that as marketplace vehicle are typically calibrated for 95 RON or occasionally 98 RON fuels, that benefits of fuels with higher RON values would not be expected to show any further benefit in terms of power or efficiency (although as described above at least one vehicle under one test cycle appeared to show an improvement). In the current study, to test the maximum efficiency benefit afforded by an increase in octane, a downsized engine was tested on a series of 4 test fuels after being calibrated specifically on each fuel.

2 TECHNICAL BACKGROUND

Octane number is a measure of a fuel's resistance to auto-ignition. Gasoline spark-ignited engines need a high octane fuel to avoid knock in contrast to diesel engines which rely on auto-ignition and so require a low octane (or high cetane number) fuel. The octane number of a fuel is measured in a special test engine known as a CFR engine that is a single cylinder test engine with variable compression ratio dating from 1928 and although the test has been progressively improved over the years, the basic engine configuration and test conditions remain the same. Tests in the early 1930s demonstrated that the knocking behaviour of fuels in vehicles of that era did not correlate with the measured Research Octane Number, therefore a new, more severe, Motor Octane Number was developed. Both methods are still in use today:

- Research Octane Number (RON) is measured at a speed of 600rpm with a specified intake air temperature of 52°C and is traditionally associated with mild to moderate driving conditions [6].
- Motor Octane Number (MON) was introduced to simulate more severe higher load conditions and uses a higher engine speed of 900rpm and a governed charge temperature of 149°C. The MON of a fuel is typically about 10 numbers lower than its RON because of the more severe test conditions, although the difference between RON and MON varies with fuel composition [7].

A fuel's octane number is determined by comparing its performance in the engine with a blend of pure compounds: iso-octane, defined to be 100 octane and n-heptane, defined to have zero octane number. Although the engine test conditions, especially the engine speed, seem far from typical of today's engines, octane number has proved a valuable measure of fuel quality up to the present and the octane requirement of even the most advanced vehicles can be described as a function of RON and MON. Fuel specifications usually set minimum requirements for both RON and MON. In most parts of the world, RON is the primary measure of gasoline octane at the point of sale. In the USA, Canada and some other countries, a different system is used where the octane measure displayed at the point of sale is the Anti-Knock Index, defined as (RON+MON)/2.

How an individual road vehicle responds to octane number depends on the details of its engine design and calibration. The 'octane requirement' of a vehicle has traditionally been determined by testing under acceleration or steady speed full load conditions, either on the road or on a chassis dynamometer. By running on a series of specially blended test fuels of progressively changing octane number, the lowest octane number that will run in the vehicle without knock can be determined. In the past, large numbers of vehicles were tested in co-operative industry programmes in Europe and the USA to build up a picture of the road vehicle fleet, so that the octane number of fuels sold could be matched to the needs of the vehicle fleet. More recently, the octane numbers are determined purely by the fuel specification and vehicles are developed to operate on them. However, a growing body of vehicle test data shows that the traditional expectation that RON correlates with mild operating conditions and MON with more severe driving no longer holds particularly in boosted downsized engines which may be representative of future vehicles [8], [9], [10].

3 PROGRAMME OVERVIEW

In the current study testing took place in two distinct stages. Firstly the engine was fitted to a test-bed, where it was calibrated and its efficiency and emission measured over a range of test points. These data were then used as inputs in a vehicle drive-cycle simulation, multiple cycles of increasing severity were calculated.

In the second segment of testing, the engine was fitted into a d/e segment vehicle, this vehicle was subsequently tested over the same drive cycles validating the simulation model.

4 ENGINE TEST-BED CALIBRATION STUDY

4.1 Engine selection

A 3-cylinder, turbocharged, direct-injection engine was chosen to undertake this study. This engine had been used in a previous studies [11], [12], [13] had shown itself to be sensitive to RON and since it used a fully flexible engine control unit (ECU) it was possible to calibrate it for best performance on each fuel.

A compression ratio of 12.2:1 was considered to be appropriate for a future looking engine with >30bar BMEP. The engine was fitted with a solenoid actuated, multi-hole direct injector and cam phasers on by intake and exhaust. The other key engine specifications are shown in table 1.

Table 1. Key engine specifications.

Cylinders	-	3
Capacity	cm^3	1199.5
Bore	mm	83
Stroke	mm	73.9
Compression Ratio	-	12.2:1
Maximum BMEP	bar	30
Peak Power (speed)	kW	120 (5000-6000 rpm)
Peak Torque (speed)	Nm	286 (1600-3500 rpm)

4.2 Test Fuels

Four fuels were selected for use in this programme, these fuels were blended to meet RON and MON targets whilst other physical properties were kept consistent wherever possible. These fuels were consistent with EN228, the European specification for forecourt gasoline.

The target RON ranged from 95, typical for regular grade gasoline in Europe, through 98, 100 to 102 the highest RON fuel currently available at an EU forecourt. Details of the test fuels are show in table 2.

Table 2. Key parameters of the 4 tests fuels.

Properties	95RON	98RON	100RON	102RON
Specific gravity	0.752	0.751	0.7509	0.7515
RON	95.4	98.3	99.9	102
MON	86.2	88.3	88.9	90.7
Olefins, vol%	2.3	4.4	3.3	6.7
Aromatics, vol%	33.1	33.1	33.1	34.2
Benzene, vol%	0.91	0.92	0.92	0.86
Oxygen, wt%	2.09	2.07	2.06	2.09
RVP, kPa	55.7	57.3	56.7	52,5
Evap. @70°C, vol%	29.5	27.1	26.4	26.4
Evap. @100°C, vol%	53.6	49.8	47.5	47.4
Evap. @150°C, vol%	94.4	94.3	93.9	94.1
LHV, GJ/T	42.2	42.3	42.05	-

4.3 Engine Calibration and Testing

To ensure best performance was achieve on each fuel, the engine was fully calibrated over a range of steady state speed and load points. Parameters optimised at each mapping point included spark-timing, cam-phasing, boost-level and lambda.

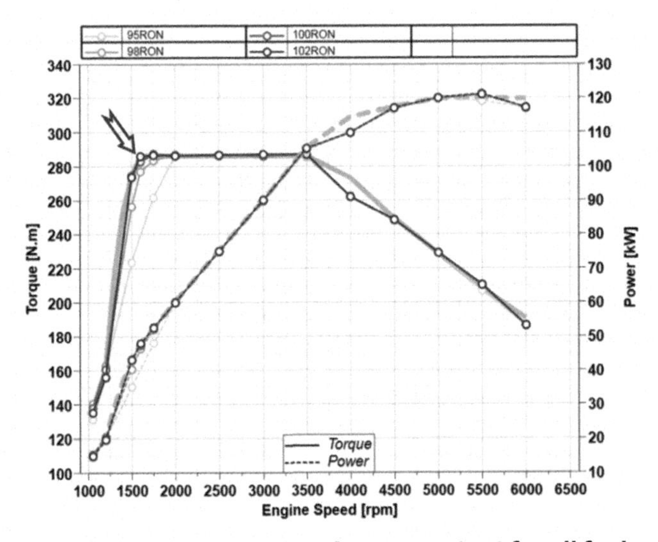

Figure 1. Chart of power and torque output for all fuels.

A requirement for this study was to ensure that similar vehicle performance was achievable on each of the test fuels. Figure 1 shows the power and torque curves for the engine with each fuel, the only noticeable deviation can be seen during low speed, high load operation, where the 95 RON fuel became knock-limited and thus was unable to achieve the same operation points.

At higher load and speed conditions, the particular care was required to ensure that the exhaust gas temperature did not exceed the material constraints of any of the exhaust system components.

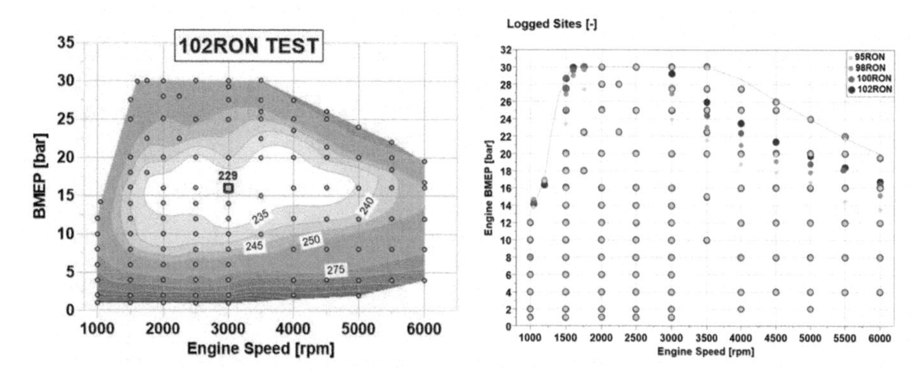

Figure 2. Speed vs Load map of a) calibration and b) logging points for all fuels.

Wherever possible the performance of each fuel was measured at the same speed and load point, as shown in the right chart of Figure 2. It is only when a significant difference in knock-susceptibility was experienced, that it became inappropriate to measure data at identical test conditions.

Measurements included key engine temperatures and pressures, together with heat-release analysis on all cylinders. Regulated emissions measurements were also recorded. Using this data, speed-load maps were created for each fuel type, which could then be used within a 1-D model to predict vehicle fuel consumption over a variety of drive cycles.

5 RESULTS AND DISCUSSION

5.1 Brake Specific Fuel Consumption (BSFC)

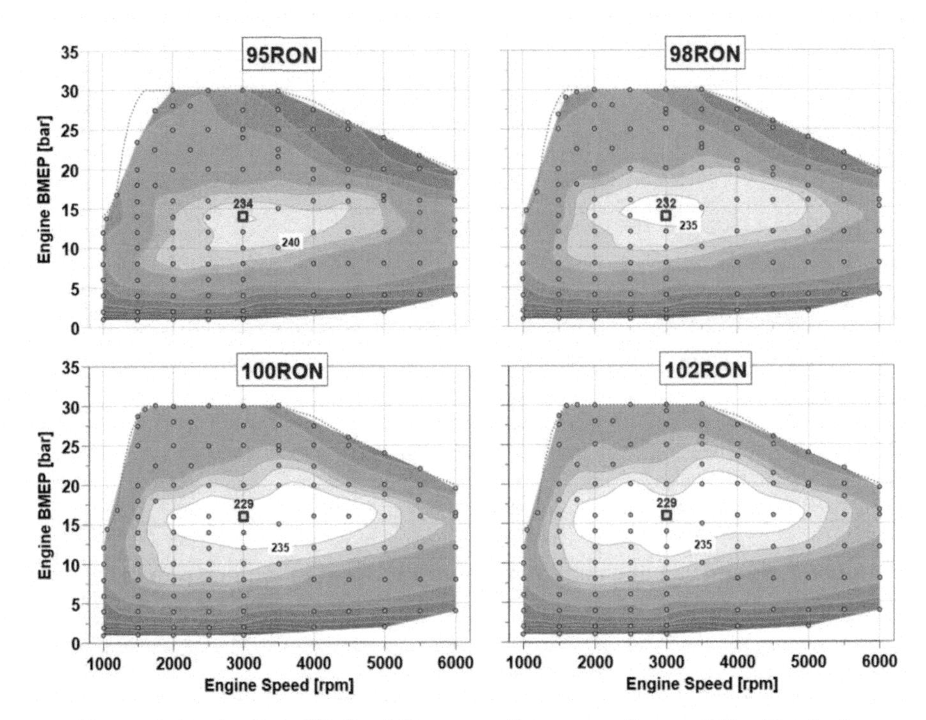

Figure 3. Brake Specific Fuel Consumption maps for each fuel tested.

The contour plots in Figure 3 demonstrated the benefit in fuel efficiency afforded by RON, as the load increases and the engine becomes more susceptible to knock, by maintaining optimum spark timing for more of the operating range, the higher RON fuels enable efficient operation over a larger portion of the range, this improvement in thermal efficiency is particularly noticeable in the size of the central area of best efficiency. On viewing the upper right portion of each chart, it is apparent that RON plays a key role in improving efficiency at high engine speeds and loads, this improvement in combustion efficiency will be discussed later.

5.2 Exhaust Gas Temperature

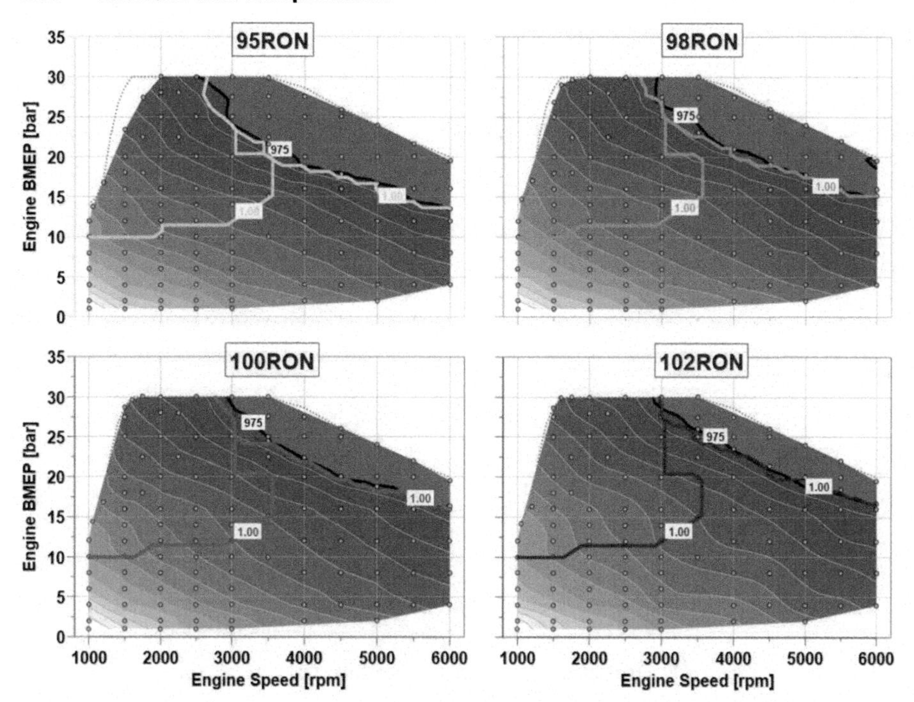

Figure 4. Exhaust gas temperature maps for each fuel.

As depicted in Figure 4, in order to protect the exhaust system components, the exhaust gas temperature profile of the engine operation on each fuel was controlled to be largely similar, deviating primarily at the higher load and speed points where lower RON fuels lead to higher exhaust gas temperatures being emitted over a larger segment of the operation range. Under these conditions, to compensate for the increase in temperature due to retarded ignition employed to avoid knocking combustion, overfuelling is employed, this point is further illustrated in Figure 5 below, where the magnitude of overfuelling required for each fuel can be clearly seen.

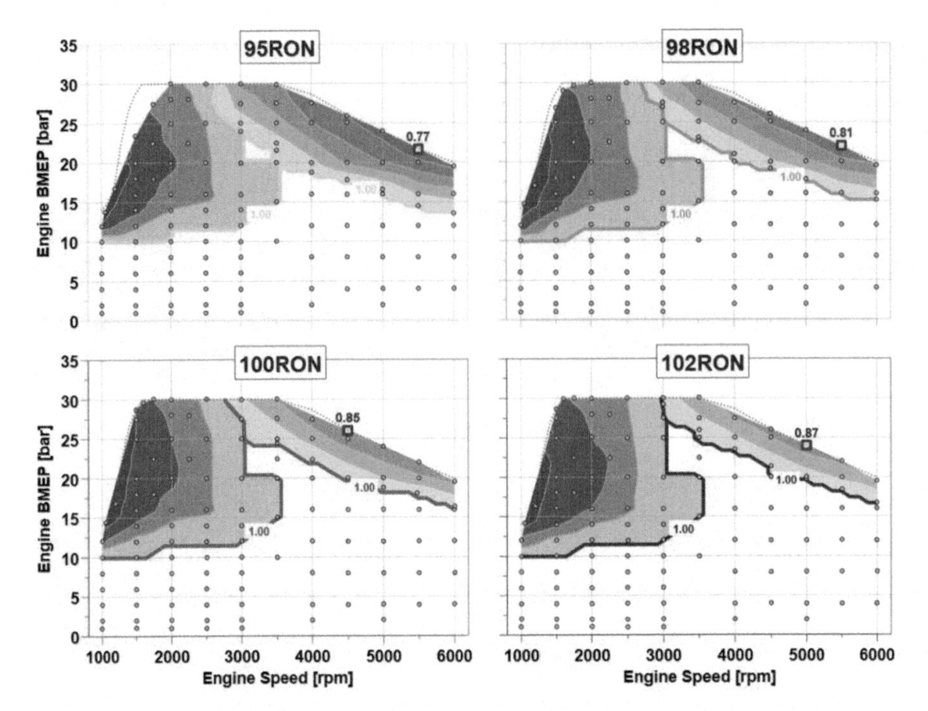

Figure 5. Depiction of exhaust lambda for each fuel.

5.3 Vehicle Modelling

A vehicle simulation was performed using GT-Drive software, this dynamic model (forward facing) model was an updated version of the model used in a previous study by Bisordi *et al* [11].

Based on a library of multi-physics elements the software enables vehicles and drivelines to be built and tested over drive-cycles for fuel consumption and pollutant emissions.

A "virtual driver" was constructed and used to generate the required system inputs such as throttle, brake, clutch and gear selection signals, to follow the time speed profiles of the various drive cycles investigated. This "virtual driver" looks one time-step ahead and calculates the torque necessary to achieve the required vehicle acceleration in order to match the requested future vehicle speed.

The calculated torque request is passed on to the engine or brake objects. In case of a torque request for the engine object, greater the fuel cut threshold, the model will look up the instantaneous fuel flow from a 3-dimensional look-up table in a quasi-static manor.

To account for transient and cold start fuelling characteristics of the real engine a number of correction tables and equations are implemented into the model.

The above method relies on the time-step to be small enough to be quasi static, usually in the order of 0.25s results in sufficiently accurate instantaneous fuel flow rates.

As the engine is turbo charged the transient time to full torque at any given engine speed is not instantaneous therefore turbo transient behaviour is also modelled via equations and look up tables based on test data. In highly dynamic drive cycles where the engine is operated transiently most of the time the above corrections have to be incorporated into the model to ensure a sufficient correlation to test data.

The inputs required for model creation combine parameters related to vehicle specifications used for driving resistance representation and powertrain data for efficiency, torque and energy flow whilst delivering the power demanded.

Vehicle specifications were either obtained via manufacturer's information or from direct measurements and were finely adjusted so road loads such as aerodynamic drag and wheel rolling resistance could be accurately represented.

In order to capture the actual losses for the vehicle under evaluation, a vehicle coast down test following 70/220/EEC guidelines was performed and the measured driving resistance curve employed later in the correlated model for the technology and fuel assessment over the drive-cycles selected.

For engine representation the GT-Drive model uses a map of measured fuel flow rate against engine speed and load. This map is obtained from dynamometer measurements taken during steady-state operation, under fully warm engine conditions. Full load and motored curves are also measured and implemented as function of accelerator position.

The transient effects such as increased fuelling during warm-up can be accurately estimated due to the repetitive nature of the NEDC. A fuel multiplier trendline is fitted to a modified Arrhenius equation in which elapsed cycle time is the main variable. This method was applied to all measurements performed on the selected fuels tested in the baseline vehicle. No appreciable differences were found between the fuel types investigated. Therefore, a single warm-up correction model for all fuel types is used. The same warm-up correlation is also used for both the WLTC and RDE cycle simulations.

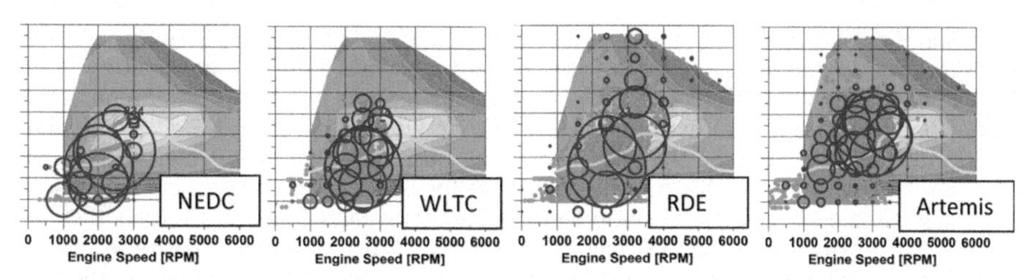

Figure 6. Fuel weighted residency maps for each drive cycle, 95RON knock-limit displayed.

The fuel-weighted speed-load residency plots in Figure 6 demonstrate the benefit that can be achieved through the use of higher octane fuels. The yellow line plotted on each chart described the knock limit of the engine, the size and number of blue circles plotted above that line giving an indication of the relative severity of each cycle, from an engine knock perspective, and therefore the potential benefit for higher RON fuels.

These benefits translate to the drive-cycle fuel efficiency for each fuel and cycle combination presented in Table 3 below.

Table 3. Simulated fuel consumption.

	95 RON	98 RON	100 RON	102 RON	95 RON	98 RON	100 RON	102 RON
Drive Cycle	**L/100km**				**% improvement vs. 95 RON**			
NEDC	7.078	7.062	7.019	6.954	-	0.22	0.83	1.75
WLTC	7.663	7.640	7.552	7.486	-	0.29	1.44	2.3
RDE	8.129	8.022	7.927	7.827	-	1.32	2.48	3.72
Artemis	8.34	8.245	8.168	8.075	-	1.14	2.06	3.17

Fuel economy benefits associated with an increase of RON from 95 to 102 of between 1.75% and 3.72% were simulated, with the lowest benefit being seen over the NEDC drive cycle, and the greatest over the chosen RDE cycle. For the NEDC cycle the engine operates at BMEP levels below the knock limit threshold for most of the cycle and therefor the effect of higher RON fuels is relatively small. The WLTP cycle is operated at slightly higher loads although the greater part of the cycle is still below the 95 RON knock limit. In addition, both the RDE cycle and the Artemis cycle operate at significantly higher loads compared to the NEDC or the WLTC cycles and, therefore, showed fuel economy improvement when higher RON fuels were used due to less overfuelling needed than for lower RON fuels. For the NEDC, WLTC and Artemis cycles, the results were also modelled with and without stop-start. The results with stop-start are shown with the benefit from using stop-start ranging from 1% for the Artemis to around 4.7% for the NEDC due to the high idle content of the latter cycle. The RDE results are shown for a shift strategy run in ECO shift mode. In addition, a SPORT shift mode was also run and for the chosen cycle there was a penalty of around 3% for all fuels.

6 VEHICLE TESTING

Following the completion of the engine test-bed testing and modelling phase, the engine was fitted within the chassis of a d/e segment car for chassis dynamometer testing. This vehicle was originally equipped with a 2.0 litre, turbocharged direct-injection engine of similar performance to the test-engine. The vehicle was tested using NEDC, WLTC and RDE simulated test cycles on the chassis dynamometer. The RDE test cycle chosen was the same as that used for the modelling exercise for direct comparison and was chosen as it represented an average cycle in terms of those available for all the fuels tested. Figure 7 shows the results for CO2 and fuel economy for the measured test cycles. Each of the results was the average of each of three repeats and the bars show the range of data round the average points. Both the WLTC and the RDE showed trends downward as RON increased with no overlap between the results from the 102RON fuel and the other fuels. The NEDC results were less clear which was consistent with the modelling and in line with the residency maps including the amount of time spend in low load versus high load conditions. The modelled fuel economy results are superimposed on the charts in Figure 8, and it can be seen that the NEDC for the modelled result at 95RON appears to follow the same trend as the others. In general the difference between the modelled and measured results was around 1.5% and below which was considered to be very good with the lowest difference in the RDE results and the biggest difference with the WLTC which was more similar to the NEDC. For all the cycle simulations particular attention was paid to such items as idle speed, road load, catalyst heating, alternating lambda and fuelling during gearshifts. In sensitivity analyses, the latter three items, in particular, were most significant in improving the comparison between the modelled and measured results. It is expected that differences between these parameters would account for the differences between the measured and modelled results for the difference cycles.

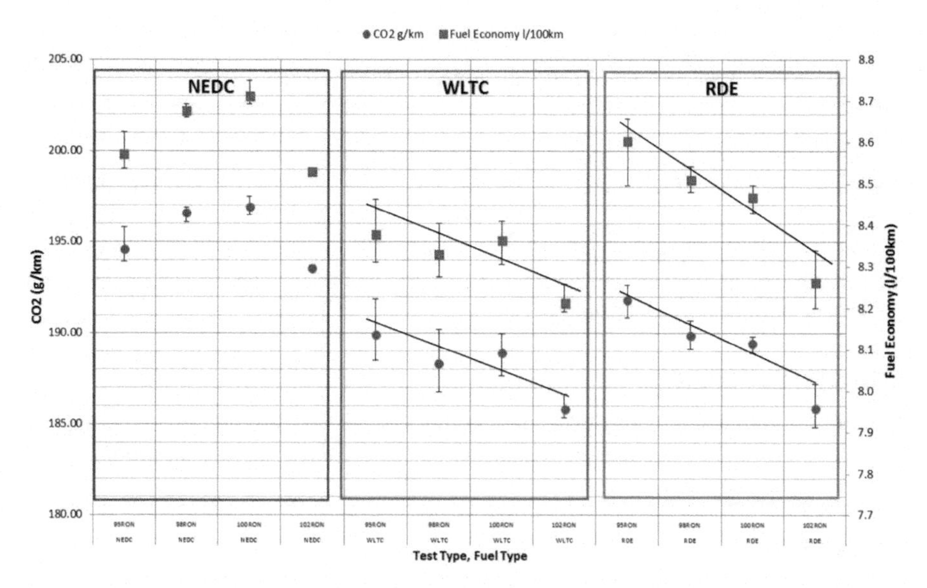

Figure 7. Vehicle results for all fuels and drive cycles.

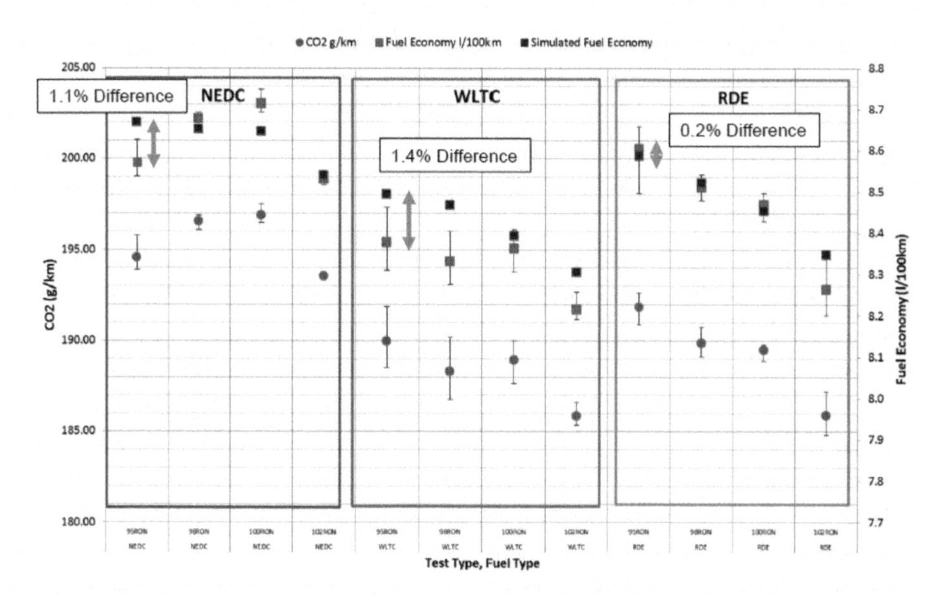

Figure 8. Comparison of modelled and measured drive cycle fuel consumption and CO2.

7 CONCLUSION

When optimised to take advantage of higher RON fuels an engine can demonstrate significant improvements in efficiency and CO2 emissions, particularly when operating under high load conditions.

During lower engine speed operation, the majority of this benefit in efficiency associated with higher RON fuels can be directly attributed to the thermodynamic improvement of earlier ignition timing. At higher engine speed, the ability to maintain advanced ignition timing also reduces the necessity to use over-fuelling as a means to protect exhaust system components from excessive gas temperature, thus providing a significant further benefit for higher RON fuels, this is more noticeable during real driving conditions.

To understand the societal impact of the future use of higher RON fuels in terms of well-to-wheels CO_2, the impact that the production of higher RON fuels has upon refinery efficiency must be understood. This topic has been studied by Concawe and will be the subject of a future publication.

ACKNOWLEDGEMENTS

The authors would like to acknowledge members of the Concawe FE/STF-20 group for their contribution to the paper as well as Mahle Powertrain Ltd, Northampton, UK and Coryton Advanced Fuels, UK for fuel blending.

REFERENCES

[1] Stradling, R. et.al, Effect of Octane on the performance of two gasoline direct injection passenger cars. SAE Technical Paper 2015-01-0767, 2015.
[2] Stradling, R. et al, Effect of Octane on Performance, Energy Consumptions and emissions of two Euro 4 passenger cars, Transportation Research Procedia (2016).
[3] Concawe Report 13/16, Phase 1: Effect of fuel octane on the performance of two Euro 4 Gasoline Passenger Cars, www.concawe.org, 2016.
[4] Williams, J, et.al. Effect of Octane on the performance of Euro 5 and Euro 6 gasoline passenger cars, SAE paper no. 2017-01-0811, 2017.
[5] Concawe Report xx/2019 (in progress).
[6] EN ISO 5164 "Petroleum products – determination of knock characteristics of motor fuels – Research method".
[7] EN ISO 5163 "Petroleum products – determination of knock characteristics of motor fuels – Motor method".
[8] Davies, T., Cracknell, R., Lovett, G., Cruff, L. et al., Fuel Effects in a Boosted DISI Engine. SAE Technical Paper 2011-01-1985, 2011.
[9] Amer, A., Babiker, H., Chang, J., Kalghatgi, G. et al, Fuel Effects on Knock in a Highly Boosted Direct Injection Spark Ignition Engine. SAE Int. J. Fuels Lubr. 5(3):1048-1065, 2012. SAE Technical Paper 2012-01-1634, 2012.
[10] Remmert, Sarah et al, Octane Appetite: The Relevance of a Lower Limit to the MON Specification in a Downsized, Highly Boosted DISI Engine. SAE Technical Paper 2014-01-2718, 2014.
[11] Bisordi, A. et al, Evaluating Synergies between Fuels and Near Term Powertrain Technologies through Vehicle Drive Cycle and Performance Simulation. SAE Paper 2012-01-0357,2012.
[12] Oudenijeweme, D. et.al. "Significant CO2 Reductions by Utilising the Synergies Between a Downsized SI Engine and Biofuels" – ImechE – Internal Combustion Engines: Performance, Fuel Economy and Emissions 2011.
[13] Oudenijeweme, D.et. al. "Downsizing and Biofuels: Synergies for Significant CO2 Reductions", Aachen Colloquium Automobile and Engine Technology 2011.

Evaluating the potential of a long carbon chain oxygenate (octanol) on soot reduction in diesel engines

I. Ruiz-Rodriguez[1], R. Cracknell[2], M. Parkes[2], T. Megaritis[1], L. Ganippa[1]

[1]Mechanical Aerospace and Civil Engineering Department, Brunel University London, Uxbridge, UK

[2]Shell Global Solutions UK, Cheshire, UK

ABSTRACT

The automotive industries are facing challenges to meet stringent CO_2 and air-quality regulations. One of the factors affecting emissions performance is the fuel's chemical composition. The presence of oxygen in a fuel offers the potential to lower soot emissions, however, these oxygenates must be compatible with existing hardware and fuels. To this end, the combustion of a long carbon-chain oxygenate, 1-octanol, was studied. It was injected at high pressures into a constant volume chamber, and a high-speed, two-colour pyrometry system was used to evaluate the spatial and temporal variations of soot and temperature. A significant reduction of soot was obtained whilst the flame temperature remained similar to that for diesel.

1 INTRODUCTION

Internal combustion engines (ICEs), in particular diesel engines, are widely used in many industrial sectors such as in agriculture, rail, marine and automotive because of their high power output, thermal efficiency, reliability, and low maintenance costs. Significant developments have been made to improve the emissions of noxious compounds such as NO_x and soot [1]. Nonetheless, regulations are getting tighter, and whilst light-duty vehicles will see increasing levels of electrification ranging from mild hybrids to battery electric vehicles (BEVs), for the heavy-duty fleet, it is difficult to envisage a complete switch in the near future because of weight and range constraints. It seems clear that ICEs will remain important in the near future, so solutions that address both legislation and demand must be sought. One such approach is to improve the in-cylinder combustion processes, as this would offer several socio-economic advantages: by making use of the current engine infrastructure, implementation costs can be reduced and; reducing in-cylinder emissions would reduce the cost and load on aftertreament systems, potentially allowing old vehicles to still operate on the roads.

In recent years, the interest for oxygenated fuels has increased due to their potential use as a drop-in fuel to help reduce emissions in diesel engine, in particular soot. In this context, a drop-in fuel is an alternative fuel to diesel that can be directly used in ICEs without major modifications to the engine. Drop-ins are normally blended with diesel at different percentages. Some oxygenates can be produced from renewable sources and even biomass, and others can be synthetically produced via microbial engineering [2-4]. One such group of oxygenates are alcohols, which are characterised by their hydroxyl functional group (-OH). The combustion of low carbon chain alcohols such as ethanol and methanol has been widely studied and characterised, and their properties are well known [5-7]. More recently, work has been done on higher carbon chain alcohols such as pentanol, hexanol, and octanol [8-10]. Higher carbon chain alcohols tend to have a higher energy content; furthermore, longer carbon chains reduce compound polarity, meaning that they are more likely to form stable blends with diesel. In an industrial context, this means that higher alcohols have the potential to be compatible with current engine hardware and would not necessitate the addition of surfactants to mix with diesel- offering a cost-effective solution. In other studies, octanol has been implemented directly in engines and has shown promising results in terms of emission performance [10, 11]. However, most studies have been done in engines, and thus optical data on its in-cylinder combustion process and soot formation are still lacking. Spatial information of how the temperature and soot distribution in spray flames change when fuelled with octanol could help to characterise its potential as a drop-in fuel for diesel engines.

This work aims to determine the suitability of octanol for engines and its effectiveness in reducing in-cylinder soot formation when compared to diesel. In order to address this, 1-octanol was injected neat (in its pure form without blending with other components) into a high-pressure, high-temperature constant volume chamber. This allowed for the decoupling of complex cylinder motions from the combustion event and for precise control of the ambient conditions. Its combustion was studied by using the two-colour pyrometry method coupled with high-speed imaging. This was done under conditions simulating an exhaust gas recirculation (EGR) environment that is typically used to control NO_x emissions- albeit normally having a trade-off on increased soot emissions. If under these conditions a fuel can produce lower soot than diesel it could potentially help address the NO_x-soot trade-off.

2 EXPERIMENTAL METHODS

2.1 Fuels

Diesel of EN590 compatibility was used as a reference fuel, with a known cetane number (CN) of 53, density of 850kg/m^3 and a lower calorific value (LCV) of ~43MJ/kg. The oxygenate studied, 1-octanol, is a linear alcohol with a carbon chain of 8 and with a hydroxyl group at one of the chain ends, as shown in Figure 1. It has a chemical formula of $C_8H_{18}O$ with a CN of 39, density of 824kg/m^3 and an LCV of ~38MJ/kg [12, 13]. From toxicity reports [14], its only known health impacts lie in the irritant category, which makes it a chemical with a relatively lower toxicity profile than diesel. This is an important fuel property, as it would make its handling relatively safe for end-users.

Figure 1. The chemical structure for 1-octanol.

2.2 The constant volume chamber

Both diesel and octanol were injected separately in their neat forms into a constant volume chamber (CVC). In these experiments, the mass of octanol injected was matched to that of diesel when injected at a duration of 1.5ms. One of the advantages of using a CVC is that the ambient conditions can be carefully controlled. For these tests, the conditions were set to an ambient temperature of ~1300K and to an ambient oxygen concentration of ~10%. This was obtained by chemically preheating the chamber with a premixed homogeneous mixture of acetylene and air, ignited by a spark plug. The combustion of the mixture would then reach high temperatures and pressures, closely monitored by a fast-response pressure transducer. Upon the decay of both temperature and pressure, the test fuel was injected at the desired conditions mentioned previously. A multi-hole common rail diesel injector was used, but only one of the sprays was imaged to maximise the camera's resolution. The injection pressure was set at 700bar to allow for a compromise between imaging area and camera speed, set at 18,000 frames per second (fps); seven combustion events were acquired for each fuel.

The schematic of the experimental set-up that was used in this work is shown in Figure 2. The window at the bottom allows for optical access for the high-speed diagnostics and the one at the side allows for additional illumination if needed for spray visualisation. For the two-colour pyrometry method (which will be described in more detail in the following subsection) the flame emission was imaged at two different wavelengths by using two narrow band-pass filters centred at 543.5nm and 670nm, having a full-width half max (FWHM) of 10nm. This was achieved by aligning a 50/50 beam splitter with a mirror directly under it, which enabled two images of the same flame to be acquired using one camera after passing through either a green or a red filter. The data was then post-processed to extract the flame from the background and to solve for soot and temperature values.

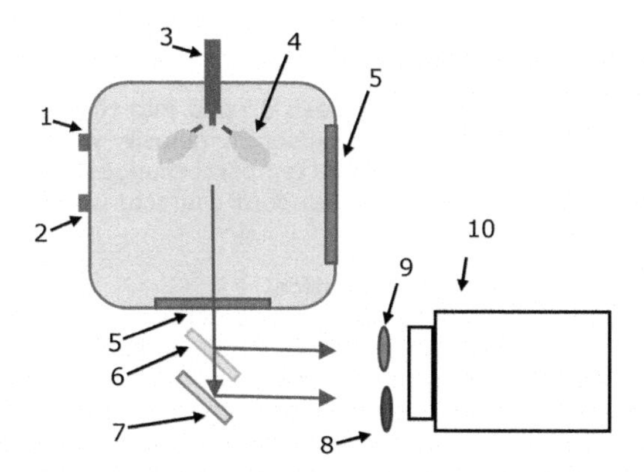

Figure 2. Experimental set-up showing the CVC and the high-speed two-colour pyrometry system. 1. Exhaust valve; 2. Gas inlet; 3. Multi-hole injector; 4. Sample flame; 5. Optical access; 6. 50/50 beam splitter; 7. Mirror; 8. Red narrow band-pass filter; 9. Green narrow band-pass filter; 10. High-speed camera.

2.3 Two-colour pyrometry

The two-colour (2C) method has been used extensively in the literature to characterise the soot and temperature distributions in flames [15-18]. The basic concept relies on the soot radiance (I_{soot}) being a multiple of the blackbody radiance (I_{BB}) and the emissivity of the body investigated (ε), which takes a value of <1 for a non-blackbody emitter, equation 1. The radiation emitted by the soot particles is dependent on both the wavelength (λ) and the temperature (T). By using Planck's laws one can further express I_{BB} in terms of λ, T and two constants, c1 and c2 known as Plank's constants, equation 2.

$$I_{Soot} = \varepsilon I_{BB} \tag{1}$$

$$I_{BB} = \varepsilon \frac{c_1}{\lambda^5 \left[\exp\left(\frac{c_2}{\lambda T}\right) - 1\right]} \tag{2}$$

Finally, the emissivity can be obtained by using semi-empirical equations expressed in terms of the soot absorption (KL factor) and λ [19]. One of the most widely used correlations for ε is that presented in equation 3, where α is the dispersion coefficient and takes a value of 1.39 in the visible wavelength range [19].

$$\varepsilon = 1 - \exp\left(-\frac{KL}{\lambda^{\alpha}}\right) \tag{3}$$

Calibration parameters are then required to relate the pixel's arbitrary units registered in the sensor of the camera to a physical parameter such as radiance. In this work, this was done by using an integrating sphere and a tungsten halogen light source. With these data, the above equations, were solved using an in-house developed numerical solver to obtain the flame temperature and KL factor for every pixel location in the flame. The sensitivity of this technique due to the choice of the detection wavelength, selection of the dispersion coefficient and other setting-specific errors have been discussed in [20, 21]. This technique provides semi-qualitative relative information about soot and temperature between fuels.

3 RESULTS AND DISCUSSIONS

The results and discussions section has been divided into three subsections. Some generic combustion characteristics based on soot luminosity will be discussed first in §3.1. Following this, in §3.2, the spatial and temporal changes in temperature will be presented. Finally, in §3.3, the spatial and temporal characterisation of the soot evolution will be discussed.

3.1 Fuel and combustion characteristics

Attempts were made in this work to report any deterioration caused to the fuelling system when injecting with neat octanol. From purely qualitative observations, there was no significant alteration of the system behaviour based on the consistent combustion behaviour. As well, the injector remained operative as there were no injection failures. After each injection, the components were visually inspected for signs of damage but none could be observed. Additional tests were also carried out,

Figure 3. Sample of qualitative soaking test carried out to identify potential material incompatibilities with neat octanol. Sample 1, stainless steel; Sample 2, steel; Sample 3, galvanised steel; Sample 5, PVC hose; Sample 6, PTFE hose. Besides a slight swell of the PVC pipe, no material alterations were observed.

where sample materials from the fuel system and CVC were soaked in neat octanol for a period of three hours. Some of the materials tested were stainless steel, steel, galvanised steel, PVC, Viton, silicone, and nitrile, as shown in Figure 3. The only effect observed in any of the materials tested was a slight swelling of the PVC hose. This indicates that for longer storage periods, PVC components might not be compatible with neat octanol. These tests show that octanol is potentially compatible with existing injection systems, and more so as a drop-in fuel at low percentages, as the concentration of octanol would be relatively lower than that tested here. The three-hour period was chosen to ensure that the fuel did not cause any break down of the high pressure chamber set-up, however, longer time periods would be required to ascertain its full compatibility characteristics. Even though the outcome is encouraging, these are only preliminary qualitative observations and a quantitative evaluation is required to fully characterise any material incompatibilities.

Regarding the combustion characteristics, two parameters were studied. First, the soot lift-off length (sLOL), Figure 4, and second, the end of (visible) combustion (EOC), Figure 5. The sLOL is the distance between the injector tip and the point at which soot has stabilised. The data in each figure represent an ensemble average of the seven injections for each fuel, with the error bars representing two standard deviations (SDs). The values were time-averaged over the stable sLOL period, which occurred from 985μs after the start of injection (aSOI) to 2217μs aSOI.

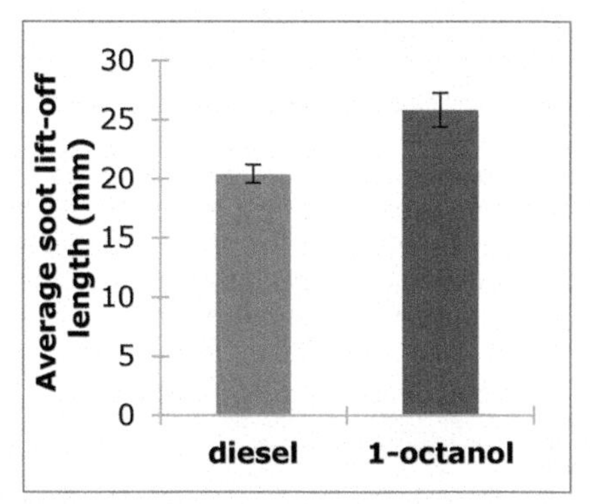

Figure 4. Soot lift-off length for both fuels averaged over the period where the sLOL was stable. This was done for the seven injections performed for each fuel and the error bars correspond to two SDs.

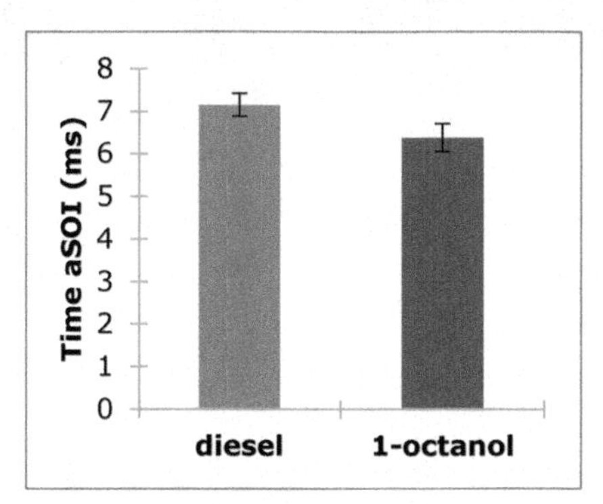

Figure 5. Point in time at which the last visible pocket of soot can be observed, referred to in the main body of the text as the end of (visible) combustion (EOC).

From Figure 4, it can be seen that for diesel, soot stabilizes on average ~20% closer to the injector tip than octanol. This indicates that spatially, the onset of soot formation is delayed for octanol to a more downstream location. It could be that either its lower reactivity or its oxygen concentration (or a combination of both) offsets the formation of soot. In practical terms, this means that soot will have less space and time to grow beyond this location before oxidation starts to control the soot development processes. This could limit the development of larger soot particles, which would also make them easier to oxidise in engines.

From Figure 5, it can be seen that soot is consumed slightly earlier for octanol than for diesel. It is interesting to note that even though 1-octanol has a lower reactivity profile than diesel, its oxidation is clearly faster, as the last pocket of soot disappeared around 700μs earlier than for diesel. In an engine, this would mean that for the same amount of injected mass of fuel, due to its faster oxidation, less soot will remain "unoxidised" when the exhaust valve opens. This can potentially reduce engine-out soot emissions as well as enhance the life of diesel particulate filters where applicable.

Though the same mass of fuel was considered in this work, as seen in §2.1, the density and energy between the fuels are also comparable and hence the outcome based on the same energy content may not be far from that presented in these results.

3.2 Spatial and temporal characterisation of the flame temperature

The spatial and temporal changes in temperature for diesel are shown in Figure 6, and those for octanol in Figure 7. For both graphs, each data point corresponds to the temperature value averaged over the detected flame, and subsequently ensemble averaged over the seven injections performed. The errors bars shown correspond to two standard deviations. The images above each graph correspond to one of the seven injections performed for each fuel, and are presented to illustrate the spatial temperature distribution throughout the flame as combustion proceeds. Each image in the sequence corresponds to one of the data points in the graphs, presented in chronological order.

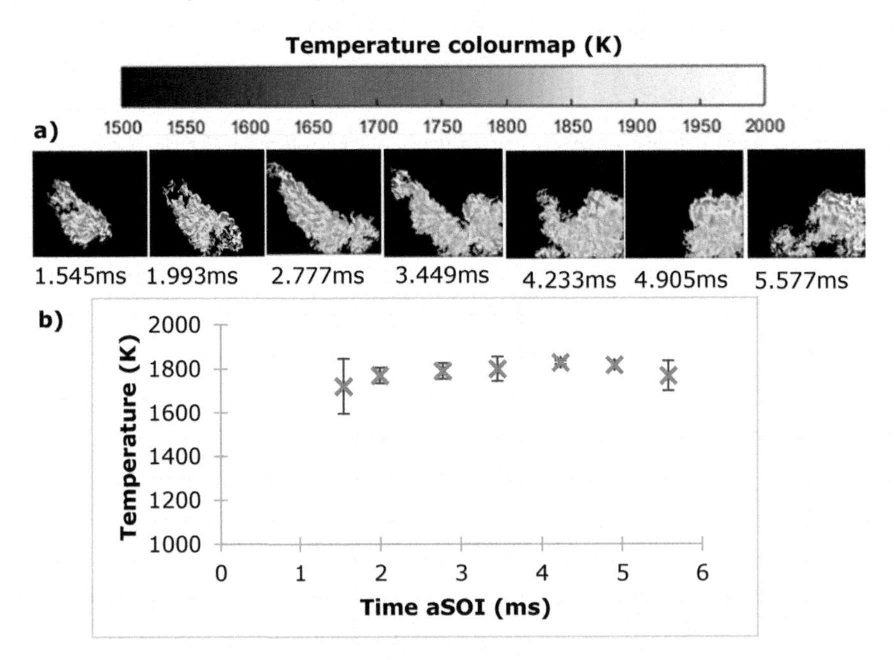

Figure 6. Temperature results obtained for diesel a) Spatial distribution of the temperature in the flame b) Temporal development of the temperature for points selected throughout the injection event.

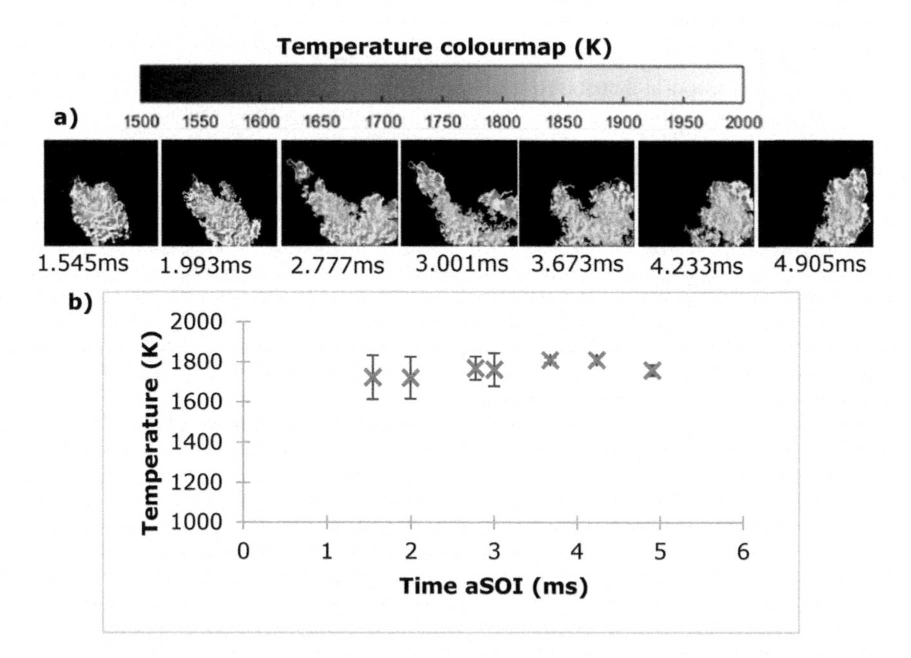

Figure 7. Temperature results obtained for octanol a) Spatial distribution of the temperature in the flame b) Temporal development of the temperature for points selected throughout the injection event.

For both fuels, the temperature was slightly lower at the SOC relative to the main combustion phase, and it then increased slightly and remained nearly constant until the combustion approached an end. Both fuels followed the expected combustion behaviour, that is, an increase in temperature during the SOC as reactions developed and then an almost constant temperature until the final reactants and products were consumed during the EOC. When looking at the average temperature, it remained between 1700K-1800K for both fuels. It could be that under the conditions studied, the similar LCV of both fuels produced this similar average flame temperature. Extrapolating these findings to an engine, this suggests that under the selected ambient conditions, octanol could produce a similar amount of thermal NO_x as that of diesel, provided the flame areas are similar. Nonetheless, this suggestion is based on the findings of this work performed in a CVC and further works on engines are needed to provide a quantitative assessment.

When looking at the spatial distribution of the flame temperature (Figures 6a and 7a), similar trends were identified for both fuels. At the SOC, there were flame regions with high and low temperature values due to the localised fuel-rich regions that had not had the time to mix and therefore burned at different temperatures depending on their mixture ratios. As combustion proceeded, these regions dissipated and the temperature distribution became more uniform except for some lower temperature patches at the tip of the flame and at the flame edges that were closer to the chamber wall. This was caused by heat transfer from the hot flame gases to the cooler CVC wall. As combustion approached an end, high temperature regions were observed at the tip of the flame, where more air was available for oxidation reactions to develop and consume the remaining pockets of fuel and soot.

3.3 Spatial and temporal characterisation of soot

The spatial and temporal distributions of soot (Figure 8 and Figure 9) have been presented in a similar manner to those for the temperature in the previous sub-section (Figure 6 and Figure 7). The KL factor represents the amount of soot as an averaged line-of-sight value. It is commonly presented in 2C pyrometry literature without units, and following this convention, it is presented as dimensionless in this work. The data points in the graphs represent averaged values over the flame and are themselves an ensemble average of the seven injections performed for each fuel. Each image also corresponds to a single shot of one of the combustion events, corresponding chronologically to the data points in the graphs.

As opposed to the similarities observed for the temperature in the previous subsection (Figure 6 and Figure 7), the soot distribution for diesel and octanol showed differences both spatially (Figure 8a and 9a) and temporally (Figure 8b and 9b). Octanol consistently showed lower KL factor values than diesel and achieved a significant reduction in peak soot of 67%. When looking at the temporal evolution of the curves for both fuels, the soot formation rate for octanol was slower, as it took longer than diesel to reach its peak KL factor value. Before the KL peak is reached, when the KL factor has a positive rate of change, soot formation dominates over oxidation. On the other hand, after the KL peak is reached and the KL factor has a negative rate of change, soot oxidation dominates over formation. The oxidation rate of octanol appeared slightly faster than that of diesel, partly because of its lower peak KL factor. However, from these observations, it seems that the effect of octanol on soot suppression is the major drive behind its soot reduction potential. Overall, this finding makes octanol an attractive drop-in fuel for engine applications because of its lower soot production rate that makes it easier to oxidise before the exhaust valve opens.

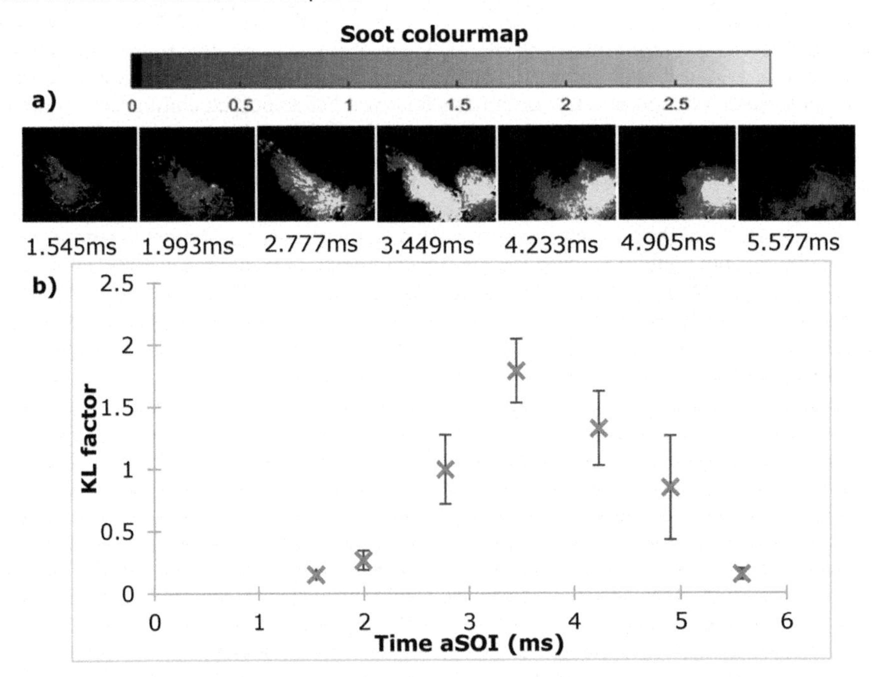

Figure 8. Soot (KL) results obtained for diesel a) Spatial distribution of the soot in the flame b) Temporal development of the soot for points selected throughout the injection event.

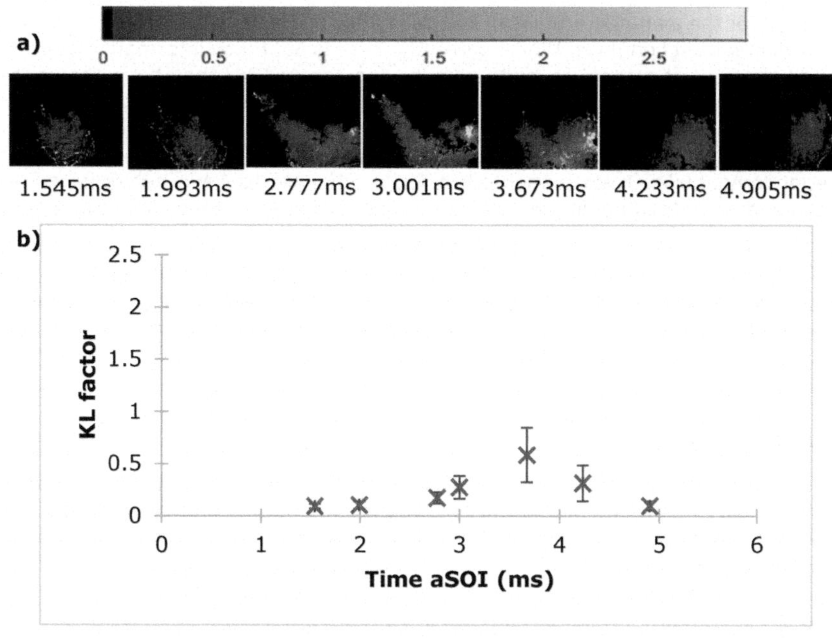

Figure 9. Soot (KL) results obtained for octanol a) Spatial distribution of the soot in the flame b) Temporal development of the soot for points selected throughout the injection event.

When looking at the soot spatial distribution images, the same conclusions can be drawn than those from the graphs: overall, octanol soots less. From the images, it can be seen that diesel shows higher KL values throughout the times considered, whereas octanol has lower KL values, which indicates a lower sooting tendency. Some interesting information could also be extracted when looking at how the soot develops spatially. That is, looking at where the high and low KL regions are in the flame and how they develop with time. For diesel, soot accumulated at the centre of the flame and towards the tip as combustion proceeded. For octanol, the higher sooting regions relative to its sooting propensity occurred -and remained- mostly at the tip. This suggests that the physio-chemical mechanisms causing a reduction in overall soot for octanol are also altering the way in which soot is distributed across the flame when compared to diesel flames. Even though at this stage it is unclear whether the soot reduction capabilities of octanol are due to the chemical effect of its hydroxyl group or due to dilution [22, 23], it is clear that octanol shows the potential to reduce soot emissions without a significant increase in flame temperature (i.e., without a significant effect on the thermal NO_x). More studies are currently being carried out by the authors to determine the extent of the effect of the hydroxyl moiety on soot reduction.

In line with the outcomes reported in this work, previous works on single-cylinder engines have also shown promising results regarding the potential of fuelling with octanol [10, 24, 25]. Under the conditions in their studies, they showed that octanol had a longer premixing time than diesel, and that this partly contributed to its low sooting levels. They have also shown a reduction in PM of up to a factor of five when fuelling with octanol, which indicates that depending on the load operation, soot suppression could be even higher than the one reported in this work.

Of interest for the manufacturers is the performance of higher carbon chain alcohols in multi-cylinder diesel engines. Whilst currently there is not much information available on the behaviour of octanol in multi-cylinder engines, in [26] they have shown that with alcohol additions of 30% to diesel and with the help of cetane improvers, a significant reduction in soot emissions is possible whilst keeping similar heat release and thermal efficiency profiles to those of diesel. Furthermore, in line with the preliminary material compatibility tests presented in §3.1, other works have not reported any major damages to their equipment when using octanol as a neat fuel or as a drop-in.

Whilst more research is required in CVCs, single, and multi-cylinder engines to understand the fundamentals behind the reduced sooting propensity of octanol, it is clear that from an industrial and application point of view, both its physical and combustion characteristics make it an attractive compound for future consideration as a drop-in fuel.

4 CONCLUSIONS

The combustion characteristics of octanol and diesel were studied by using high-speed two-colour pyrometry, which allowed to elucidate the spatial and temporal data for both soot and temperature at every flame location. The soaking test performed with octanol revealed no operational or physical alterations of the injection and pumping systems, but it was observed that octanol did cause a slight degree of swelling of the PVC hose.

The spatial and temporal changes of the temperature and soot in the spray flame were also analysed. Both fuels showed a similar average flame temperature throughout the combustion event as well as similar temperature distributions under the conditions studied. Nonetheless, large differences were observed between the fuels for the sooting propensity, with octanol consistently showing lower sooting values. Octanol also had a slower soot formation rate, which in an engine would facilitate its prompt oxidation. It was also found that for octanol, the soot only accumulated at the flame tip, whereas for diesel it was more evenly distributed throughout the flame. From our work, it can be concluded that octanol offers the potential to reduce engine-out soot emissions, which would also help reduce the load on the particular filters.

ACKNOWLEDGEMENTS

This work was financially supported by the EPSRC.

NOTATION

ε	Emissivity (-)
λ	Wavelength (nm)
2C	Two-colour
BEV	Battery electric vehicle
CN	Cetane number
CVC	Constant volume chamber
EGR	Exhaust gas recirculation
Fps	Frames per second
FWHM	Full width at half maximum
EOC	End of combustion
I_{soot}	Soot radiance (W/sr m2 nm)
I_{BB}	Blackbody radiance (W/sr m2 nm)
ICE	Internal combustion engine
KL	Soot optical thickness (-)
LCV	Lower calorific value (MJ/kg)
sLOL	Soot lift-off length
SOC	Start of combustion
SOI	Start of injection
T	Temperature (K)

REFERENCE

[1] Williams, M., Minjares, R.: A technical summary of Euro 6/VI vehicle emission standards. ICCT, Int. Counc. Clean Transp. 1-12 (2016).

[2] Julis, J., Leitner, W.: Synthesis of 1-octanol and 1,1-dioctyl ether from biomass-derived platform chemicals. Angew. Chemie - Int. Ed. 51, 8615-8619 (2012). 10.1002/anie.201203669

[3] Stepan, E., Enascuta, C.E., Oprescu, E.E., Radu, E., Vasilievici, G., Radu, A., Stoica, R., Velea, S., Nicolescu, A., Lavric, V.: A versatile method for obtaining new oxygenated fuel components from biomass. Ind. Crops Prod. 113, 288-297 (2018). 10.1016/j.indcrop.2018.01.059.

[4] Dekishima, Y., Lan, E., Shen, C., Cho, K., Liao, J.: Extending carbon chain length of 1-butanol pathway for 1-hexanol synthesis from glucose by engineered Escherichia coli. JACS. 133, 11399-11401 (2011).

[5] Leach, F., Stone, R., Davy, M., Richardson, D.: Comparing the effect of different oxygenate components on PN emissions from GDI engines. In: IMechE ICE (2015).

[6] Jamrozik, A., Tutak, W., Gnatowska, R., Nowak, Ł.: Comparative Analysis of the Combustion Stability of Diesel-Methanol and Diesel-Ethanol in a Dual Fuel Engine. Energies. 12, 971 (2019). 10.3390/en12060971.

[7] Çelebi, Y., Aydın, H.: An overview on the light alcohol fuels in diesel engines. Fuel. 236, 890-911 (2019). 10.1016/j.fuel.2018.08.138.

[8] De Poures, M.V., Sathiyagnanam, A.P., Rana, D., Rajesh Kumar, B., Saravanan, S.: 1-Hexanol as a sustainable biofuel in DI diesel engines and its effect on

combustion and emissions under the influence of injection timing and exhaust gas recirculation (EGR). Appl. Therm. Eng. 113, 1505-1513 (2017). 10.1016/j.applthermaleng.2016.11.164.
[9] Yilmaz, N., Atmanli, A.: Experimental assessment of a diesel engine fueled with diesel-biodiesel-1-pentanol blends. Fuel. 191, 190-197 (2017). 10.1016/j.fuel.2016.11.065.
[10] Kerschgens, B., Cai, L., Pitsch, H., Heuser, B., Pischinger, S.: Di-n-buthylether, n-octanol, and n-octane as fuel candidates for diesel engine combustion. Combust. Flame. 163, 66-78 (2016). 10.1016/j.combustflame.2015.09.001.
[11] Heuser, B., Jakob, M., Kremer, F., Pischinger, S., Kerschgens, B., Pitsch, H.: Tailor-Made Fuels from Biomass: Influence of Molecular Structures on the Exhaust Gas Emissions of Compression Ignition Engines. (2013). 10.4271/2013-36-0571.
[12] Sigma Aldrich: 1-Octanol Safety Data Sheet, (2015).
[13] Yanowitz, J., Ratcliff, M.., McCormick, R.., Taylor, J.., Murhpy, M..: Compendium of Experimental Cetane Numbers. (2017).
[14] European Chemical Agency: Octanol Dossier, https://echa.europa.eu/substance-information/-/substanceinfo/100.003.561.
[15] Matsui, Y., Kamimoto, T., Matsuoka, S.: A Study on the Time and Space Resolved Measurement of Flame Temperature and Soot Concentration in a D. I. Diesel Engine by the Two-Color Method. SAE Int. (1979).
[16] Choi, C.Y., Reitz, R.D.: Experimental study on the effects of oxygenated fuel blends and multiple injection strategies on DI diesel engine emissions. Fuel. 78, 1303-1317 (1999). doi:10.1016/S0016-2361(99)00058-7.
[17] Tree, D.R., Svensson, K.I.: Soot processes in compression ignition engines. Prog. Energy Combust. Sci. 33, 272-309 (2007). 10.1016/j.pecs.2006.03.002.
[18] Jing, W., Roberts, W., Fang, T.: Comparison of Soot Formation For Diesel and Jet-A in a Constant Volume Combustion Chamber Using Two- Color Pyrometry. SAE Tech. Pap. 2014-01-1251. C, (2014). 10.4271/2014-01-1251.Copyright.
[19] Hottel, H.C., Broughton, F.P.: Determination of True Temperature and Total Radiation from Luminous Gas Flames: Use of Special Two-Color Optical Pyrometer. Ind. Eng. Chem. - Anal. Ed. 4, 166-175 (1932). 10.1021/ac50078a004.
[20] Payri, F., Pastor, J. V., García, J.M., Pastor, J.M.: Contribution to the application of two-colour imaging to diesel combustion. Meas. Sci. Technol. 18, 2579-2598 (2007). 10.1088/0957-0233/18/8/034.
[21] Di Stasio, S., Massoli, P.: Influence of the soot property uncertainties in temperature and volume-fraction measurements by two-colour pyrometry. Meas. Sci. Technol. 5, 1453-1465 (1994). 10.1088/0957-0233/5/12/006.
[22] Das, D.D., McEnally, C.S., Kwan, T.A., Zimmerman, J.B., Cannella, W.J., Mueller, C. J., Pfefferle, L.D.: Sooting tendencies of diesel fuels, jet fuels, and their surrogates in diffusion flames. Fuel. 197, 445-458 (2017). 10.1016/j.fuel.2017.01.099.
[23] Lemaire, R., Lapalme, D., Seers, P.: Analysis of the sooting propensity of C-4 and C-5 oxygenates: Comparison of sooting indexes issued from laser-based experiments and group additivity approaches. Combust. Flame. 162, 3140-3155 (2015). 10.1016/j.combustflame.2015.03.018.
[24] Heuser, B., Mauermann, P., Wankhade, R., Kremer, F., Pischinger, S.: Combustion and emission behavior of linear C^8-oxygenates. Int. J. Engine Res. 16, 627-638 (2015). 10.1177/1468087415594951.
[25] Klein, D., Pischinger, S.: Laser-Induced Incandescence Measurements of Tailor-Made Fuels in an Optical Single-Cylinder Diesel Engine. SAE Int. J. Engines. 10, (2017). 10.4271/2017-01-0711.
[26] Zhang, T., Jacobson, L., Björkholtz, C., Munch, K., Denbratt, I.: Effect of using butanol and octanol isomers on engine performance of steady state and cold start ability in different types of Diesel engines. Fuel. 184, 708-717 (2016). 10.1016/j.fuel.2016.07.046.

SESSION 6: INTERNAL COMBUSTION ENGINES

Evaluation and optimization of the scavenging system of a 2-storke poppet valve diesel engine

Chenxi Wang[1], Yiqiang Pei[1], Jing Qin[2], Yan Zhang[3], Wei Liu[3], Shuyong Zhang[3]

[1]State Key Laboratory of Engines, Tianjin University, Tianjin, China
[2]Internal Combustion Engine Research Institute, Tianjin University, Tianjin, China
[3]China North Engine Research Institute, Tianjin, China

ABSTRACT

Conventional approach of 4-stroke engines to down-sizing using direct injection and turbocharging has several limitations, such as maximum cylinder pressure, rapid heat release rate and so on. 2-stroke engines can produce higher power densities compared to a 4-stroke cycle counterpart because of faster combustion frequency. To expand engine's operation ranges, 2/4-stroke switchable multi-cylinder gasoline engine was proposed by Ricardo. Considering advantages of 2/4-stroke engine, same concept was used in a single-cylinder diesel engine in this paper aiming to obtain high output power. This 2/4-stroke engine was built on a traditional 4-stroke diesel engine by using a switchable camshaft system. Although high power could be achieved, only 63.9% trapping ratio and 79% scavenging efficiency was put out on 2-stroke mode due to inadequate exchange process. In this paper, a method to measure cycle-resolved scavenging efficiency based on CO_2 concentration is proposed. Combining experiment with CFD simulation to optimize scavenging process and engine performance. It is found that under the limitation of head height, the optimization of port structure only has 0.8-2% increasing of scavenging efficiency and 2.6-3.6% increasing of trapping ratio. And the effect of structure weaker with speed increasing. In the prototype engine, intake and exhaust valves open at the same time, which lead exhaust gas backflow, affects the scavenging process badly. By optimizing the 100 ° CA valve overlap can guarantee the best trapping ratio and scavenging efficiency. Finally, trapping ratio of 73.3% and scavenging efficiency of 84.3% enable the engine to achieve 111kW power when the air flow rate of 661kg/h, 14.3% lower than the air flow rate of the original 771kg/h. 14.3% lower than the air flow rate of the original 771kg/h.

Keywords: 2/4-stroke diesel engine, scavenging efficiency, trapping ratio, optimization, CFD simulation

1 INTRODUCTION

With the increasingly requirements of higher thermal efficiency and engine power density, many approaches have been carried out on diesel engines, such as down-sizing, super-high injection pressure, thermal management and powertrain hybridization etc., However, conventional approach of 4-stroke engines to down-sizing using direct injection and turbocharging has a number of limitations such as maximum cylinder pressure, rapid heat release rate and so on [1]. 2-stroke engines can produce higher power densities compared to a 4-stroke cycle counterpart because of faster combustion frequency , but this advantage is often offset by drawbacks regarding gaseous emissions and engine components durability. The modern down-sized 4-stroke engine design can greatly benefit from this attribute of the 2-stroke cycle [2,3]. To maximize engine's operation over wider speed and load ranges, 2/4-stroke switchable multi-cylinder gasoline engine was proposed by Ricardo [4]. It provides a new idea to increase the fuel economy potential of downsizing still further. By combining 2 and 4-stroke operation in the same engine it could harness the benefits of each of the operating cycles but avoid their problems. A GDI engine with two stage boosting system and electro-hydraulic valvetrain that has the potential to improve fuel consumption by up to 30% compared to a conventional gasoline engine. Another merit of 2/4-stroke engine is the convenience of HCCI combustion concept application, and a lot of research have been carried out in Burnel University [5-9].

However, there is no corresponding research on diesel engine at present, considering advantages of 2/4-stroke GDI engine, to improve the torque and power of diesel engine to meet the requirements, same concept was used in a single-cylinder diesel engine in this paper aiming to obtain high output power. In this study, a fundamental study of power increasing was carried out on a 2/4-storke diesel engine, which based on a traditional 4-stroke poppet valve diesel engine, through using switchable camshaft system changing valves timing to achieve 2-stroke mode.

Owing to the structure of original 4-stroke intake/exhaust port and large valve overlap when 2-stroke mode, the prototype engine has relatively poor scavenging process and low trapping ratio because of the air short-circuiting comparing with traditional 2-stroke engines. So, in this study intake/exhaust port structures and valves timing were optimized for the best engine performance.

Unfortunately, it is difficult to get accurate scavenging efficiency of 2-stroke engine directly by experiments. In 1970s, Jante proposed an experimental method to measure the short-circuiting by using the distribution of the gas velocity measured by the pitot tubes first in 2-stroke engines [10], which can be used to deduce trapping and scavenging efficiency. In 1999, Heywood proposed a new method to calculate scavenging efficiency during firing 2-stroke engine operation by measuring CO_2 concentrations in cylinder [11], and how to get the accurate CO_2 concentration is the obstacle of this method until 2012, a CO_2 fast-response non-dispersive infrared (NDIR) analyzer from Cambustion Company was used by Brunel University in the exhaust port to calculate cycle-resolved short-circuiting rate [12]. That gave a good solution for scavenging efficiency measurement, which was used in this study.

Other method to give trapping and scavenging efficiency of a 2-stroke engine is simulation. Both 1D and 3D software could give this answer [13-14]. Hence, a scavenging experiment which based on the method of Heywood and computational fluid dynamics simulation were used in this paper to measure realistic scavenging efficiency and optimize fresh charge motion to provide best engine performance.

2 EXPERIMENTAL SETUP AND SIMULATION SETUP

2.1 Principle of the scavenging efficiency measurement

Scavenging efficiency η_{sc} is a measure of the extent the exhaust gas residuals have been replaced in the cylinder with fresh charge, and is defined by Equation 1.

$$\text{Scavenging efficiency} = \frac{\text{Fresh air quality in the cylinder after scavenging}}{\text{Total mass of gas in the cylinder after scavenging}} \tag{1}$$

$$\text{Trapping Ratio} = \frac{\text{Mass of delivered fresh air retained in cylinder}}{\text{Total mass of fresh charge}} \tag{2}$$

The mass of CO_2 in the cylinder at the pre-combustion stage is the sum of CO_2 in the residual exhaust gas and fresh air, ignore the minimal difference in molar mass of air, residual exhaust gas and their mixed gases:

$$[co_2]_u=[co_2]_b(1-\eta_{sc})+[co_2]_a\eta_{sc} \tag{3}$$

$$\eta_{sc}=\frac{[co_2]_b-[co_2]_u}{[co_2]_b-[co_2]_a} \tag{4}$$

Where

$[co_2]_b$ = CO_2 concentration in the cylinder after combustion

$[co_2]_u$ = CO_2 concentration in the cylinder before combustion

$[co_2]_a$ = CO_2 concentration in the intake port

In order to measure instantaneous CO_2 concentration, a sampling probe was installed in cylinder from the cylinder pressure measuring hole. As shown in Figure 1, a fast-response CO_2 analyzer type NDIR500 from Cambustion Ltd was used to measure the instantaneous CO_2 concentration through the sampling probe. The fast response time is approximately 5 ms, which means that the CO_2 volume fraction measurement in a single cycle can be achieved. The NDIR analyzer had an accuracy to within 2% by calibrating with different concentrations of CO_2 gas.

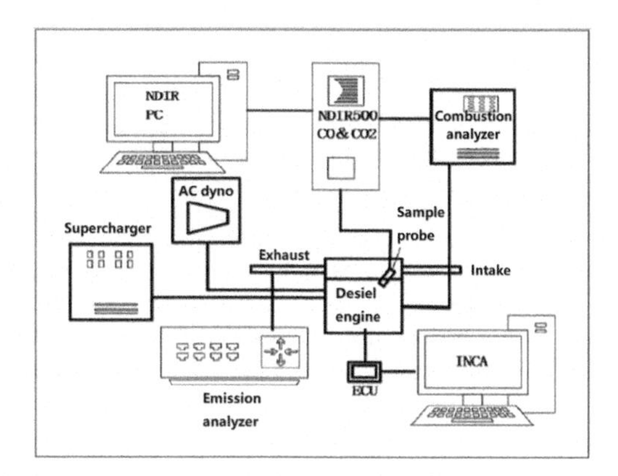

Figure 1. 2/4-stroke single-cylinder engine.

2.2 Baseline design of 2/4 diesel engine

The prototype engine used in this work is a single-cylinder research engine. It can be operated in either 2-stroke mode or 4-stroke mode by switch different camshafts. An external boosted system was connected to the intake system through a buffer tank that supplied constant compressed air. The intake air mass flow rate was measured with a mass flow meter installed in the engine's intake system and intake/exhaust temperatures measurement by k-type thermocouple. The Kistler piezoelectric pressure sensor was used to measure the in-cylinder transient pressure, and the Kistler combustion analyzer was used for data acquisition and combustion process analysis. The specifications of this engine are shown in Table 1.

Table 1. Specifications of prototype engine.

Parameter	Value
Bore (mm)	110
stroke (mm)	110
Connecting rod (mm)	183
Geometric compression ratio	13.5
Displacement (L)	1.04
Cylinder head	2 intake valves, 2 exhaust valves
Fuel	Diesel
Intake/exhaust valve opening (°CA)	130/268
Intake/exhaust valve closing (°CA)	130/268

2.3 Simulation methods

In this study, 1D simulation by GT-Power was used to evaluate the engine performance and scavenging process mainly output power, trapping ratio and scavenging efficiency, meanwhile obtain boundary conditions and initial conditions for later port structure design in 3D simulation.

Because the engine still in development, only 2000 r/min of data have been tested so far, and the goal of this engine is to obtain high power at 3000 r/min. In order to predict 3000 r/min operating condition, the predicting combustion mode "EngCylComDI-Pulse" was used. Four combustion factors of this model: Entrainment Rate Multiplier, Diffusion Combustion Rate Multiplier, Ignition Delay Multiplier and premix Combustion Rate Multiplier were derived from experimental data at 2000 r/min. The computational l was shown in Figure 2. And the DOE function in GT-power will design orthogonal experiments for different variables according to the settings of user, and the cases that meet the requirements can be selected according to the requirements.

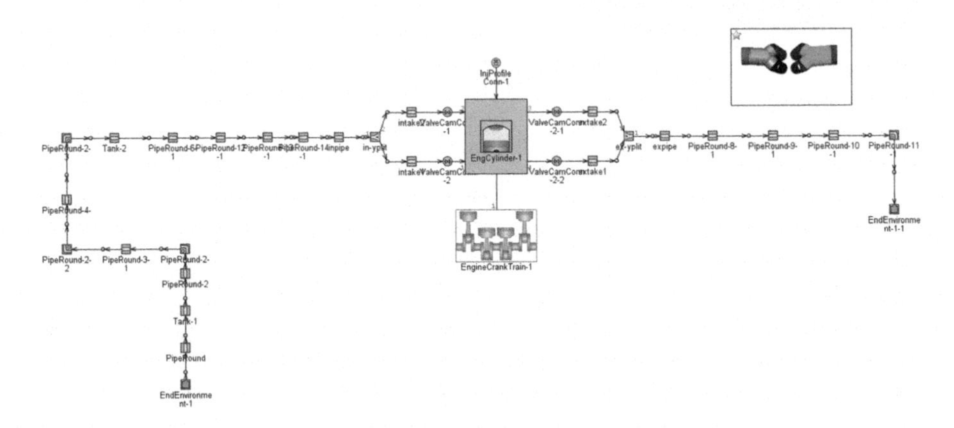

Figure 2. GT-power model.

In order to analyses the prototype engine's defects in scavenging process intuitively and estimate modified port structures, this study used CFD simulation. As shown in Figure 3, the model including the fluid field of the intake/exhaust ports and the combustion chamber. The average mesh size was 2 mm, and the regions surrounding the valve seat and head were refined. The corresponding simulation conditions, including important parameters, initial and boundary conditions, are provided by GT-Power. The initial conditions and boundary conditions were provided by 1D simulation shown in Table 2.

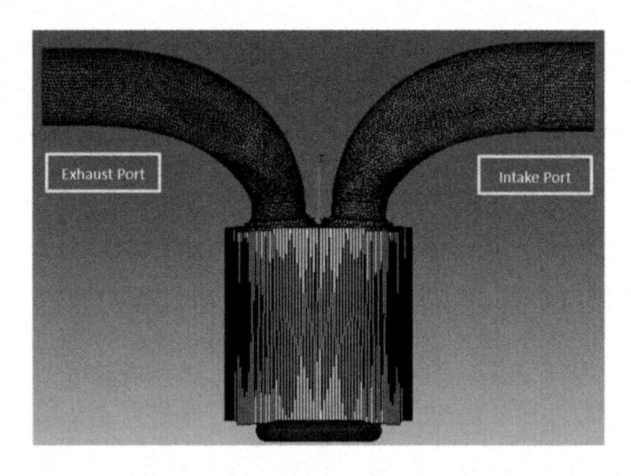

Figure 3. Mesh generation and boundary division in converge.

Table 2. Simulation condition.

Parameters	**2000r/ min@50kW**	**2000r/ min@44kW**	**2000r/ min@34kW**
Fuel mass	124.4 mg	107 mg	84 mg
Exhaust valve duration	139°CA ATDC	139°CA ATDC	139°CA ATDC
Exhaust valve lift	8.03 mm	8.03 mm	8.03 mm
Exhaust valve opening	130° CA ATDC	130° CA ATDC	130° CA ATDC
Intake valve duration	139°CA ATDC	139°CA ATDC	139°CA ATDC
Intake valve lift	8.08 mm	8.08 mm	8.08 mm
Intake valve opening	130° CA ATDC	130° CA ATDC	130° CA ATDC
Initial conditions @130°CA			
Cylinder temperature	1477 K	1329 K	1226 K
Cylinder pressure	11.56 bar	10.39 bar	9.67 bar
Intake temperature	340 K	335 K	332 K
Intake pressure	4.99 bar	4.91 bar	4.66 bar
Exhaust temperature	631 K	657 K	661 K

3 RESULT AND DISCUSSION

3.1 Scavenging efficiency experimental data process

As shown in Figure 4(a), 150 cycles' CO_2 transient concentration need be record when engine is in steady operation at the 2000 r/min experimental condition. And when calculation and analysis, to prevent engine's cyclical fluctuation, the final scavenging efficiency is the average value after removing 25 maximum and 25 minimum values.

The CO_2 concentration reached minimum due to the entry of fresh charge, then it remains stable after the intake and exhaust valves are closed before the combustion begins, $[CO_2]u$ is the concentration of carbon dioxide at this time. Then, the CO_2 concentration increases instantaneously to maximum because of in-cylinder combustion, $[CO_2]b$ is the concentration of carbon dioxide at this time. Therefore, based on the measurements $[CO_2]b$、$[CO_2]u$, and taking $[CO_2]a$ as 0.04%, the single-cycle scavenging efficiency can be calculated according to the Equation 4.

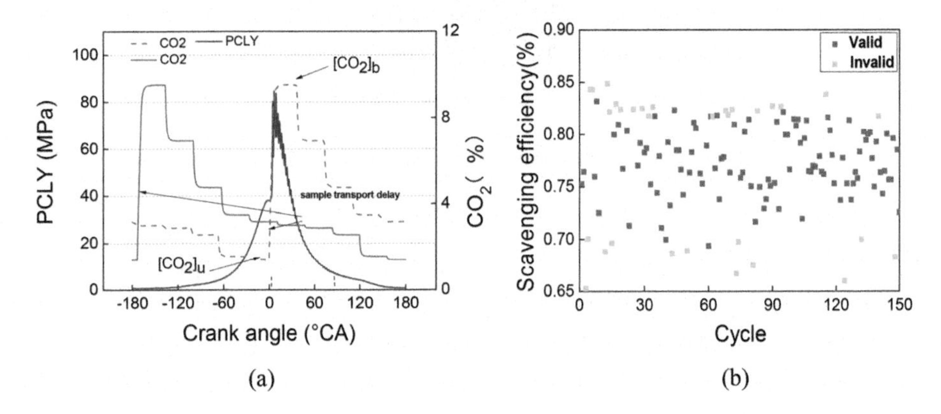

Figure 4. In-cylinder CO_2 transient concentration and scavenging efficiency.

3.2 Evaluation of prototype engine

3.2.1 *Model calibration*

By comparing experimental data and simulation result in Table 3, it can be seen that the relative error of power, torque, intake mass flow and fuel consumption under each load are 1.74%, 1.62%, 1.75% and 1.3%, respectively. Therefore, the simulation model is considered suitable to predict the actual performance parameters of the 2-stroke diesel engine.

Table 3. Calibration of GT predict model.

Operating condition	2000 r/min51 kW		2000 r/min44 kW		2000 r/min34 kW	
	Experiment	Simulation	Experiment	Simulation	Experiment	Simulation
Power (kW)	50.8	51.7	44	45	34.1	34.5
Torque (N·m)	242	246.5	210.2	214.6	162.7	164.2
Air flow rate (kg/h)	408.7	411.6	414.2	405	419.8	410
Maximum Pressure (bar)	158.6	156.6	149.1	150	140	144.6
BSFC (g/kWh)	294.6	289.1	290.5	285.7	302	301.1

As shown in Figure 5, 3D results show that the scavenging efficiency at 50kW , 44 kW and 34 kW operating condition are 81.2%, 74.16% and 72.45%. Meanwhile, the experimental data is 79.8%, 75.26% and 71.48%. The scavenging efficiency was measured at different operating conditions at 2000 r/min. It can be seen that, at the same speed, with the increase of the intake pressure form 4.66 bar to 4.99 bar. The air flow rate was increased, more and stronger airflow improved the scavenging efficiency. And the error of simulation and experiment are below 2%. It is in good agreement with the experiment results, so it is believed that 3D simulation can predict the scavenging efficiency in the cylinder correctly.

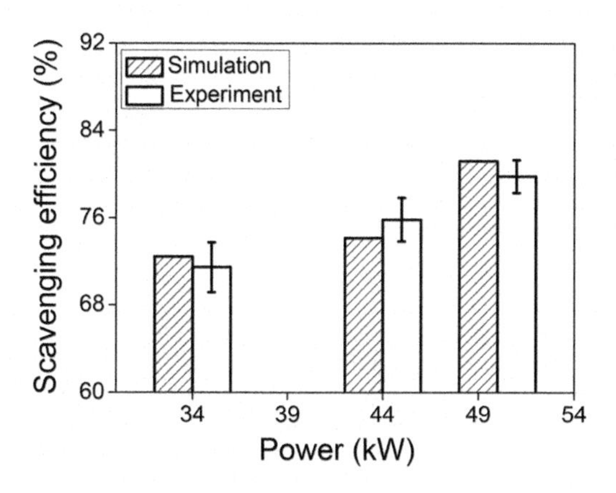

Figure 5. Calibration of converge model.

3.2.2 *Evaluation of prototype engine*

According to the prediction, the maximum output power of prototype engine is 83.7 kW at 3000 r/min under the current intake flow rate ability as shown in Table 4. Theoretically, the mass of intake air needs to be increased to 771 kg/h so that the power could reaches 111 kW due to the low trapping rate and scavenging efficiency. But, in order to match supercharger, the air flow rate need reduce to a reasonable range. Others information can also be obtained from the Table 4 that higher air flow rate elevates the scavenging efficiency, that leads to more fresh air thus less residual gas stay in the cylinder. However, which means more short circuit of fresh air with less trapping ratio at the meanwhile. That indicates that improving the capture ability of fresh air in cylinder is also one of the important means to increase output of engine's power, which requires the optimization design of the gas exchange system. In order to meet maximize power output and other constraints reequipments, the optimizing of high trapping ratio and scavenging efficiency is necessary. This study will optimize the structure of the engine from both the intake and exhaust, and the phase design of inlet charge system and valve timing system.

Table 4. Prediction of 3000 r/min operating condition.

Case	3000 r/ min@83.7 kW	3000 r/ min@111 kW	Limitation
Fuel mass (mg/cycle)	124.4	160	-
λ	2.1	2.1	
SOI (°CA)	-13	-13	-
IVO (°CA)	130	130	-
EVO (°CA)	130	130	-
Differential pressure (bar)	5	7	
Power (kW)	83.7	111.1	110
Air flow rate (kg/h)	685.4	771.2	700
Max pressure (bar)	178.5	207.5	220
BSFC (g/kWh)	267.5	259.3	260
Trapping ratio (%)	65.7	63.9	-
Scavenging efficiency (%)	76.3	79	-

As shown in Figure 6, through the analysis of the mass flow rate of the intake, we can see that the reason that why the scavenging effect of the prototype engine is poor. The intake and exhaust valves are opened at the same time, high cylinder pressure let a part of exhaust gas backflow to the intake port, which wastes part of the intake process and reduces the effective intake time. The actual charge in the cylinder did not reach the expected effect.

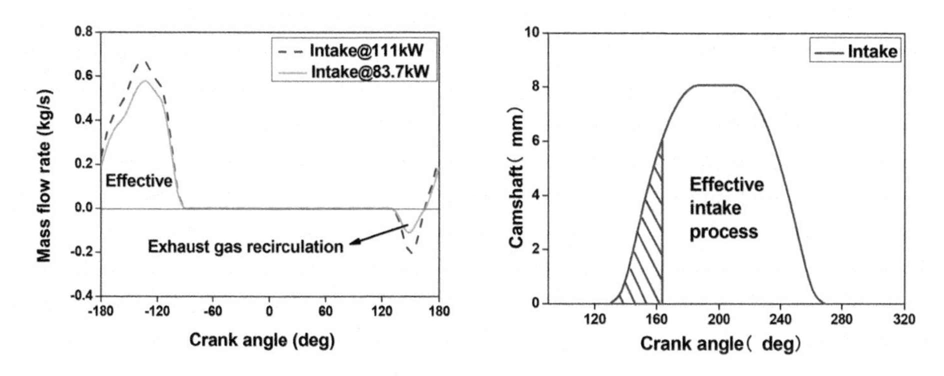

Figure 6. Intake mass flow rate with varied valve timing.

3.3 Intake and exhaust port optimization

As present in Figure 7, (a), (b) and (c) are air speed and CO_2 concentration field when scavenging process. When adopting the current valve timing and the structure of intake and exhaust ports, the prototype engine has the following problems in the scavenging process:

(1) Short-circuiting rate : In Figure 7(a) region 1 and 5, more fresh air escape from intake ports to exhaust ports, which make trapping ratio decrease at scavenging beginning.
(2) Backflow of exhaust gas: due to the simultaneous opening timing of the intake and exhaust valves, the differential pressure between the cylinder and the intake ports is large, resulting in exhaust gas backflow and affecting the scavenging process.
(3) Poor exhaust: there is turbulence in the exhaust ports at region 4, which makes exhaust gas cannot be discharged smoothly
(4) Weak airflow movement in the cylinder: as Figure 7(c), a scavenging dead zone with high CO_2 concentration is formed in the cylinder at the end of scavenging. Therefore, the intake flow in zone 2 and 3 should be strengthened to form a strong reverse roll flow so that fresh air can scavenge the cylinder more fully.

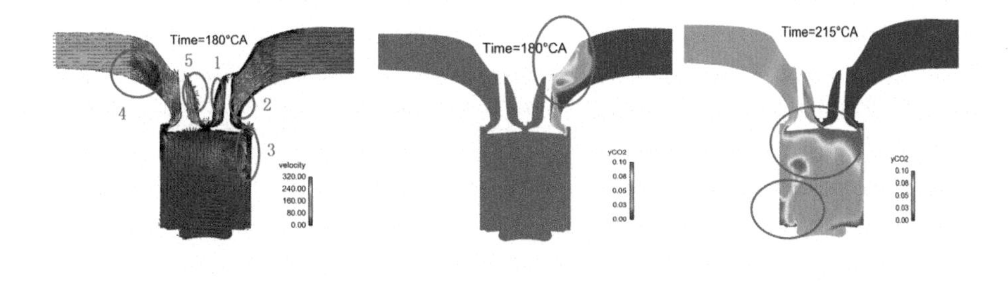

(a) Speed field b) CO_2 concentration field (c) CO_2 concentration field

Figure 7. In-cylinder flow field of prototype engine.

3.3.1 ***Structure optimization***

To solve the above problems, reasonable changes were made on the intake and exhaust port structures. As shown in Figure 8, a total of 4 cases were designed and calculated by 3D simulation:

(1) case1: Raise the intake ports in order to increase the flow in region 2 and 3

(2) case2: Narrow the intake ports

(3) case3: Enlarge valve seat rim chamfering

(4) case4: Change the bifurcation of intake and exhaust ports

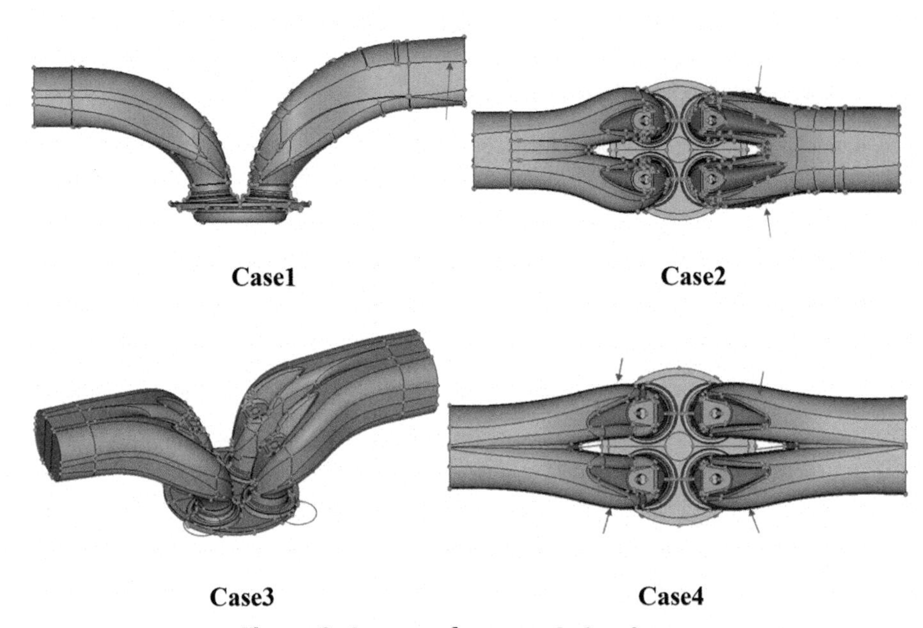

Figure 8. 4 cases of new port structures.

As shown in Figure 9, the different case of scavenging system was simulated at 2000r/min @50kW and 3000r/min @83.7kW working conditions. The results showed that little modification which base on the 4-stroke intake and exhaust ports had only a small impact on scavenging efficiency, and the influence of the intake/exhaust port structure gradually weakened with the increase of engine speed.

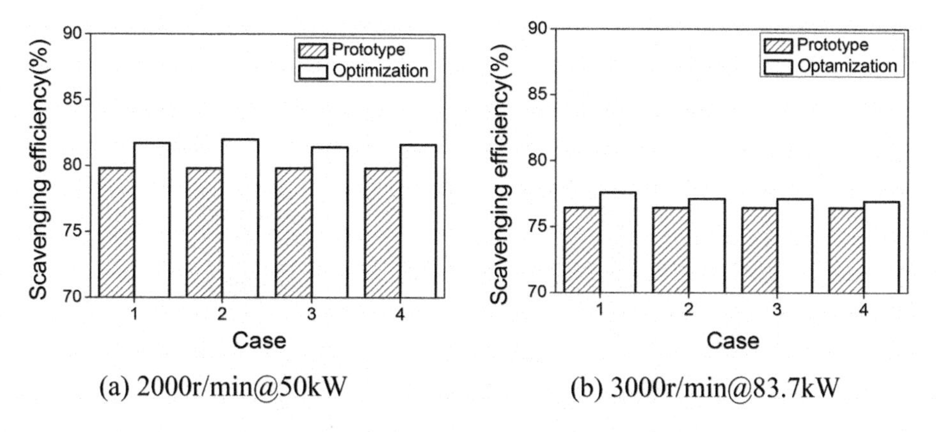

Figure 9. Scavenging efficiency of four ports.

As shown in Figure 10, modification of intake and exhaust structure has better effect at lower speed to trapping ratio, and the difference of port structure gradually weakened with the increase of engine speed. But different with scavenging efficiency, trapping ratio increases as engine speed increases.

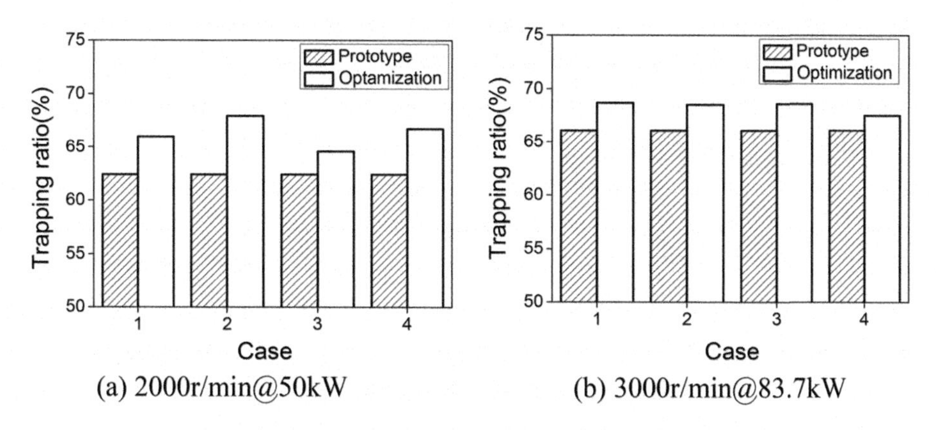

Figure 10. Scavenging efficiency of four ports.

As the best case, case1 is taken as an example to analyze the flow field in the cylinder. As present in Figure 11, after the intake port is raised, the air velocity at the intake port is significantly accelerated, and the guidance of the intake makes more airflow move along the cylinder wall, so that the flow velocity is faster and reversed tumble is stronger. The speed was slightly increased but not significantly changed compared to the 3000 r/min condition.

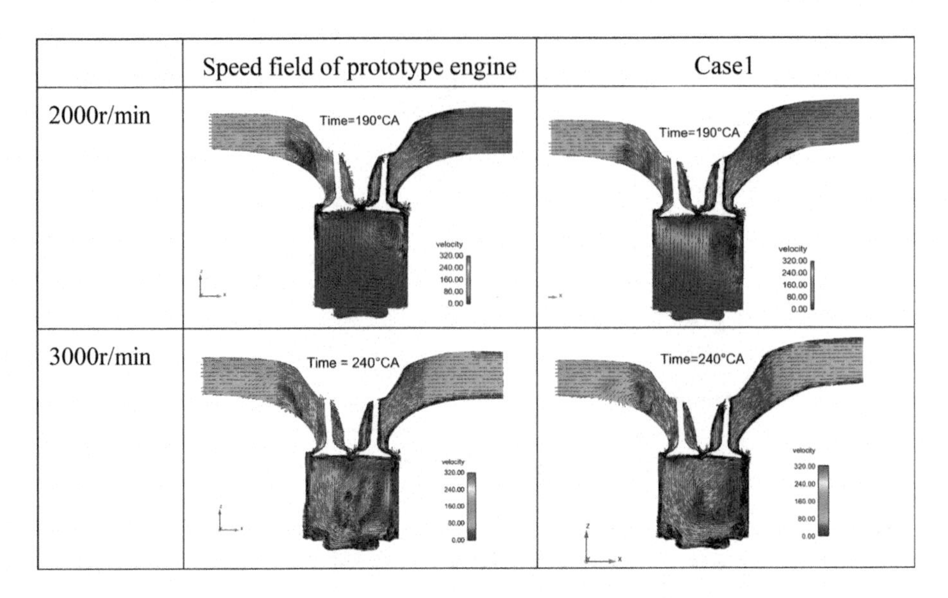

Figure 11. In-cylinder speed field.

As present in Figure 12, when the intake port is raised, higher tumble intensity produces better scavenging effect. Lead to less CO_2 residue at the bottom of the cylinder, reducing the existence of scavenging dead region. Compared with 3000r/min, low engine speed has more sufficient scavenging time to scavenge exhaust gas. At high speed and high air flow rate, the improvement of that changing intake and exhaust ports structure is not so significant.

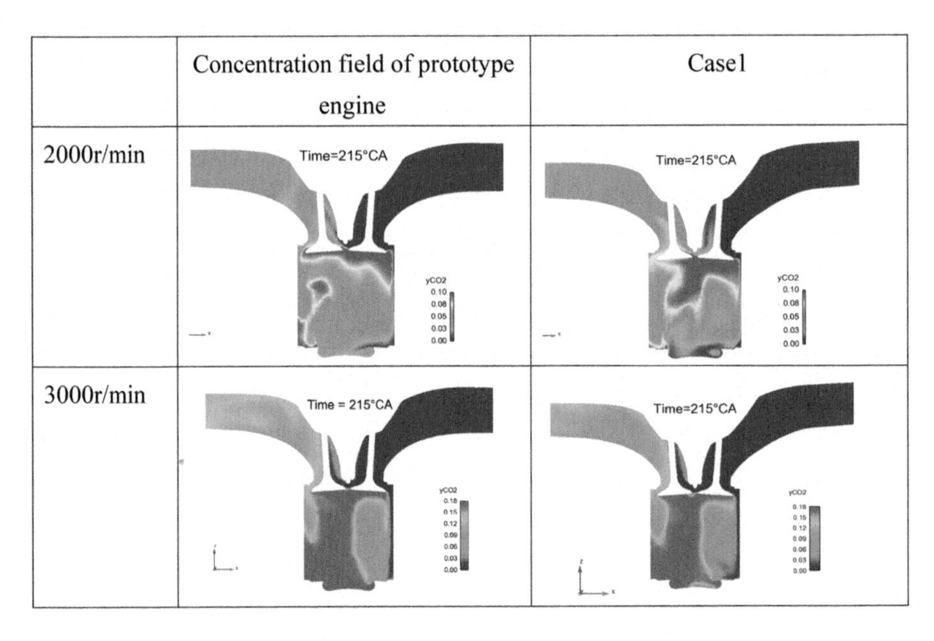

Figure 12. In-cylinder CO_2 concentration field.

3.4 VVT optimization

According to the above analysis, the influence of intake and exhaust ports on scavenging efficiency and trapping ratio is weak under the limit of 4-stroke intake and exhaust ports. And it is weaker under large load and high air flow rate. Therefore, more effective optimization is attempted through the valve phase.

3.4.1 *The influence of the single variable on scavenging efficiency and trapping ratio*

In this study, 7 factors that may affect scavenging efficiency and trapping ratio: intake valve opening (IVO), exhaust valve opening (EVO), intake valve duration (ID), exhaust valve duration (ED), differential pressure, injection fuel mass and start of injection (SOI) was simulated to study its sensitivity.

To provide sensitive and directional guidance for DOE optimization. a univariate analysis was conducted. For example, as shown in Figure 13, with the delayed opening of the intake valve, more fresh gas is retained in the cylinder under the same air flow rate, and the backflow of exhaust gas is reduced because exhaust valve opens earlier relatively. So, the trapping ratio and scavenging efficiency are improved. The increase of fresh air improves the power of combustion and reduces the fuel consumption.

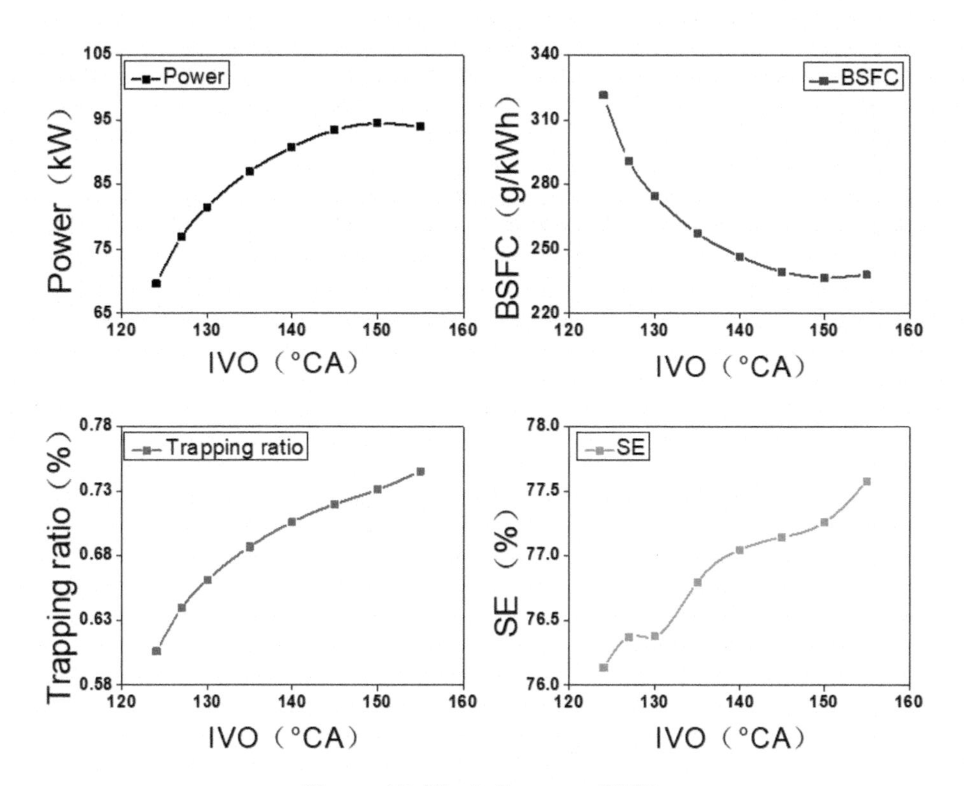

Figure 13. The influence of IVO.

As shown in Table 5, seven single variables that may affect the scavenging process of two-stroke engine was studied to get sensitivity analysis. The upward arrows indicate that this factor has a greater impact.

Table 5. Sensitivity study.

Parameter	Fuel mass	SOI	Differential pressure	IVO	EVO	ID	ED
Power	↑	↓	↑	↑	↓	↓	↓
BSFC	↑	↓	↑	↑	↑	↓	↓
Max pressure	↓	↑	↑	↑	↑	↓	↓
TR	↓	↓	↑	↑	↑	↓	↑
SE	↓	↓	↑	↑	↑	↓	↑

3.4.2 *DOE (design of experiments) optimization*

From above analysis, injection fuel mass, start of injection, differential pressure, intake valve opening time, exhaust valve opening time and exhaust valve duration will have more obvious effect on engine performance and scavenging process. Though the DOE function, the combination optimization of the above parameters is carried out.

As shown in Table 6, the case that meets all the limitation and has the highest power is finally selected. The change of intake and exhaust is shown in the Figure 15. The intake valves open 18°CA later and the exhaust valves starts 20°CA earlier than prototype engine. The injection fuel mass enable to increase to 160 mg/cycle, because higher trapping ratio. The power of the diesel engine can achieve 111 kW only needs 661 kg/h air flow rate. At this operating condition, the trapping ratio from 63.9% at prototype engine increase to 73.3%, and the scavenging sufficiency increases 79% to 84.3%.

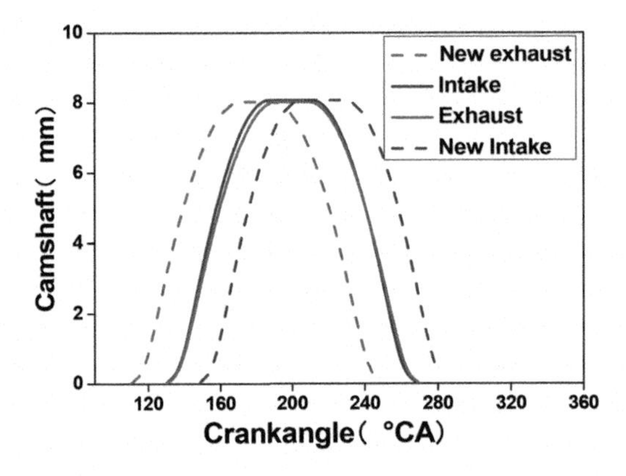

Figure 14. Change of intake and exhaust valves open timing.

Table 6. Optimization case.

Case	3000 r/min@111 kW (Prototype engine)	3000 r/min@111 kW (Optimization)	Limitation
Fuel mass (mg/cycle)	160	160	-
λ	2.1	2.1	
SOI (°CA)	-13	-13	-
IVO (°CA)	130	148	-
EVO (°CA)	130	110	-
Differential pressure (bar)	7	4	
Power (kW)	111.1	111.6	110
Air flow rate (kg/h)	771.2	661.6	700
Max pressure (bar)	207.5	189.8	220
BSFC (g/kWh)	259.3	258.1	260
Trapping ratio (%)	63.9	73.3	-
Scavenging efficiency (%)	79	84.3	-

As shown in Figure 15, the optimized valve timing eliminates the original exhaust gas backflow, more fresh charge enables to enter the cylinder smoothly, utilizing the intake process effectively. Due to the improvement of trapping ratio and scavenging efficiency, the engine can obtain more sufficient air at a lower intake pressure and air flow rate.

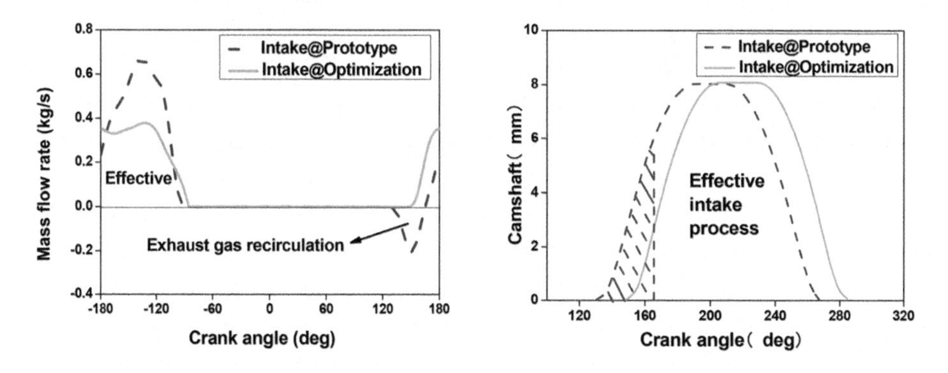

Figure 15. Intake mass flow rate with varied valve timing.

As present in Figure 16, the exhaust valves open earlier than prototype after the optimization by DOE, and a large amount of exhaust gas discharged from the exhaust ports first, so that the flow in intake ports can be smoother. All those reasons make engine achieve a stronger scavenging process. As shown in Figure 17, the exhaust backflow in engine intake ports is significantly reduced after DOE optimization, and exhaust gas in cylinder is significantly reduced after intake and exhaust valves are closed.

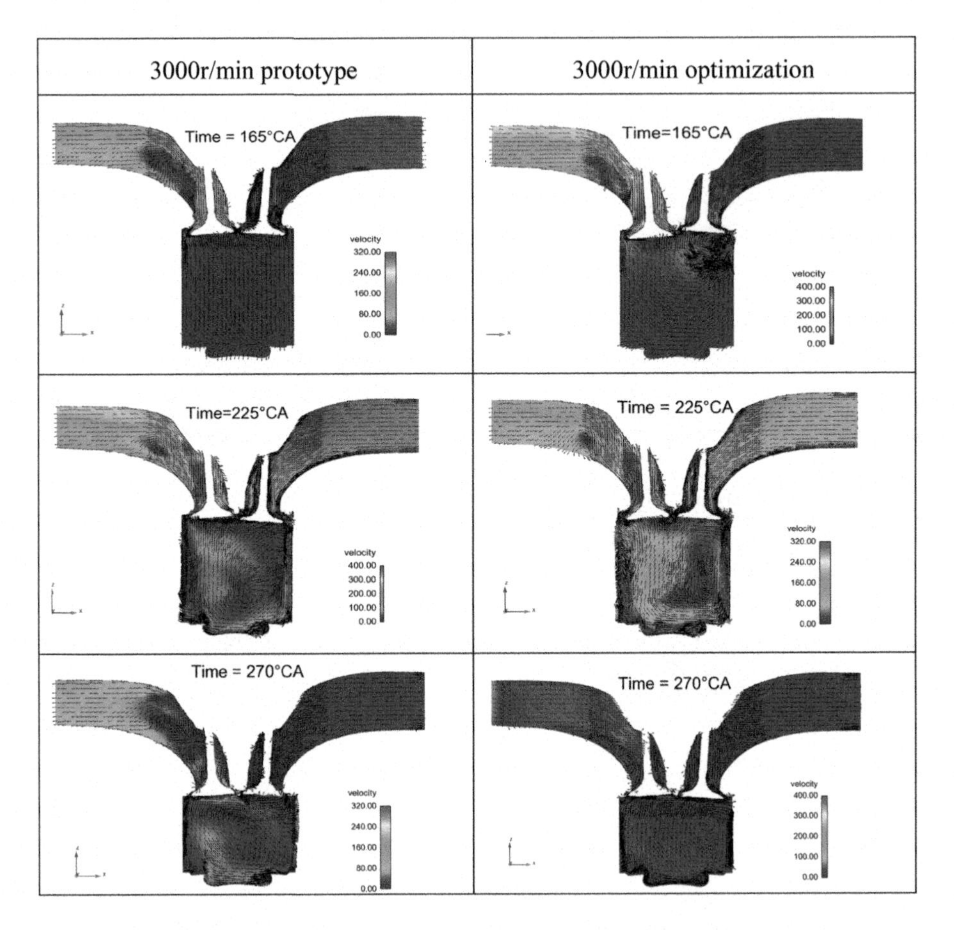

Figure 16. Speed field comparison of prototype engine and DOE optimization.

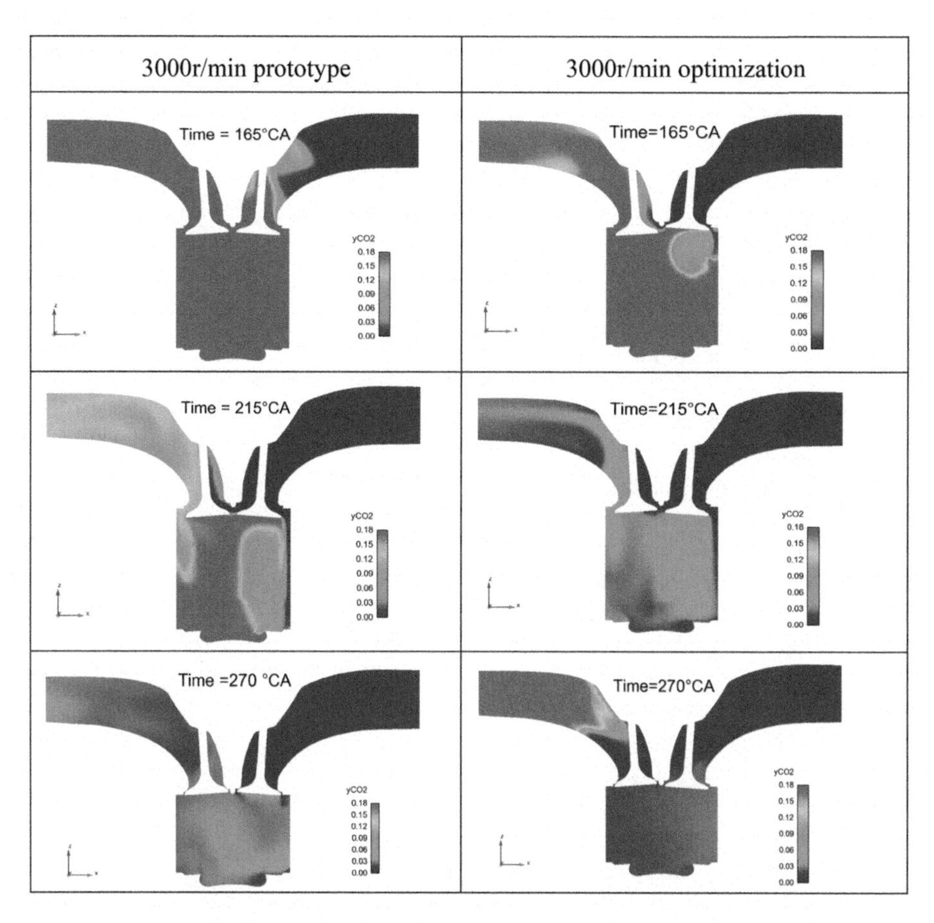

Figure 17. Concentration field comparison of prototype engine and DOE optimization.

4 CONCLUSIONS

In this study, 1D GT-power and 3D Converge simulations were adopted to evaluate scavenging process of 2/4-stroke engine, in order to obtain more realistic scavenging process, this study used an accurate method to measure scavenging efficiency of 2-stroke engine. 7 important effect factors were investigated in detail. Finally, achieved 111kW at 3000r/min optimization when 661kg/h air flow rate, 14.3% lower than the air flow rate of the original 771kg/h. The conclusions are summarized as follows:

Thorough transient CO_2 concentration of pre-combustion and post-combustion, cycle-resolved scavenging efficiency can calculate in cylinder.

In order to prevent adverse effects on the 4-stroke mode and the limitation of engine head, only minor structural changes have been made on intake/exhaust ports. the optimization of port structure only has 0.8-2% increasing of scavenging efficiency and 2.6-3.6% increasing of trapping ratio. And the effect of structure weaker with speed increasing.

In the prototype engine, intake and exhaust valves open at the same time, which lead exhaust gas backflow, affects the scavenging process badly. By optimizing the 100 ° CA valve overlap can guarantee the best trapping ratio and scavenging efficiency. Finally, trapping ratio of 73.3% and scavenging efficiency of 84.3% enable the engine to achieve 111kW power when the air flow rate of 661kg/h,

Abbreviations

1D one dimensional
3D three dimensional
CFD computational fluid dynamics
IVO intake valve open
IVC intake valve close
EVO exhaust valve open
EVC exhaust valve close
ID intake valve duration
ED exhaust valve duration
SOI start of injection
DOE design of experiments
IMEP indicated mean effective pressure
TDC top dead center
BDC bottom dead center
BSFC brake specific fuel consumption
η_{sc} scavenging efficiency

REFERENCES

[1] Rebhan D I M, Stokes J. Two-stroke/four-stroke multicylinder gasoline engine for downsizing applications[J]. MTZ worldwide, 2009, 70(4):40-45.
[2] Lucas D Pugnali1, Rui Chen. Feasibility study of Operating 2-stroke Miller Cycles on a 4-stroke Platform through Variable Valve Train. SAE Technical paper 2015-01-1974.
[3] Dalla Nora M, Lanzanova T, Zhang Y, et al. Engine Downsizing through Two-Stroke Operation in a Four-Valve GDI Engine. SAE Technical Paper 2016-01-0674.
[4] Osborne R J, Stokes J, Lake T H Development of a Two-Stroke/Four-Stroke Switching Gasoline Engine - The 2/4SIGHT Concept[J]. 2005.
[5] Zhang Y. Effects of Injection Timing on CAI Operation in a 2/4-Stroke Switchable GDI Engine[J]. Sae International Journal of Engines, 2003, 5(2):250-5.
[6] Y Zhang, H Zhao, M Ojapah. Experiment and Analysis of a Direct Injection Gasoline Engine Operating with 2-stroke and 4-stroke Cycles of Spark Ignition and Controlled Auto-Ignition Combustion. SAE Technical Paper 2011-01-1174.
[7] Y Zhang, M Ojapah, Alasdair Cairns. 2-Stroke CAI Combustion Operation in a GDI Engine with Poppet Valves. SAE Paper 2012-01-1118.
[8] Ojapah M, Zhao H, Zhang Y. Effects of Ethanol on Part-Load Performance and Emissions Analysis of SI Combustion with EIVC and Throttled Operation and CAI Combustion. SAE Technical Paper 2014-01-1611.
[9] Yan Zhang, Macklini DallaNora and Hua Zhao. Investigation of Valve Timings on Lean Boost CAI Operation in a Two-stroke Poppet Valve DI Engine. SAE Paper 2015-01-1794.
[10] Jante A. Scavenging and other problems of two-stroke spark-ignition engines. SAE paper 680468, 1968.

[11] Heywood JB and Eran S. The two-stroke cycle engine: its development, operation, and design. London: Taylor & Francis, 1999.
[12] Zhang Y, Zhao H. Measurement of short-circuiting and its effect on the controlled autoignition or homogeneous charge compression ignition combustion in a two-stroke poppet valve engine[J]. Proceedings of the Institution of Mechanical Engineers, Part D: Journal of Automobile Engineering, 2012, 226(8):1110-1118.
[13] Rival D, Ciccarelli G. Evaluation of Scavenging Performance in a Novel Two-Stroke GDI Engine[J]. Proceedings of the National Academy of Sciences of the United States of America, 2006, 103(1):81-86.
[14] Wang X, Ma J, Zhao H. Evaluations of Scavenge Port Designs for a Boosted Uniflow Scavenged Direct Injection Gasoline (BUSDIG) Engine by 3D CFD Simulations. SAE Technical Paper 2016-01-1049.

Application of a rotary expander as an energy recovery system for a modern Wankel engine

G. Vorraro, R. Islam, M. Turner, J.W.G. Turner

University of Bath, UK

ABSTRACT

A Wankel rotary engine produced by Advanced Innovative Engineering (AIE) UK Ltd with a capacity of 225cc is investigated, employing a rotary expander as an energy recovery system, aiming at maximizing the expansion of the gas and improving the overall efficiency of this propulsive system. In this configuration, the expander is placed alongside to the engine and receives exhaust gas from the engine by means of a short and straight connecting pipe. Unlike the engine, the expander has a Wankel-type 1:2 configuration. A thorough description of the machine, including the main mechanical, kinematic and port-timing parameters are reported, in addition to details of the test rig and experimental methodologies adopted. Finally, this work presents the experimental results obtained from the engine equipped with the expander and compares those results with the engine baseline performance.

1 INTRODUCTION AND BACKGROUND

1.1 The Wankel engine as a range extender

The Wankel rotary engine has in recent years become the focus of engineers once more as a potential range extender. Mazda, the only automotive vehicle OEM (original equipment manufacturer) to persist with the Wankel rotary engine up until 2012, announced the reintroduction of the Wankel engine as a range extender as part of their 'Sustainable Zoom-Zoom 2030' development program [1].

Wankel engines in theory offer several advantages over reciprocating piston engines as range extenders even if they are notoriously known for their poor fuel economy. They typically offer exceptional NVH characteristics and, as a direct result of this and a combustion event every 360 degrees (for a 3-flank rotor) they also demonstrate good specific power. Subsequently, due to their low mass owing to a lack of reciprocating parts, it has been demonstrated that Wankel engines are highly suitable as range extenders for hybrid vehicles application [2]. There also have an additional advantage over reciprocating engines in their low inertia, which makes engine switch-on and –off events potentially less intrusive in a series hybrid.

1.2 Engineering challenges of Wankel engines

While the unusual geometry of the Wankel rotary is the root cause of its beneficial characteristics of low NVH and high specific power, unfortunately it also presents some unique challenges. More detailed explanation the geometry of the Wankel engine and its operation can be found in [3-6].

1.2.1 *Compression ratio*

In a rotary engine the compression ratio is dictated by the ratio between the generating radius **R** and the eccentricity **e** as referred in Figure 1, often referred to as the **K**-factor. Engines with high **K**-factors are capable of very high compression ratios. Conversely a low **K**-factor results in low compression and difficulties in sealing the rotor effectively. In practice Wankel engines are often designed with higher **K**-factors and in order to limit the compression ratio to levels suitable for the planned fuel, a recess is often addedto the rotor face which plays a significant part in the combustion process; this recess is also necessary to transfer charge in the combustion chamber across the minor axis of the housing when the rotor passes minimum volume (for ease of terminology this will be referred to as TDC in this paper, although clearly there is no such kinematic condition in a Wankel engine).

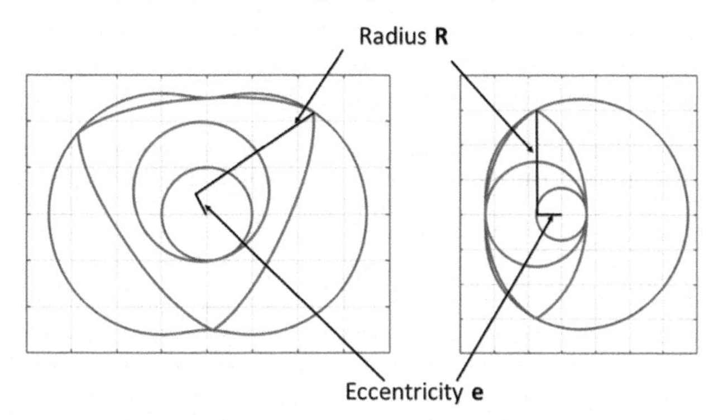

Figure 1. Schematic of the 3:2 and 2:1 trochoidal configurations showing the radius R and shaft eccentricity e.

1.2.2 *Combustion characteristics and flame speeds*

Wankel engine combustion chambers, by their nature have constantly varying geometry, with the available shape moving with the rotor flank as it travels around the housing. As a result the placement of the spark plug is critical to flame propagation. Figure provides an example of flame front propagation with respect to rotor motion. As engine speed increases in many cases the flame front is unable to keep up with the leading apex seal and a pocket of unburnt charge often escapes when the expansion phase completes. In addition the advancing 'trailing' apex seal creates an area of localised high pressure which pushes back against the flame front creating a second area of unburned charge.

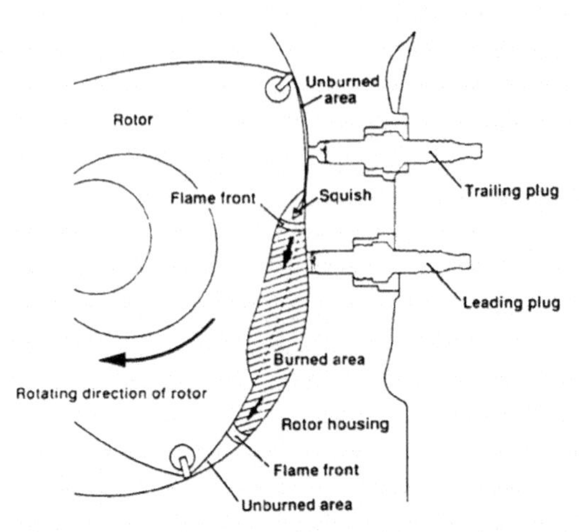

Figure 2. Flame propagation within a Wankel rotary engine with an example of multiple spark plugs arranged in a leading-trailing configuration, reproduced from [7].

To combat this Mazda employed two sequential spark plugs employing different timing in the housings of their production engines in an attempt to improve combustion characteristics, even going as far as introducing a third trailing spark plug in the R26B race engine [7].

The geometry and volume of the rotor recess also plays a significant part in the combustion process [6] and this coupled with the variable chamber shape and volume can lead to high NOx and hydrocarbon emissions.

1.2.3 *Managing port overlap*

A further factor impacting hydrocarbon emissions is again down to the Wankel geometry, specifically the location of the intake and exhaust ports. In a typical Wankel engine running on gasoline a trapped compression ratio in the region of 8 or 10:1 is often targeted. As a result significant port overlap is often encountered owing to the physical location of the intake and exhaust port to achieve the desired compression ratio along with an acceptable expansion ratio. This overlap is most significant when both the intake and exhaust port are located peripherally on the housing (as in Figure), although this overlap can be mitigated through relocating both ports to the side housing, as Mazda did with their RENESIS engine [8].

By relocating the ports in this way Mazda were able to reduce their HC emissions by around 30-50% in comparison to the previous generation Wankel engine which had a side intake port and peripheral exhaust port [8].

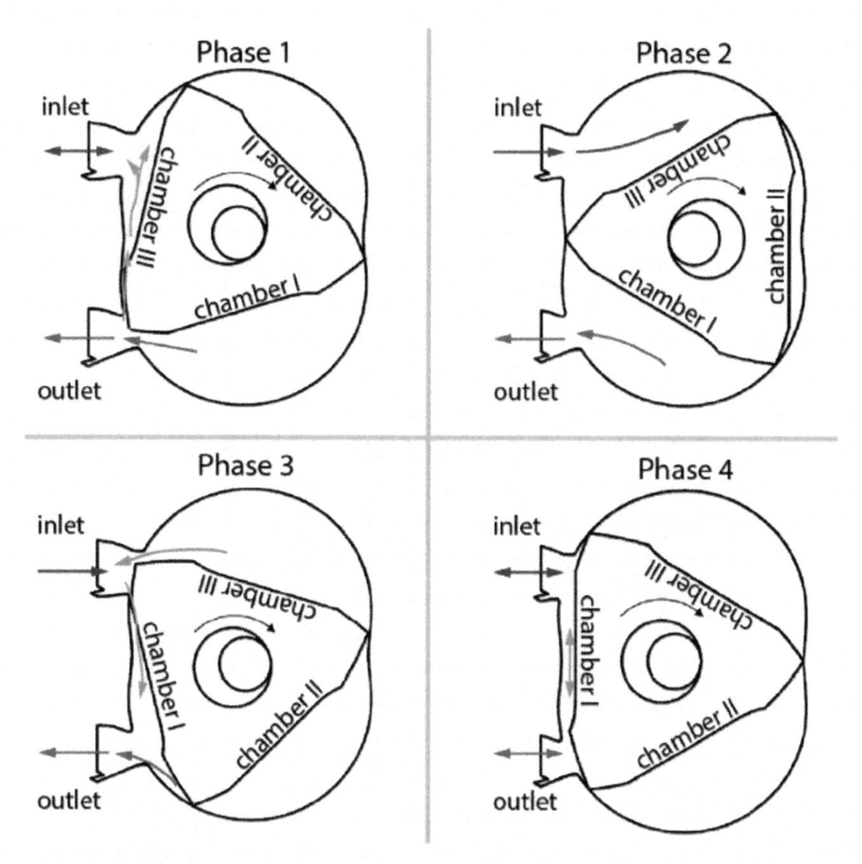

Figure 3. Four phases of gas exchange within a Wankel engine, intake highlighted in blue, exhaust in red and overlap in orange, reproduced from [9].

1.2.4 *Hydrocarbon emissions*

As a consequence of the unusual combustion chamber shape, port overlap and relatively low compression ratios, Wankel engines suffer from high hydrocarbon emissions which can be made more significant as a result of maintaining sufficient lubrication at the apex seals. As previously mentioned, this challenge became significant enough that since 2012 no automotive manufacturer has sold a vehicle fitted with a rotary engine [10]. Mitigating HC emissions in the face of these challenges remains the crux of Wankel engine development. Further work is ongoing within the research group on improving Wankel rotary engine emissions with a current focus around optimising the air-fuel ratio control [11].

1.3 Exhaust gas energy recovery

A further by-product of the unusual Wankel geometry is the relatively low thermal efficiency, in comparison to a reciprocating piston 4-stroke running on gasoline. Since most of the energy is lost to exhaust gas energy and so there is significant potential for the implementation of exhaust energy recovery systems.

1.3.1 *Turbomachinery*

Also if alternative boosting systems using exhaust energy have been proposed by other authors [12], most typical exhaust energy recovery systems revolve around the implementation of some kind of turbomachinery, which utilise a turbine driven by the exhaust gas to compress the intake air. It is well known from the technical literature that the matching between the engine and a turbocharger is a delicate task in order to assure the adequate engine performance and avoid the dangerous operating condition such as the surge [13,14]. One of the best example of this application with a Wankel engine was Mazda's 20B engine that used a combination of an intake port located on the side housing and a peripheral exhaust port supplemented with a sequential twin-turbo system, as shown in Figure .

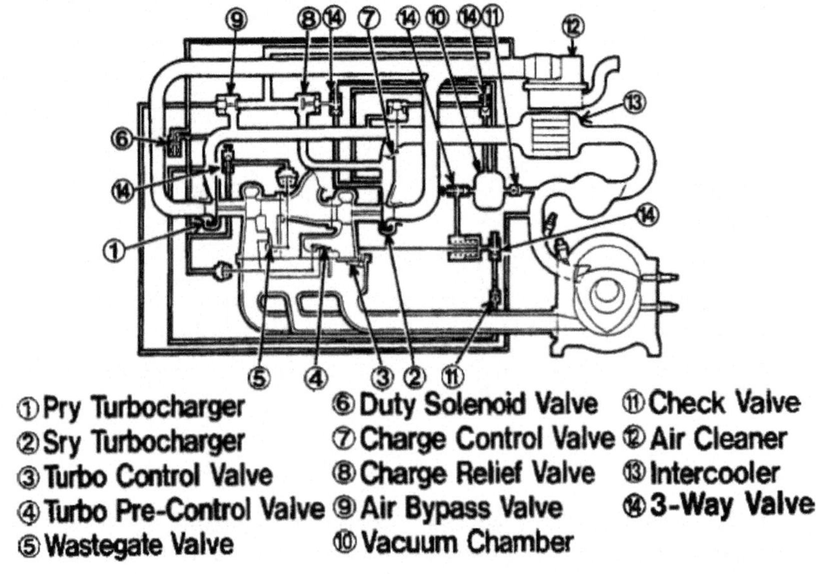

Figure 4. Mazda 20B-RE sequential twin turbo rotary engine schematic, reproduced from [15].

1.3.2 *Sequential two-stage rotary engine*

Rolls-Royce previously investigated operating a rotary engine on the diesel cycle which required a compression ratio that typically the rotary engine struggled to support. Their approach was effectively to combine two rotary engines with differing geometries; one smaller rotor handled high pressure compression and expansion, while a larger rotor provided initial low pressure compression and secondary exhaust expansion. It is worth highlighting that the two rotors were mechanically linked at a 1:1 ratio to the same output shaft [6]. Because of its general configuration, apparent in Figure, the engine was colloquially known as the "Cottage Loaf". While it was capable of delivering sufficient compression for effective diesel combustion, as reported previously it faced some challenges as a result of its unusual combustion chamber geometry.

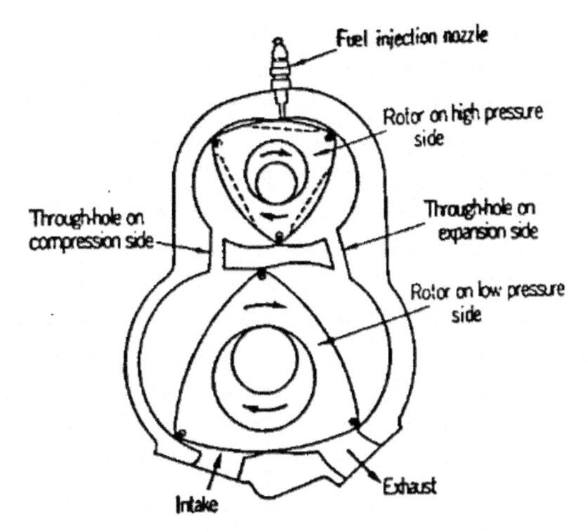

Figure 5. Rolls-Royce 'Cottage Loaf' two-stage rotary engine, reproduced from [6].

1.3.3 *Rotary expander*

The Rolls-Royce Cottage Loaf engine was unusual in that it utilised a pair of 3-flank rotors to improve the expansion and compression ratio, with the first larger rotor supporting both the initial compression phase and the secondary expansion phase. As such the engine was a double compression-expansion engine (DCEE). Recently rotary engine specialists have investigated combining a 3-flank rotor coupled with a second 2-flank rotor to greatly increase the expansion ratio of the unit [16]. Advanced Innovative Engineering (AIE UK Ltd) are one such rotary engine manufacturer who have licenced this technology and are currently developing the technology for sale [17], the detail of which will be reported in a later section.

1.4 Experimental activities overview

The focus of this investigation deals with the characterisation of the 225CS rotary engine developed by AIE, and a comparison of performance when run with a rotary expander. The experimental activities are divided into the following broad categories:

- Phase 1 - Full characterisation of the 225CS in standard trim
- Phase 2 - Full characterisation of the 225CS with the expander fitted
- Phase 3 - Optimisation of the engine control strategy to improve performance and emissions

The key performance and characterisation metrics are detailed in Table 1.

Table 1. Key performance metrics.

Metric	Nomenclature	Units
Brake Specific fuel consumption	BSFC	g/kWh
Brake mean effective pressure	BMEP	bar
Power	P	kW
Torque	T	Nm
Hydrocarbon	HC	ppm
Nitrogen Oxide	NO_X	ppm
Oxygen	O_2	ppm
Carbon Dioxide	CO_2	ppm
Carbon monoxide	CO	ppm

Once fully characterised, the impact of oil type and consumption on the performance metrics will be established. Finally research into Wankel engine combustion chamber pressures will be extended to include the expander [18].

2 WANKEL ENGINE AND EXPANDER CONFIGURATION

2.1 Overview of the engine and expander configurations

The most attractive peculiarity of Wankel engines is undoubtedly the geometrical configuration of both the housing and the rotor, in addition to the kinematics developed by the rotor, the eccentric shaft and central mechanism. In this section two configurations will be described. From a geometrical point of view, the curves that define both the rotor and the housing belong to the trochoidal family. Specifically, the rotor derives from a hypotrochoid while the housing from an epitrochoid. Nevertheless, as reported in [5], the Belerman-Morley theorem ensures that trochoidal curves can be generated in different ways, such as the Wankel engine housing itself which can be generated both as an epitrochoid or a peritrochoid.

The general mathematical equations for the trochoidal curves are parametric and can lead to different rotor-housing configurations. While the mathematical derivation and complete classification is outside of the scope of this work and extensively reported in the classic textbooks [3,5,6], an example of the aforementioned configurations is shown in Figure .

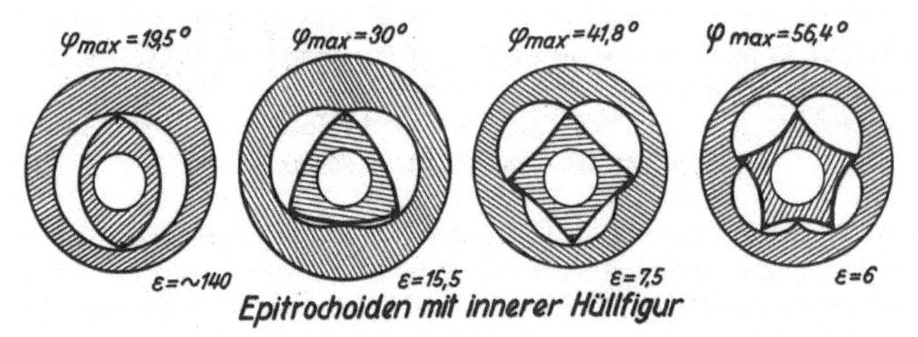

Figure 6. Different configuration of the trochoidal housing and rotor, from [3].

Some of the most important parameters for a trochoidal configuration are represented by the

- generating radius **R**
- crankshaft eccentricity **e**
- **m** parameter that defines the trochoidal configuration (1:2, 2:3, etc.) [3,5,6]
- **K** factor defined as **R/e**

Specifically, choosing **m**=2,3,...,n will define the 1:2,2:3,...,(n-1)/n configurations. One of the most important consequences of the **K** factor for a rotary engine and expander is that it affects the compression ratio of the machine: the higher the **K** factor the higher the compression ratio. Also, the real compression ratio is different from the theoretical one since it is affected by the

- parallel displacement of the housing curve defined by the **a** parameter.
- parallel displacement of the rotor curve defined by the **a'** parameter.
- presence of the rotor recesses.

Both the parallel displacements **a** and **a'** are introduced in order to change the line of contact between housing and rotor apices and create vital clearances between the rotor and the housing [6] while the recesses are used on engines for flame propagation and hence combustion improvement. A summary of these parameters for the engine and expander presented in this work are reported in the following Table 2.

The mathematical relations for the 2:3 configuration are well developed and extensively reported in the aforementioned textbooks [5,6]: for these configurations it is possible to compute all the main parameters such as the swept volume and surfaces in relation to the eccentric shaft angle. The mathematical relations for the generic (n-1)/n configuration have been developed by the authors but they are outside of the scope of this paper and will be presented in a future work. Nevertheless Figure presents the non-dimensional swept volumes for the 1:2 and 2:3 configurations involved in this paper, which will be of primary importance in establishing the timing between the two machines and for the comparison of the performance. Also, the swept volume for different **K** factors will be reported in the figures for the aforementioned configurations. It is apparent from the figures that the engine and expander cycles have different durations, i.e. 1080° and 720° of the eccentric shafts respectively. In addition, it must be mentioned that due to the two lobes of the housing the engine presents two different geometrical TDCs at 0° and 540° while the expander has only one TDC at 0° of the eccentric shaft. Finally, both the machines investigated in this work present peripheral ports that are clearly visible in the Figure 8b. In the same figure the rotors are represented with the actual phase, i.e. with the expander rotor at TDC and the inlet port fully open while the engine rotor has one of the apices at the leading edge of the exhaust port ready for the exhaust phase. Further details about the rotor timing will be reported in a subsequent section.

Table 2. Geometrical parameters of the engine and expander.

Definition	Abbreviation	Engine	Expander	Units
Generating Radius	R	69.5	73	mm
Eccentricity	e	11.6	17	mm
Width of Rotor Housing	B	51.941	63	mm
Rotor Offset	a'	1.5	0	mm
Housing Offset	a	2	0.9	mm
No. of Rotors		1		
Total Displacement		225	653	cc
Mass (excluding ancillaries)		10	10.5	kg
Geometric Compression Ratio		9.6:1	30.2	-

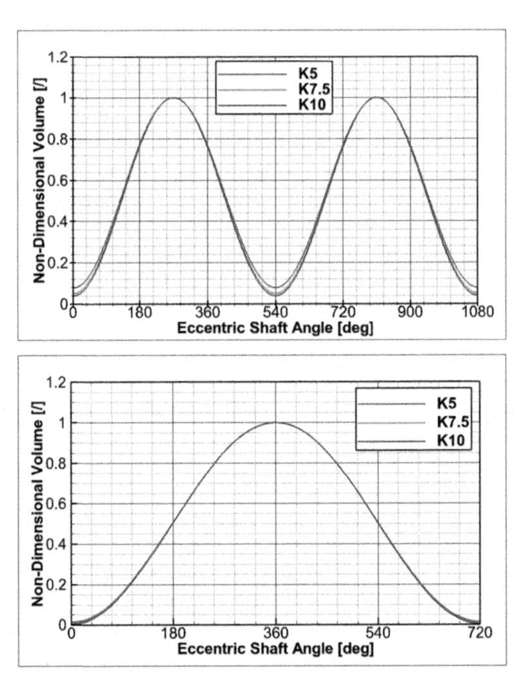

Figure 7. Non-dimensional swept volume for a Wankel engine (7a, top) and Wankel expander (7b, bottom) for different K factor values.

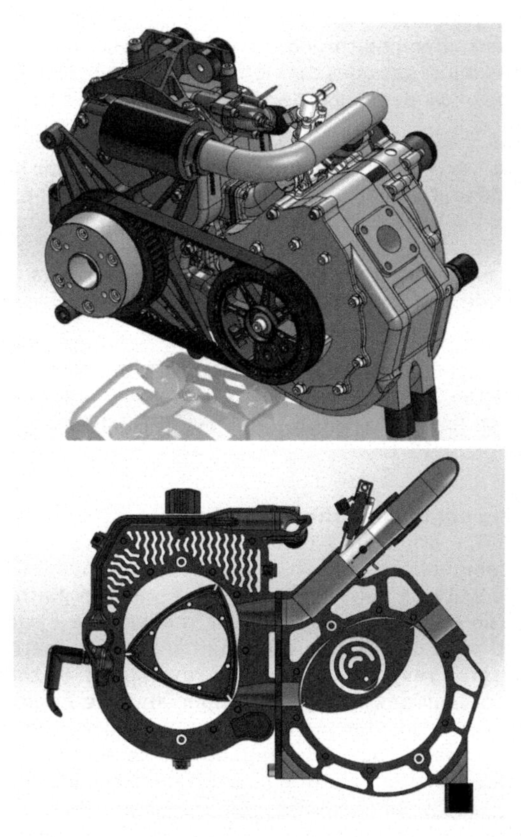

Figure 8. Engine and expander assembly (8a, top) and section drawing of the assembly (8b, bottom).

2.2 Overview of the AIE 225CS engine and the expander

The Advanced Innovative Engineering 225CS Wankel engine has been largely described together with the most important parameters and current numerical models used to predict its performance in previous works [4,17,18,19]. AIE UK also developed a compact exhaust energy recovery system in the form of a rotary type expander to be connected to the engine mentioned above. As shown in Figure 8, the engine and the expander appear as a compact single machine and are mechanically connected by a pulley and belt system with a speed ratio of 1:1. A subsequent section will explain in more detail how the port and rotor timing works. From the section drawing reported in Figure 8b it is possible to distinguish the two different configurations for the engine and the expander: the engine has the classic 2-lobe housing and 3-flank rotor (hence "2:3"), while the expander is of a 1:2 type. As previously stated, the expander rotor is positioned at its TDC, therefore a flank is facing the minimum volume and the opposite the maximum one: even visually it is possible to appreciate the large difference between the two volumes which lead to high compression/expansion ratios, in agreement with the theory presented in the previous section. Also in the same figure it is apparent that as shown both rotors rotate in the counter-clockwise direction and the fresh charge reaches the engine through the upper intake manifold. It is also clear that the path of the exhaust gas goes through the engine exhaust port to the expander inlet and finally to the expander exhaust and into the exhaust system.

Both the engine and the expander are water-cooled and can be connected to separate cooling circuits or in series, sharing the same external heat exchanger. In this way the engine outlet cooling port is connected to the inlet port of the expander. On the other hand, for safety reasons, each machine receives lubricating oil from two different circuits, including separate pumps and tanks. In both the cases the oil is metered through the journal bearings and subsequently reaches the rotor side seals, the side flanks and the apex seals in contact with the housing.

The engine and the expander are both equipped with the patented SPARCS used to reject the heat from the rotor to the cooling fluid through internal heat exchangers. A general overview on how the SPARCS (Self-Pressurising-Air Rotor Cooling System) works is reported in [4,17,18,19].

2.3 Timing of the engine and the expander

The mechanical and thermo-fluid dynamic connection and timing of the two machines is realizable because both the machines have intake and exhaust events with a periodicity of 360°. This can be explained mathematically relying on the fact that, for each rotor flank, a thermodynamic cycle is completed in a complete revolution of the rotor. Also, due to the kinematics of the internal mechanism, each rotor revolution needs **m** revolutions of the eccentric shaft, i.e. two and three revolutions for the expander and the engine respectively, as shown in Figure 8. Each rotor having a number of flanks equal to **m**, it is apparent that the ratio between the cycle angular displacement and the number of flanks gives a periodicity of events that is equivalent to 360° for both machines.

The area and port timing for each configuration related to its own eccentric shaft displacement are reported in Figure 9. In order to maximize the expansion phase, the two rotors and the ports are phased so that at the incipient exhaust of the engine, the gas will find the inlet port of the expander fully open with the expander rotor at TDC and ready to over-expand the working fluid. Furthermore when the expander rotor is at TDC the engine rotor is at 740° (apex at exhaust port leading edge) and hence a phase of 740° exists between the two rotors relative to the expander TDC. More interesting are the superimposed engine exhaust and expander inlet phases reported in Figure 10, where the timing of the rotors is taken into account. As can be seen the two phases have similar starting points and durations with the engine exhaust phase almost entirely covering the expander intake phase, which lead to the maximum over-expansion.

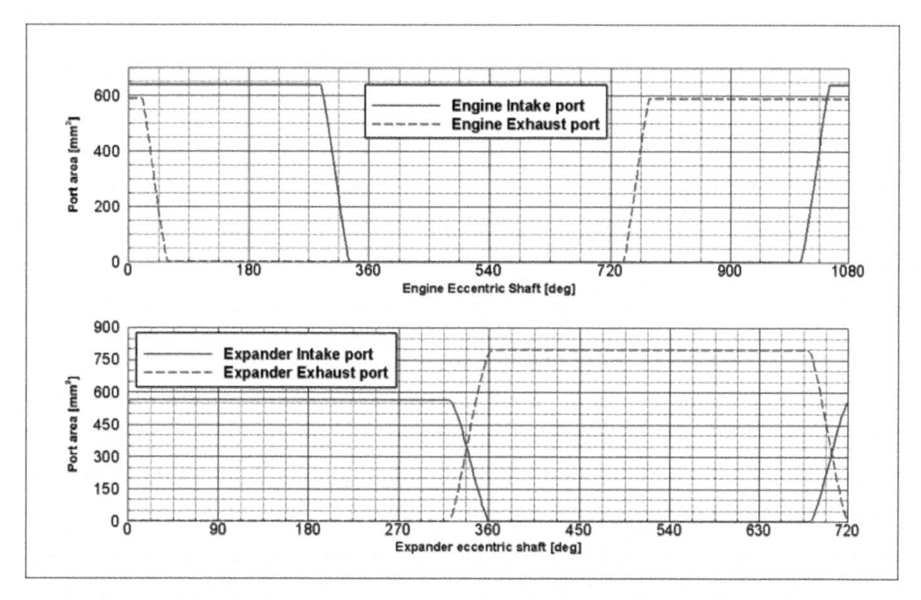

Figure 9. Intake and exhaust port timing for the engine and the expander.

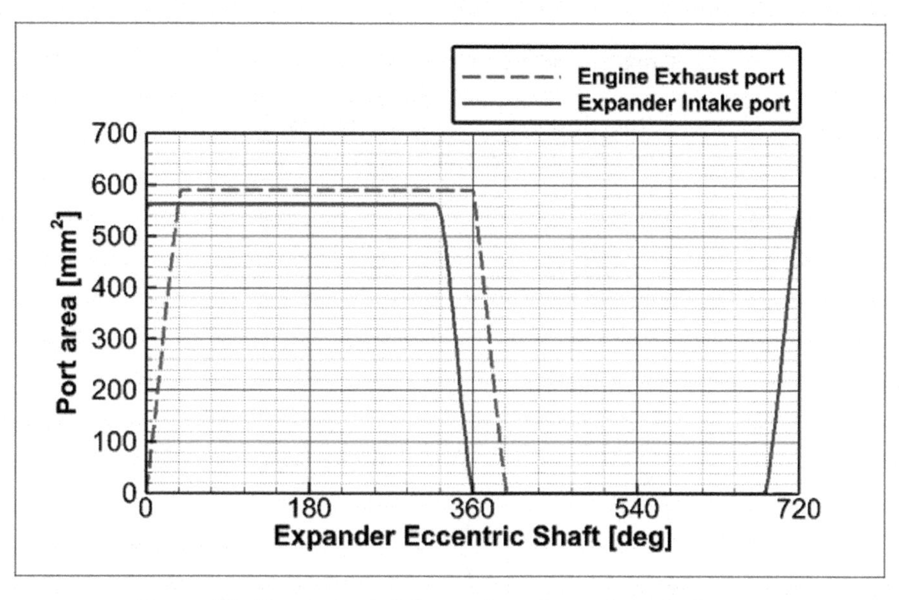

Figure 10. Expander intake and engine exhaust phases.

For the sake of completeness, Figure 11 shows the volume variation of both the machines in relation to the expander eccentric shaft displacement and taking into account only one flank for each rotor and the actual timing between the rotors.

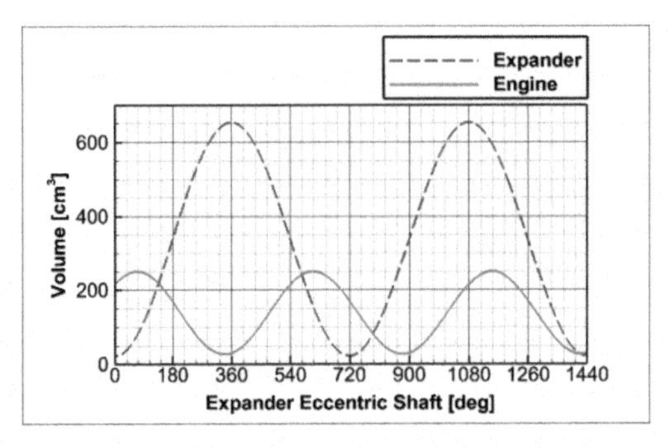

Figure 11. Engine and expander swept volumes.

As mentioned above, the computation of the total volume and the real values for compression and expansion ratios must take into account the additional volumes introduced by all the geometries that are not considered in an ideal model of the trochoidal configuration, such as recesses, parallel displacement of the housing and the rotor, additional ducts and volumes, etc. This is the case when the engine is connected to the expander: the engine exhaust and expander inlet duct behaves like an additional and instantaneously-introduced clearance volume. Therefore it must be included in the total volume formed by the two machines expanding simultaneously the same mass of exhaust gas. Table 3 reports all the values related to the clearance volumes used in the computation of the total instantaneous volume shown in Figure 12.

Table 3. Additional clearance volumes.

Description	Volume [cc]
Engine rotor recesses	7.53
Engine exhaust duct	19.6
Expander inlet duct	9.78
Connection flange	3.56

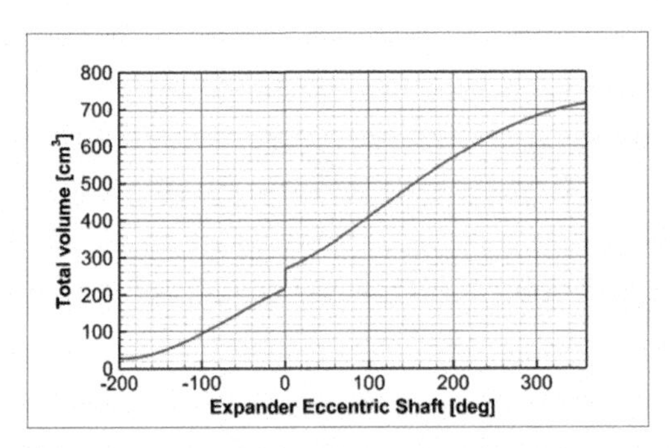

Figure 12. Total volume of the engine, expander and additional clearance volume.

The Figure 12 reports the minimum and maximum values for the volume of the two machines working together, which lead to a total effective expansion ratio of 27.4:1 and the opportunity to achieve an extended Miller cycle. Indeed, because the basic cycle is a constant-volume one, and there is so much expansion, there is an argument that the entire system is in fact operating on the Humphrey cycle which has the first three phases (isentropic compression, constat-volume heat addition and isentropic expansion) in common with the Otto cycle and differs from it only for the final constant-presure heat rejection thus increasing the cycle work. It is also apparent from this that the expander will over-expand the gas under ambient pressure if the pressure at the start of the exhaust phase is not high enough, resulting in over-expansion and thus negative work at the expander eccentric shaft when the expander exhaust port opens. Such a case will be reported in the section related to the analysis of the performance, where at low speed the BSFC of the compound machine is larger that the single engine one.

Finally, some decisive advantages of Wankel-type geometries versus typical 4-stroke reciprocating engine machinery in this application become clear from the above discussion: [1] the gas transfer ducts are much smaller and shorter to the benefit of reduced heat loss; [2] there are reduced pressure losses within the cycle due to the absence of poppet valves; and [3] there are lower friction losses because of the Wankel design itself (disregarding any extra valve drive mechanisms). Consequently the parasitic losses which can negate any theoretical benefit of the application of expander cylinders to reciprocating engines are significantly mitigated. The practical benefit and experimental proof of this will be discussed later on.

3 TEST RIG AND INSTRUMENTATION

Performance, efficiency and emissions testing of the baseline and expander version of engine are both based on the same rig facility which uses an AC current dynamometer as described in the earlier published work [18].

Several transducers are employed to measure different parameters such as temperatures, pressures, flow rate, torque, speed etc. (Table 4). K-type thermocouples are installed to monitor coolant inlet and outlet, air inlet and exhaust temperatures. Additionally, more K-type thermocouples are installed in the expander to measure expander its SPARCS air in and out and coolant and exhaust temperatures between the main engine and the expander unit. K-type thermocouples have been chosen for their large operating range with acceptable level of accuracy.

Table 4. Instrumentation for baseline and expander engines.

Parameter	Transducer Type	Operating Range	Accuracy
All temperatures	K type thermocouples	-180 to 1350°C	±0.75%
Fuel mass flow rate	Emerson Coriolis Micro motion CMF010M	93.5 kg/hr nominal	Less than 0.1% above 2 kg/hr
Coolant volume flow rate	Krohne Optiflux 5000 sensor	Maximum 36 L/min	Less than ±0.15%
Ambient air temperature and relative humidity	Omega HX94AV	0 to 100°C, 3 to 95% RH	±0.6°C, ±2.5% for 20-80% RH and rest ±3.1%
Fuel Pressure	GE Druck UNIK 5000 piezo resistive pressure transducer	0 to 6 barG	±0.2%
Torque	HBM T40B torque flange	0 to 100 Nm	Less than or equal to 1% above 5 Nm
RPM	Encoder embedded in the AC motor	0 to 8500 rpm	±0.01%

All the analogue and digital signals are collected by a CADET data acquisition system from Sierra CP and then processed using their respective calibrations to deliver the real-time measurement values. The system is capable of handling high speed data sampling for transient signals as well as low speed sampling for quasi-steady signals. The latter was used in this investigation due to the nature of the tests. Moreover, the system is equipped with a CAN node able to collect data from the standard CAN bus. This allows the user to compare the data collected by the ECU with the ones mentioned above.

All the important operations of the engines like fuel injection timing and metering, spark timing, closed-loop lambda control, oil and coolant flow metering etc. are controlled by an ECU supplied by one of the other project partners within ADAPT, GEMS UK. The ECU model EM80 collects the measurement data of all the basic engine parameters - coolant and air inlet temperatures, inlet manifold pressure (MAP), rotational speed etc. - and execute the implemented strategies to control the engine. Several OEM transducers (Table 5) are adopted to monitor vital engine parameters and the ECU uses different 2D or 3D look up tables to implement the strategies. The GEMS ECU is configurable in real time when connected with its proprietary software GWv4 and the strategies can be manipulated to meet the performance and emission requirements.

Table 5. OEM transducers used on the engines.

Parameter	Transducer Type
Air inlet temperature	KA NTC1 Thermistor
Coolant out temperature	KA NTC1 Thermistor
Rotor air inlet temperature	KA NTC1 Thermistor
Inlet manifold pressure	KA ASLAB series pressure transducer 0-1.2 bara
SPARCS pressure	KA ASL series pressure transducer 0-10 barg
Crank position	KA SPC hall effect speed sensor
Lambda	Bosch LSU wide band lambda sensor
Knock	Bosch piezoelectric knock sensor
Throttle position	KA SPC hall effect speed sensor

The analysis of the engine's exhaust emission was one of most important focus of this study together with the overall performance evaluation of both the machines. The measurement of the emission gases such as CO_2, O_2, CO, HC and NOx would facilitate this research work to assess the current state of Wankel engine emission performance and ways to improve them to meet current emission regulations. A Horiba MEXA 7000 series gas analyser performed the measurement of above mentioned gases as well as calculating the operational AFR using the carbon balance method.

4 EXPERIMENTAL RESULTS AND DISCUSSION

4.1 Performance, fuel economy and emission of baseline 225CS

The characterisation of performance, efficiency and emission of 225CS baseline engine was undertaken for a speed range of 3000 to 6500 rpm with a full range of load conditions. Test data were collected under quasi-steady state conditions for coolant temperature of around 60°C. The fuel supply was adjusted to achieve a lambda value of 1. A map of baseline engine fuel efficiency with its performance is illustrated by the BSFC contour plot in Figure 13.

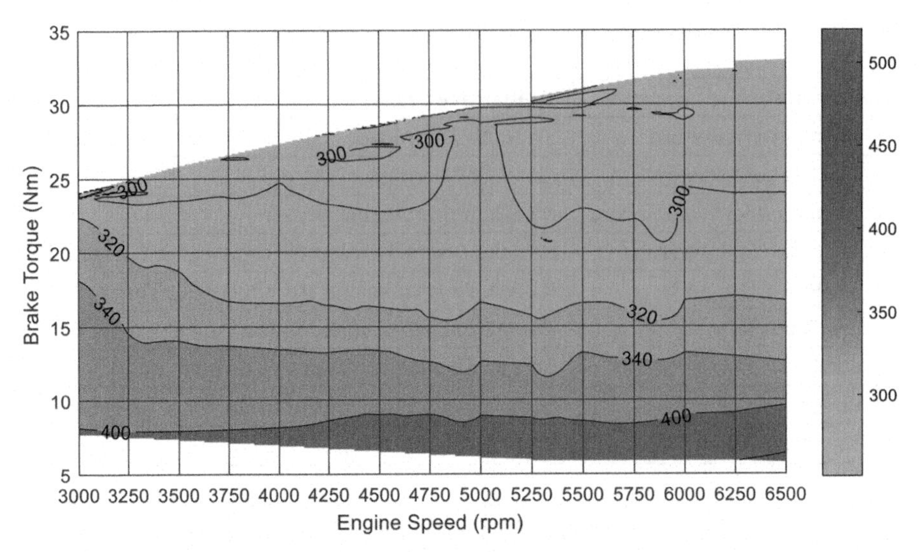

Figure 13. BSFC map (in g/kWh) with relation to engine speed and brake torque for baseline engine.

The brake load for the 225CS baseline engine increased with increasing speed and a maximum brake torque and power of 33.5 Nm and 22.8 kW respectively were observed at an engine speed of 6500 rpm. The specific fuel consumption achieved by the engine reduced with increasing engine load across the operating speed range as can be seen from the plot. The BSFC for full-load conditions was in the range of 290-300 g/kWh making this Wankel engine most efficient at high load and speed conditions. Earlier aircraft rotary engines of 1980s reported a BSFC of around 240-255 g/kWh [20] whereas modern rotary engine for automotive application could achieve 250 g/kWh [21].

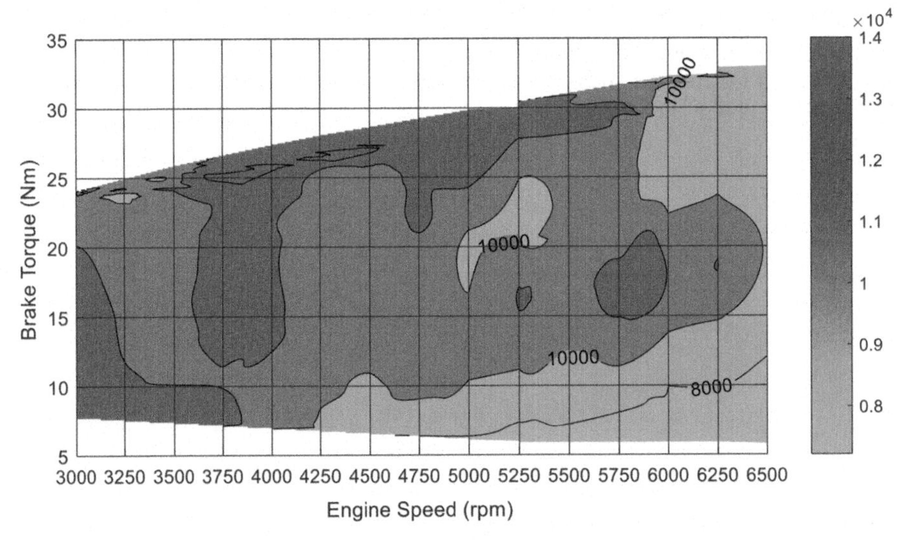

Figure 14. Map of unburnt hydrocarbon emissions with relation to engine speed and brake torque for baseline engine.

As already mentioned, one of the biggest downsides of the Wankel engine is the high amount of unburnt hydrocarbon emission in the exhaust. The total hydrocarbon (THC] emission was analysed using the MEXA gas analyser along with the other emission gases CO_2, O_2, CO and NOx. In this experimental study, raw untreated exhaust emission from each engine configuration were collected and analysed for direct comparison of their emission performance. As reported in Figure 14, the THC emissions were in range of 11000-16000 ppm at full load for the speed range between 3750 and 5500 rpm. For higher speeds, the emissions were slightly lower but only as far as 9000 ppm. These values are also significantly higher than the values reported by Mazda in their RENESIS engine; at 2000 rpm the RENESIS engine had HC emissions in the range of 800-1200 ppm [22], achieved partly through the adoption of side intake and exhaust ports.

As the engine was operating under closed loop lambda 1 control, the high amount of unburnt HC also resulted in a significant amount of oxygen in the exhaust which was in range of 1.6-2% under full load operations. Another important indicator of poor combustion performance is exhaust CO and values as high as 1% was observed.

4.2 Assessment of expander engine

The performance of the 225CS engine equipped with the expander was assessed against the baseline to identify the potential energy recovery through the expansion of exhaust gases. Tests were undertaken for a speed range of 3000 to 5000 rpm under a variety of load conditions. During the investigation the coolant temperature and fuel strategy were kept similar to the baseline tests for ease of direct comparison. A significant improvement in brake torque over the baseline was observed for the expander engine as presented in Figure 15. A minimum 13% increase in brake torque was identified for low speed operation to a maximum of 30% for higher speed of 4750 rpm. This benefit in performance was not only recorded for high load but also across different load ranges.

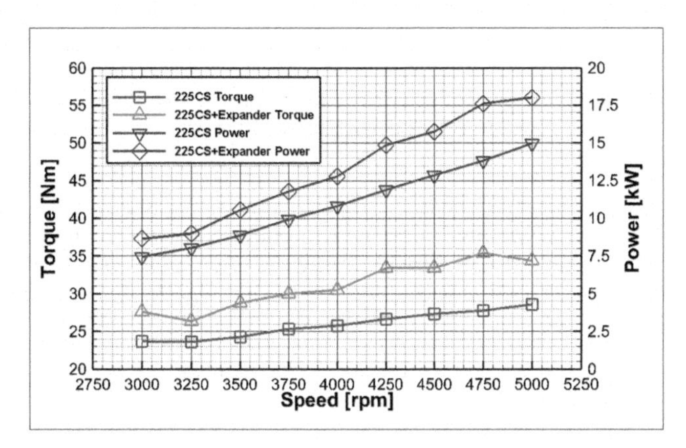

Figure 15. Comparison of performance of baseline and expander engine at wide open throttle conditions.

The improvement in brake performance can be linked directly to the increased expansion of the exhaust gas and consequent energy recovery in the expander unit. A significant reduction in exhaust temperature was observed in conjunction with the increase in brake torque. As an example case, for 5000 rpm under full load operation the exhaust gas leaving the expander was around 500°C compared to a very high temperature of 900°C for the baseline engine. The two-stage Cottage Loaf engine developed by Rolls-Royce provided a concept of similar technology and they also reported an increase in power output through two-stage expansion [23].

The true benefit of the expander unit can be identified by the assessment of fuel economy. The BSFC of the expander engine was compared with the baseline to analyse the advantage of the inclusion of the expander unit as illustrated in the Figure 16. For the speed range of 3750 to 5000 rpm, an approximate 8% improvement in minimum BSFC was recorded under wide open throttle condition. The benefit was more pronounced at higher speeds and a maximum 12% reduction from baseline BSFC was observed at a speed of 4750 rpm.

However, lower speeds yielded a higher BSFC compared to the baseline. Here the gain in brake torque was offset by comparatively higher fuel consumption resulting in higher BSFC. The expander engine had around 12-16% higher air consumption in the speed range investigated and consequently higher fuel intake was attained due to the adjustment in fuel supply to achieve a lambda value of 1.

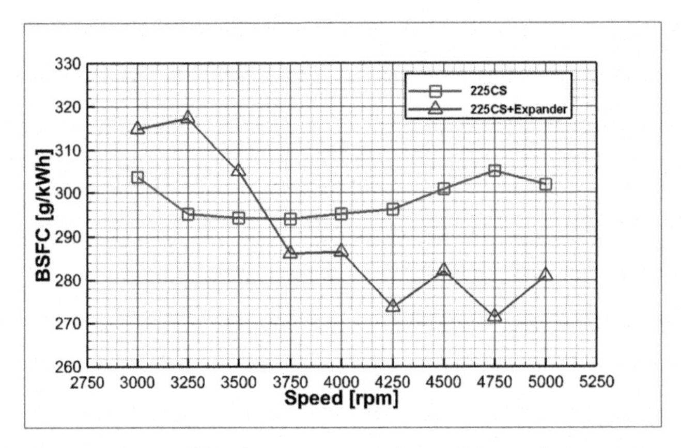

Figure 16. Comparison of fuel economy of baseline and expander engine at wide open throttle conditions.

As stated in the previous subsection, the HC emissions are one of the biggest concerns of Wankel engines and necessary mitigation steps are sought by engine manufacturers to meet the emissions requirements [22]. The expander has a great potential to perform as a thermal reactor due to the condition of the exhaust gas leaving the main engine. Thermal reactors were long used in Wankel rotary engines to reduce HC emissions as described by Yamamoto [6]. High temperatures and high oxygen concentration in the exhaust provides the conditions for additional oxidation reactions.

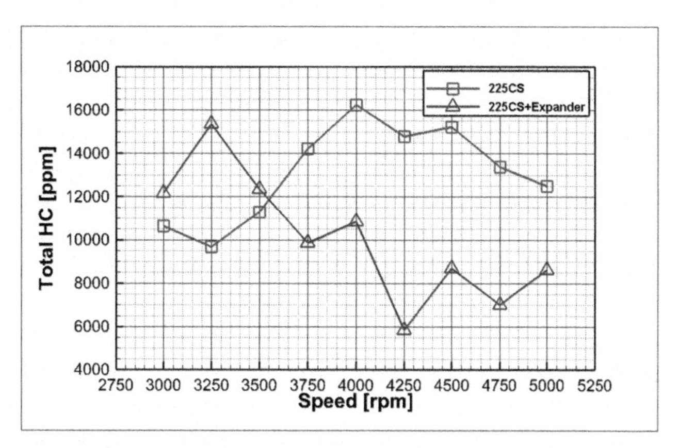

Figure 17. Comparison of hydrocarbon emissions of baseline and expander engine at wide open throttle conditions.

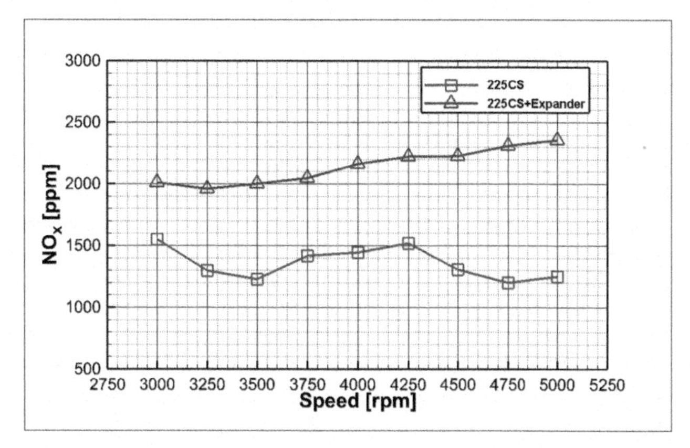

Figure 18. Comparison of NOx emissions of baseline and expander engine at wide open throttle conditions.

The expander engine proved to be not only beneficial in improving the fuel economy but also effective in reducing unburnt hydrocarbons in the exhaust. Secondary oxidation reactions were identified in the expander unit as illustrated by the reduction in HC emissions in Figure 17. A significant reduction in Total HC emission was observed for the speed range of 3750 to 5000 rpm under wide open throttle conditions. The minimum Total HC content in the exhaust was around 6000 ppm for the above conditions which was around 60% lower than that for baseline. Further improvement in Total HC, around 4000 ppm, was recorded for a lower load of 33 Nm at 4250 rpm. The research and development work by Mazda on their RENESIS engine [22] implemented a secondary air supply system to facilitate the secondary combustion process.

However, higher THC emissions were observed for speeds lower than 3750 rpm as can be seen in Figure 17. The reason could be identified as higher fuel consumption without gaining proportionate brake torque during the exhaust expansion.

The reduction in THC was matched by a reduction in oxygen concentration in the exhaust where part of it was consumed during the expansion and secondary oxidation process. The oxygen level diminished to a level of 0.5-1% which further provides evidence of secondary combustion in the expander. The additional combustion process also helped in reducing CO emission during high load and speed operation. CO level in the range of 0.4 to 0.6% was observed during this investigation. Another important emission product is NOx which was higher for the improved integrated unit as shown in Figure 18. Higher air consumption along with higher brake torque (and possible higher peak gas temperature) can be suggested as reasons behind it.

5 CONCLUSIONS AND FURTHER WORK

A rotary expander for exhaust energy recovery has been analysed in this work. The expander, designed and produced by Advanced Innovative Engineering UK Ltd, has been tested together with a Wankel engine from the same company at the University of Bath within the ADAPT-IPT project funded by the Advanced Propulsion Centre and Innovate UK. With the current analysis it has been demonstrated that the rotary expander is able to improve the performance of the single Wankel engine in terms of torque, power, efficiency and emissions. In detail, from a direct comparison in the operating range from 3000 to 5000 rpm a maximum torque increment of 30% has been observed together with a 12% reduction of the BSFC at 4750 rpm. Unfortunately the emissions figures are still high but the expander has been show to work as a thermal reactor and promote secondary combustions that lead to THC reduction in addition to a lower concentration of oxygen in the exhaust gases. Conversely, all these improvements come at the price of the increased weight and complexity of the compound machine that needs further ancillaries such as separate lubrication circuits and increased cooling requirements. Better packaging could be obtained by using a single eccentric shaft for both the machines and side ports, placing the expander parallel to the engine in a way similar to a Wankel multi-rotor configuration with the aim of reducing the weight and increasing the power-to-weight ratio. Further work will include the 1D and 3D modelling of both the machines working together in order to clarify all the phenomena related to combustion, and the mass and heat transfer between the working chambers. Finally, an expander equipped with fast pressure transducers will be tested over an extended operating range in order to collect the pressure traces from the working chambers and validate the results obtained from the numerical models.

AKNOWLEDGEMENTS

The authors wish to express their grateful thanks to Innovate UK and Advanced Propulsion Centre UK for their support and funding for this research. Thanks also go to the University of Bath's partners in the ADAPT-IPT project, namely Westfield Cars, Advanced Innovative Engineering (UK) Ltd, GEMS, and Saietta.

DEFINITIONS AND ABBREVIATIONS

1D	–	One-dimensional
3D	–	Three-dimensional
AFR	–	Air-fuel ratio
BSFC	–	Break specific fuel consumption
CAN	–	Controller area network
CO	–	Carbon monoxide
CO_2	–	Carbon dioxide
ECU	–	Engine control unit
HC	–	Hydrocarbon
MAP	–	Inlet manifold pressure
NO_X	–	Nitrogen Oxide
NVH	–	Noise, Vibration and Harshness
rpm	–	Revolution per minute
SPARCS	–	Self pressurized air-rotor cooling system
TDC	–	Top dead centre
THC	–	Total hydrocarbons

REFERENCES

[1] Mazda EU Press Office – *Mazda rotary engine to return as EV range-extender* – Published 02/10/2018 – accessed 08/05/2019 - https://www.mazda-press.com/eu/news/2018/mazda-rotary-engine-to-return-as-ev-range-extender-/.

[2] Turner, M., Turner, J., Vorraro, G., "Mass benefit analysis of 4-Stroke and Wankel range extenders in an electric vehicle over a defined drive cycle with respect to vehicle range and fuel consumption", SAE Technical Paper 2019-01-1282, 2019, https://doi.org/10.4271/2019-01-1282.

[3] Norbye, J. P., *The Wankel engine – design, development, applications* (Philadelphia: Chilton Book Co., 1971).

[4] Peden, M., Turner, M., Turner, J., and Bailey, N., "Comparison of 1-D Modelling Approaches for Wankel Engine Performance Simulation and Initial Study of the Direct Injection Limitations", SAE Technical Paper 2018-01-1452, 2018, https://doi.org/10.4271/2018-01-1452.

[5] Ansdale, R.F., *The Wankel RC Engine* (London: Lliffe Books Ltd, 1968).

[6] Yamamoto, K., *Rotary Engine* (Hiroshima: Sankaido Co. Ltd, 1981).

[7] Shimizu, R., Tadokoro, T., Nakanishi, T., Funamoto, J., *Mazda 4-rotor rotary engine for the Le Mans 24-hour endurance race* (SAE Technical Paper: 920309) *Michigan, Detroit*: 1992.

[8] Ohkubo, M., Tashima, S., Shimizu, R., Fuse, S. et al., "Developed Technologies of the New Rotary Engine (RENESIS)," SAE Technical Paper 2004-01-1790, 2004, https://doi.org/10.4271/2004-01-1790.

[9] Spreitzer, J., Zahradnik, F., and Geringer, B., *"Implementation of a Rotary Engine (Wankel Engine) in a CFD Simulation Tool with Special Emphasis on Combustion and Flow Phenomena,"* SAE Technical Paper 2015-01-0382, 2015, doi:10.4271/2015-01-0382.

[10] European Environment Agency, *"Monitoring of CO2 Emissions from Passenger Cars - Regulation (EC) No. 443/2009,"* European Environment Agency, 23 April, accessed July 09, 2018, https://www.eea.europa.eu/data-andmaps/data/co2-cars-emission–14.
[11] Chen, AS, Herrmann, G, Na, J, Turner, M, Vorraro, G and Brace, C 2018, Non-linear Observer-Based Air-Fuel Ratio Control for Port Fuel Injected Wankel Engines. in *2018 UKACC 12th International Conference on Control (CONTROL 2018): Proceedings of a meeting held 5–7 September 2018, Sheffield, United Kingdom*. Institute of Electrical and Electronics Engineers (IEEE), pp. 224–229. https://doi.org/10.1109/CONTROL.2018.8516842.
[12] Romagnoli, A., Vorraro, G., Rajoo, S., Copelan, C., Martinez-Botas, R., "Characterization of a supercharger as boosting and turbo-expansion device in sequential multi-stage systems", Energy Conversion and Management 136, 2017, 127–141.
[13] Bontempo, R., Cardone, M., Manna,M., Vorraro,G.,"A statistical approach to the analysis of the surge phenomenon", Energy 124, 2017, 502–509.
[14] Bontempo, R., Cardone, M., Manna,M., Vorraro,G., "Steady and unsteady experimental analysis of a turbocharger for automotive applications", Energy Conversion and Management 99, 2015, 72–80.
[15] Tashima, S., Taqdokoro, T., Okimoto, H., and Niwa, Y., "Development of Sequential Twin Turbo System for Rotary Engine," SAE Technical Paper 910624, 1991, https://doi.org/10.4271/910624.
[16] Garside, D., "Rotary Piston Internal Combustion Engine power unit," Patent Publication No. WO/2009/115768, September 24, 2009.
[17] Advanced Innovative Engineering *"AIE sign exclusive licence agreement to use patented engine technology"* Published 31st May 2015 – Accessed – 29th May 2019. https://www.aieuk.com/aie-sign-exclusive-licence-agreement-use-patented-engine-technology/.
[18] Vorraro, G., Turner, M., and Turner, J., "Testing of a Modern Wankel Rotary Engine - Part I: Experimental Plan, Development of the Software Tools and Measurement Systems," SAE Technical Paper 2019-01-0075, 2019, https://doi.org/10.4271/2019-01-0075.
[19] Bailey, N., Louthan, L., "The Compact SPARCS Wankel Rotary Engine", SAE J 2015.
[20] Mount, R.E., and LeBouff, G.A., "Advanced Stratified Charge Rotary Engine Design," SAE paper 890324, SAE, Warrendale, PA, 1989.
[21] Hubmann, C., Friedl, H., Gruber, S. and Foxhall, N., Single Cylinder 25kW Range Extender: Development for Lowest Vibrations and Compact Design Based on Existing Production Parts SAE Technical Paper 2015-32–0740.
[22] Ohkubo, M., Tashima, S., Shimizu, R., Fuse, S. et al., "Developed Technologies of the New Rotary Engine (RENESIS)," SAE Technical Paper 2004-01-1790, 2004, https://doi.org/10.4271/2004-01-1790.
[23] Feller, F. (1970). The 2-Stage Rotary Engine—A New Concept in Diesel Power. *Proceedings of the Institution of Mechanical Engineers*, *185*(1), 139–158. https://doi.org/10.1243/PIME_PROC_1970_185_022_02.

SESSION 7: SIMULATION OF INTERNAL COMBUSTION ENGINES

One-dimensional simulation of the pressurized motoring method: Friction, blow-by, temperatures and heat transfer analysis

C. Caruana[1], M. Farrugia[1], G. Sammut[2], E. Pipitone[3]

[1]Mechanical Engineering, University of Malta, Malta
[2]Jaguar & Land Rover Ltd., UK
[3]Department of Engineering, University of Palermo, Italy

ABSTRACT

Mechanical friction in internal combustion engines is nowadays largely optimized as a result of long years of research in this field. The ever increasing demand for better performance and less emissions however still puts mechanical friction as one of the areas in which further improvements can be done. This requirement calls for more accurate one-dimensional simulation models, along with reliable and robust experimental data to support such models. At University of Malta, a Pressurized Motoring test rig was developed and reported in SAE paper 2018-01-0121. Such method has proved to be relatively accurate in determining the mechanical friction of an internal combustion engine. As compared to the conventional Indicating technique, Morse test and Breakdown test, the Pressurized Motoring method offers the robustness that the uncertainty in the FMEP, as a result of error propagation is kept low. Furthermore, amendments to the traditional Pressurized Motoring technique were also done and presented in SAE paper 2019-01-0930, where Argon was used in place of Air to raise the bulk in-cylinder peak temperature to a value similar to what is found in a fired engine. In this proposal for publication, the authors now discuss a simple one-dimensional simulation model whose results are compared with those obtained and reported in SAE paper 2019-24-0141. Such work is aimed to test traditional friction and heat transfer models in one-dimensional software whilst identifying potential optimizations that can be done to the developed Pressurized Motoring test rig for its data to be more appealing to the one-dimensional researcher.

1 INTRODUCTION

Mechanical friction and heat transfer measurements are known to be fundamental to engine research and development, especially in an era when the internal combustion engine is facing stringent requirements to be met. At University of Malta, a Pressurized Motoring setup was built and tested (1) in which the exhausted gas of the motored engine is rerouted back to the intake side via a shunt pipe. This setup proved to allow for robust friction data to be obtained, due to mitigation of error propagation. Apart from the shunt pipe, which is relatively simple to realize, such method also proved to require no extra apparatus than what is usually available in a conventional engine test cell. The use of the shunt pipe minimizes greatly the quantity of gas required to be supplied for achieving peak in-cylinder pressures similar to that in fired engines, in fact a conventional shop floor compressor was found to be more than adequate (1). In a recent research (2), the engine was also operated on Argon, instead of Air, in order to address one of the main limitations of any motoring testing method, i.e. the low in-cylinder temperatures. The main aim of this complete research project was to provide the one-dimensional researcher with reliable experimental data to compare against traditional friction and heat transfer models as used in one-dimensional simulation software. This publication presents a simple one-dimensional model of the pressurized motoring setup used in (1) (2) (3) which was calibrated against the experimental results presented in (3).

1.1 Engine geometry

The specifications for the engine in the one-dimensional model were made to match those of the engine used in (1) (2) (3) and presented in Table 1. The engine is an under-square with the pistons having three rings. The piston bowl shape has a relatively deep dish with a protruding sphere as shown in Figure 1. The pistons are also jet-cooled. The cylinder head chamber is flat with two valves per cylinder. The log-style intake manifold is shown in Figure 2 while the exhaust manifold is shown in Figure 3. The OEM exhaust manifold has an EGR port which in the pressurized motoring configuration was instead used for regulated gas supply. The intake manifold is made of cast Aluminum, whereas the exhaust manifold is made of cast Iron. The intake port in the cylinder head has a curved nature to impose swirl to the gas induced. The exhaust port has only an approximate $90°C$ short bend. All intake and exhaust ports are identical.

Figure 1. OEM Piston.

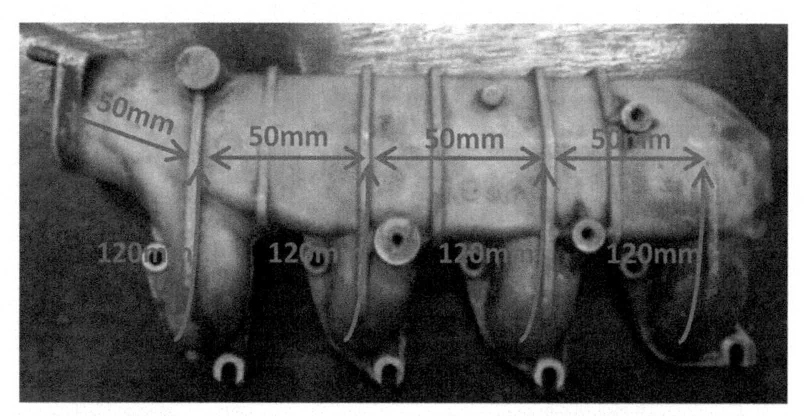

Intake Manifold:
Runner Diameter: 35mm
Plenum Cross-Sectional Area: 70mm x 50mm
Intake Manifold Entry Diameter: 40mm

Cylinder Head Intake Side:
Port Length: 120mm
Port Diameter: 35mm to 30mm

Figure 2. OEM Intake Manifold.

Exhaust Manifold:
Runner Diameter: 30mm
Exhaust Manifold Exit Diameter: 45mm
EGR Diameter: 25mm
EGR Length: 80mm

Cylinder Head Exhaust Side:
Port Length: 95mm
Port Diameter: 30mm

Figure 3. OEM Exhaust Manifold.

Table 1. Engine Specifications.

Make and Model	Peugeot 306 2.0HDi
Year of Manufacture	2000
Number of Strokes	4
Number of Cylinders	4
Valvetrain	8 Valve, OHC
Compression Ratio	18:1
Engine Displacement [cc]	1997
Bore [mm]	85
Stroke [mm]	88
Connecting Rod Length [mm]	145
Intake Valve Diameter [mm]	35.6
Exhaust Valve Diameter [mm]	33.8
Intake Valve Opens (1mm lift)	170 CAD BBDC intake
Intake Valve Closes (1mm lift)	20 CAD ATDC compression
Exhaust Valve Opens (1mm lift)	45 CAD BBDC expansion
Exhaust Valve Closes (1mm lift)	10 CAD BTDC exhaust
Intake Max. Valve Lift [mm]	9.6
Exhaust Max. Valve Lift [mm]	9.7

2 BUILDING THE ONE-DIMENSIONAL MODEL

In this research, Ricardo WAVE was used as the one-dimensional simulation software. The one-dimensional model discussed was initially developed by Camilleri (4) some years back when the engine was still used in the fired mode. During that time, the engine was coupled to a water-brake dynamometer in the aim of developing a programmable engine management for common rail diesel engines (5). The model was now modified to suit the new setup configuration of the Pressurized Motoring. The canvas for the one-dimensional model is shown in Figure 4.

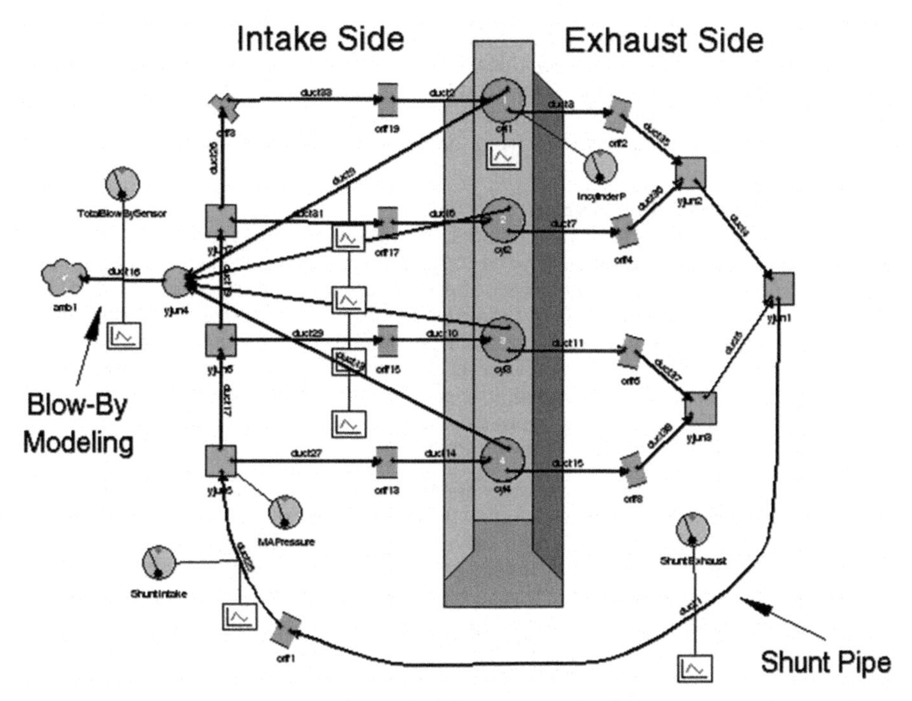

Figure 4. One-Dimensional Model Canvas.

In building the one-dimensional model, all relevant lengths and diameters for the ports, manifolds and shunt pipe were entered as measured on the real designs. The ports were assigned a discharge coefficient of unity. Their bend angle, as well as friction coefficient were switched off, since their inefficiency was accounted for in the valve flow coefficients, which in this study were experimentally obtained. Valve lift profiles were also measured and made available by Camilleri (4). Experimental data for the manifold temperatures was not available; however experimental coolant temperatures were measured and used in the model. The wall temperature for the manifolds was set to around $15^{\circ}C$ lower than that of the coolant for the particular setpoint considered.

The sub-models used in this one-dimensional model were primarily two; that for heat transfer and that for mechanical friction. Ricardo WAVE presents three alternatives to heat transfer estimation; Annand's model (6), Woschni's model (7) and Colburn's model. Apart from these, WAVE allows for the IRIS and a user-defined heat transfer model. In this study, the model by Annand and Woschni were used separately as discussed later in the paper.

The only friction model offered by Ricardo WAVE is that based on the Chen-Flynn correlation (8). Such model consists of four terms representing the friction contribution due to engine speed, engine load, windage and accessories. The results from this correlation are discussed in the results section.

Additional sub-models like the conduction model and the thermocouple model were also considered and used, but preferred to turn them off for the final version of the model as their results were questionable.

Parameters like the thermodynamic loss angle are known to be sensitive to blow-by flow (9), even though its effect is minimal compared to heat transfer. Since experimental data for blow-by flow was available, a blow-by model was devised as suggested by WAVE knowledge centre (8). The model consists of a third valve on each cylinder of the orifice type. Each of these valves were connected to a large volume, representing the crankcase and exhausted to atmosphere, represented by an 'Ambient' element. The diameter of the orifice valve was varied until the blow-by flow on each setpoint matched that obtained from the experimental sessions (1).

2.1 Cylinder head flow testing

In support of developing the WAVE model, an experimental test session was performed on an in-house developed flow-bench in the aim of determining the valves' discharge and flow coefficients. The cylinder head used in this testing session was of the same make and model of the 2.0HDi engine used on the Pressurized Motoring Setup and the one-dimensional model. All tests carried out on the flow-bench were done at 28" of water (i.e. 7kPa) below atmospheric conditions for pull through configuration (port to cylinder), and above atmospheric conditions for blow through configuration (cylinder to port). Temporary bellmouth using modeling clay were manually formed at entrance as is common practice in flow bench testing.

When set up on the flow-bench, the cylinder head was in the condition as dismantled from the running engine, i.e. not cleaned. It was discovered that since the particular engine has an EGR system, the intake valves contained heavy soot deposits on their back end as seen in Figure 5. The cylinder head was flow tested in this condition and the flow coefficients were determined in this state. Later, the valves were thoroughly cleaned and flow tested again. Figure 6 shows the mass flow comparison between two clean intake valves and the same valves before cleaning. It is clearly visible that when clean, the two valves have relatively similar flow behaviors, whereas before cleaning, the flow was marginally lower for valve 3 and considerably lower for valve 2, dependent on the amount and shape of the deposited soot. The exhaust valves were also checked for soot deposition, but it was noted that the amount of soot present was not abnormal to that usually found on exhaust valves. All exhaust valves were cleaned prior flow testing.

Figure 5. Soot present due to EGR on intake valve, cylinder two.

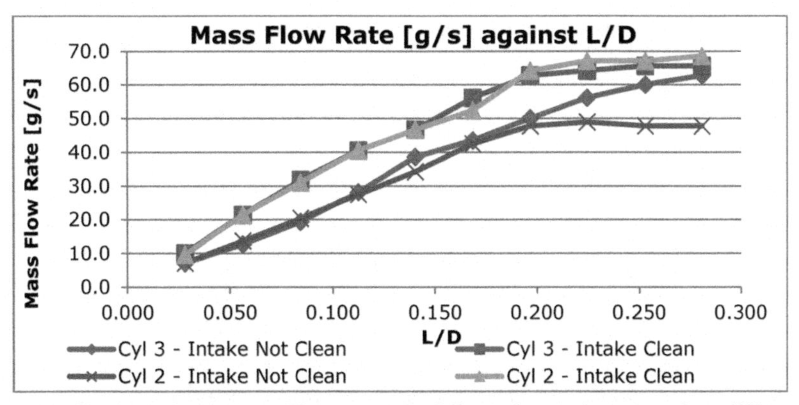

Figure 6. The Graph of Mass Flow Rate [g/s] against L/D, showing difference between clean and sooted intake valves.

The second aim of this flow testing session was to determine the restriction that both the intake and exhaust manifolds had on the mass flow. Figure 7 shows that the intake manifold was not a major restrictor at low lifts, but did restrict the flow at higher lift values. The exhaust manifold on the other hand seems to have aided the flow through the exhaust valve at lower lifts but restricted it at higher lifts.

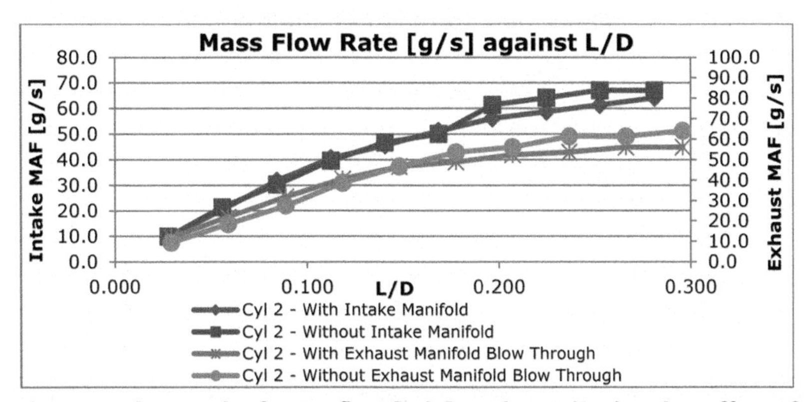

Figure 7. The graph of mass flow [g/s] against L/D showing effect of manifolds.

The exhaust valves were tested with two flow configurations; 'blow through' (i.e. from the cylinder to the port) and 'pull through' (i.e. from the port into the cylinder). The two results were compared and shown in Figure 8. In usual flow bench practice the exhaust valve is tested with a blow through setup to mimic better the actual flow direction in the engine. To follow such practice, in the one-dimension simulation software, the flow coefficient obtained from the blow through test data was used for the exhaust.

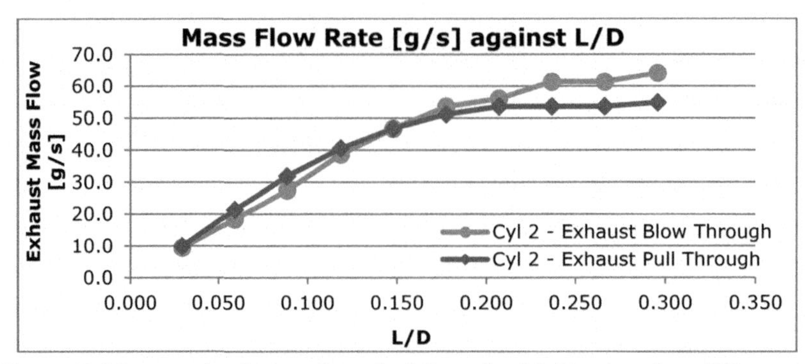

Figure 8. The graph of mass flow [g/s] against L/D, showing difference between blow through and pull through.

The 2.0HDi cylinder head ports are known to be relatively identical between cylinders. On the other hand however, the OEM exhaust manifold of the 2.0HDi is known to be unsymmetrical with respect to the manifold outlet. Due to this, it was deemed necessary to flow test each port with the manifold attached to determine the flow characteristic of each runner. Figure 9 shows that the second exhaust runner, which coincidently is the shortest and straightest of the four seems to flow in excess of 7g/s more than the least flowing runner, i.e. that of cylinder 3.

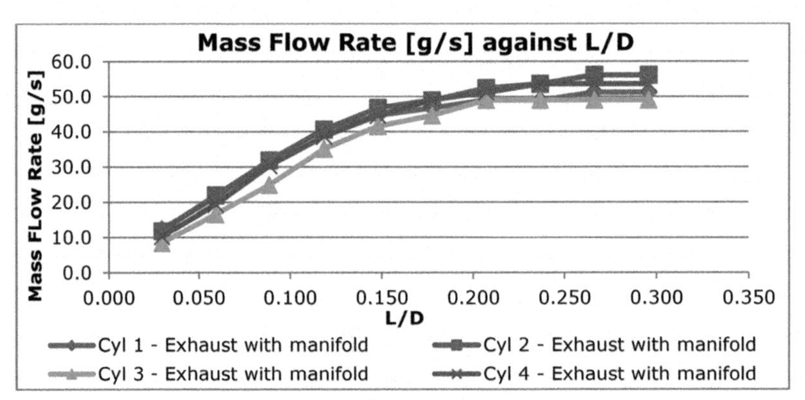

Figure 9. The graph of mass flow rate [g/s] against L/D showing difference between each exhaust runner.

2.2 Verification of flow coefficients through 1D simulation

In obtaining the flow coefficients from the mass flow experimental data, three equations were used and reported below in 1, 2 and 3. Equation 1, derived from Bernoulli's equation assumes an incompressible flow. Both equation 2 from Heywood (10) and equation 3 from Ricardo Knowledge Centre (8) assume a compressible flow. Equation 4 gives the discharge coefficient and is given by Ricardo Knowledge Centre (8). Both the coefficient of flow and coefficient of discharge can be inputted in WAVE. During runtime, the software will automatically detect whether the information given is C_f or C_d, based on the first point inputted in the array.

$$C_f = \frac{\dot{m}}{A_{valve} N_v \sqrt{2\rho\, \Delta P}} \tag{1}$$

$$C_f = \frac{\dot{m}}{A_{valve} N_v P_{up} \frac{1}{\sqrt{RT_{up}}} \left(\frac{P_{atm}-\Delta P}{P_{up}}\right)^{\frac{1}{\gamma}} \left(\frac{2\gamma}{\gamma-1}\right)^{\frac{1}{2}} \left[1 - \left(\frac{P_{atm}-\Delta P}{P_{up}}\right)^{\frac{\gamma-1}{\gamma}}\right]^{\frac{1}{2}}} \tag{2}$$

$$C_f = \frac{4A_{eff}}{\pi D^2} \tag{3}$$

$$C_d = \frac{A_{eff}}{\pi DL} \tag{4}$$

Where

$$A_{eff} = \frac{\dot{V}}{\sqrt{2\left(\frac{\gamma}{\gamma-1}\right) RT_{up}\left[1 - \left(\frac{P}{P_{up}}\right)^{\frac{\gamma-1}{\gamma}}\right]}} \tag{5}$$

To verify the flow coefficient calculation using equation 3, a steady-state model was built in Ricardo WAVE, using an orifice valve with configurable flow coefficient as input (11). This simulated the poppet valve in the experimental flow-bench setup having different flow coefficients at different lifts. The bore-tube adaptor used and the cylinder head port were also modeled, but switching off their friction and discharge coefficient models for the reason mentioned in the previous section. In the one-dimensional steady-state simulation, the flow coefficients obtained from experimental testing and using equation 3 were input as individual cases for different lift values and the mass flow was obtained in return. From this relatively simple check it transpired that the mass flow obtained using 1D simulation was within 2% of that obtained experimentally, as shown in Figure 10. This demonstrates that the flow coefficients calculated were correctly obtained.

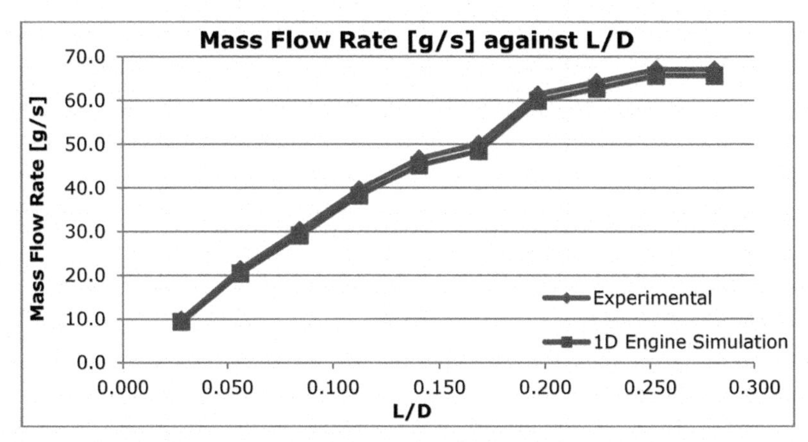

Figure 10. The graph of mass flow [g/s] against L/D comparing the experimental and simulated mass flow on intake valve.

3 RESULTS

In calibrating the pressurized motored one-dimensional model, the experimental data presented in (3) were used. The test matrix for such experimental data entailed engine speeds of 1400rpm, 2000rpm, 2500rpm and 3000rpm with a peak in-cylinder pressure of 84bar and gas compositions of gamma (c_p/c_v) 1.4, 1.5, 1.6 and 1.67, where 1.4 represents that of air and 1.67 represents that of pure argon. For the purpose of the one-dimensional modeling conducted in this study, all setpoints related to air were considered successfully, however the simulation software used did not support Argon and hence experimental data for gamma of 1.5, 1.6 and 1.67 could not be modeled.

To calibrate the one-dimensional model, the data as obtained from experiments were input as constants for the different setpoints tested. Such information includes the engine speed, manifold absolute pressure, shunt pipe intake and exhaust temperatures, valve lifts, valve flow coefficients and the engine geometry. Results like Peak In-cylinder Pressure, Gross IMEP, PMEP, Net IMEP and FMEP were then obtained and compared to the experimental data presented in (3). With such strategy, similar results were obtained, however the discrepancy was still deemed too large for any qualitative deduction to be made.

One initial problem which was troublesome regarded the configuration with which the manifold pressure was being imposed in the one-dimensional simulation. Initially, the manifold pressurization was imposed by connecting an ambient element with forced pressure and temperature equal to the measured MAP and shunt pipe temperature respectively to the exhaust collector y-junction. Results showed that the manifold pressure being imposed on the one-dimensional simulation was resulting in very large deviations in the peak in-cylinder pressure. Such issue was traced down to the ambient condition which was imposing too strict of a boundary condition. To solve such problem the ambient element was removed, and instead the manifold pressure was imposed as an initial pressure condition to the shunt pipe (acting as an initial reservoir) with an additional extra pressure required to fill all ducts with the experimentally obtained MAP, whilst allowing also some flow as blow-by. The initial extra pressure in the shunt pipe was adjusted until the final condition in all manifolds matched that obtained from the experimental MAP sensor to within ± 0.001bar. Later on in this work, such process was automated by using a PID controller responsible of varying the area of an orifice which connected an ambient element to the y-junction of the exhaust collector. With such configuration, better control of the MAP was obtained, whilst still retaining the system detached from the forced ambient element through the controlled orifice.

To understand further the discrepancy between simulation and experimental results, the trapped mass at IVC, obtained on each setpoint from the simulation was compared to that calculated using the ideal gas law equation on the respective experimental data. The trapped mass at IVC was matched to be within 10%. The in-cylinder pressure curve against crank angle was then obtained from the one-dimensional model for each setpoint and plotted against the relevant $p-\theta$ curve from the corresponding experimental data. To be able to compare the two traces, the experimental pressure curve had to be pegged on the intake stroke to the trace obtained from the one-dimensional software. When pegging was attempted, it was seen that an acceptable match was evident between the wave nature of the simulation in-cylinder pressure intake stroke, and that obtained experimentally. Also, the simulation exhaust stroke in-cylinder pressure seemed to show well the recompression on the exhaust displacement phase, visible also in the experimental data. The phasing and magnitude of this recompression however were slightly different. To rectify the issue, an attempt was made to assign 0.3mm of valve lash to each of the intake and exhaust valves. On this amendment, the in-cylinder pressure curve on the intake and exhaust strokes obtained from the simulation matched even better in its wave nature to that obtained from the experimental data. This was noted on all the setpoints tested and hence pegging could be now done not just at one crank angle, but on the whole intake stroke. Figure 11 shows the intake stroke comparison between that obtained from the simulation and the experimental data for two of the setpoints. It should be noted that the simulation model gives the possibility of pegging the experimental data to the in-cylinder pressure generated. Randolph (12) in his publication mentions method 8 of pegging to the ideal gas law equation, but stated that such method does not take into consideration the heat and blow-by losses. However it can now be noted that such issue is catered for when pegging to the simulation in-cylinder pressure.

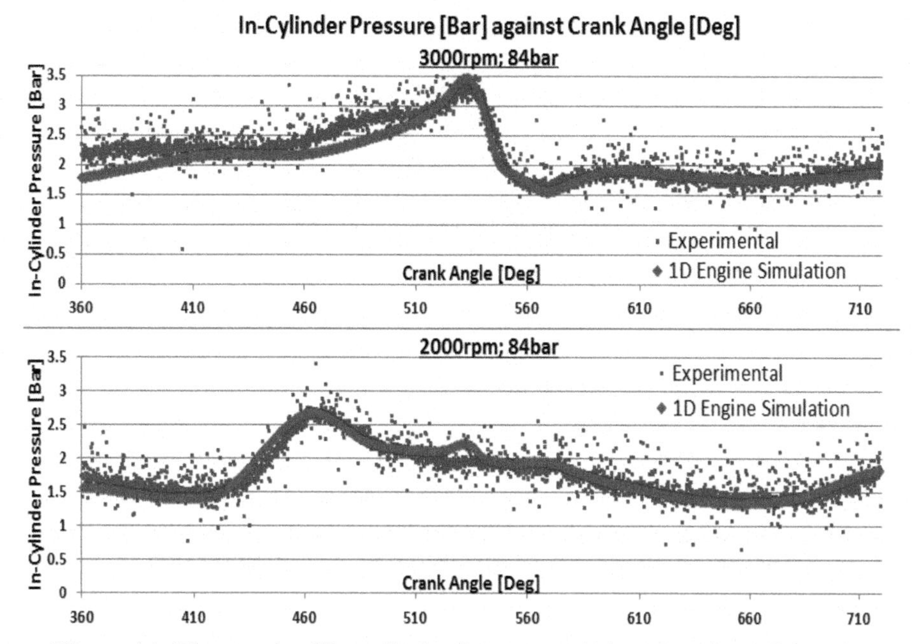

Figure 11. The graph of In-cylinder Pressure against Crank Angle on the intake and exhaust strokes for two different setpoints.

After pegging the traces on the whole intake strokes, the full 720° simulation in-cylinder pressure cycle was compared to that from the corresponding experimental data. It was shown that some discrepancy on the peak in-cylinder pressure was evident. The heat transfer multiplier on closed valve for the heat transfer correlation were tweaked in the hope of getting a match on the peak in-cylinder pressure, however it was seen that when the peak pressure was successfully matched, the in-cylinder pressure during the compression stroke for the simulation fell below that of the experimental pressure trace by a significant amount as seen in Figure 12a. A similar observation was noted on the expansion stroke. Such discrepancy on the compression and expansion strokes led to a 38% discrepancy in the gross IMEP on the setpoint of 3000rpm; 84bar between the simulation and experimental cases. To rectify this issue, the cylinder head, piston and liner temperatures were shifted to higher temperatures than the coolant temperature obtained experimentally by different amounts dependent on the engine speed setpoint, but not higher than the oil temperature of 80°*C*, as imposed experimentally. The higher the engine speed, the higher the temperature required due to a lesser time for heat to flow out of the system. Such amendment seemed to give a better all-around match between the simulation and experimental in-cylinder pressure traces as shown in Figure 12b. Consequently the gross IMEP showed a better match as well. Figure 13 shows the overall comparison between the simulation and experimental pressure traces on the same 3000rpm; 84bar. With such better comparison, the discrepancy in the Gross IMEP on the same setpoint of 3000rpm; 84bar was around 9%.

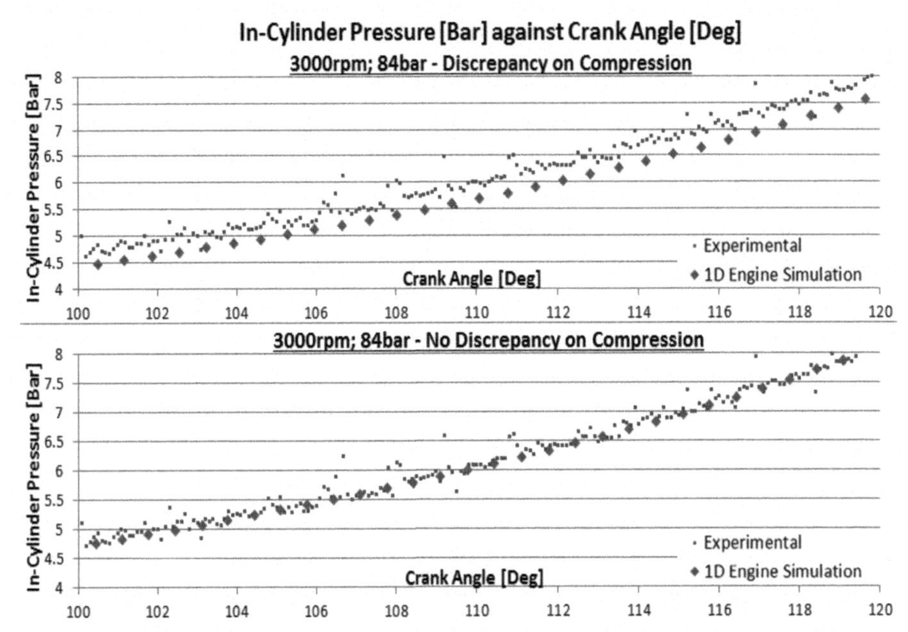

Figure 12. The graph of in-cylinder pressure against crank angle: a) Discrepancy on compression, b) No discrepancy on compression.

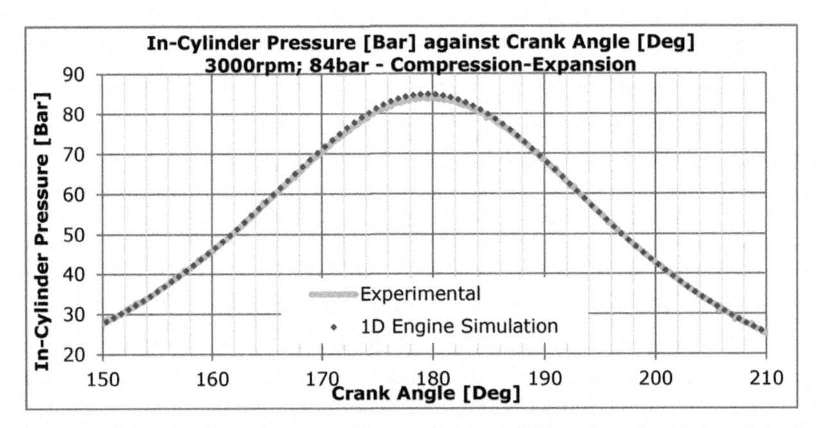

Figure 13. The graph of in-cylinder pressure against crank angle, showing comparison between experiment and simulation data on compression and expansion strokes.

Figure 14 to Figure 17 compare the results obtained from the experiments as reported in (3) to those obtained from the one-dimensional model using Annand's and Woschni's heat transfer model with intake scavenging (8). It can be seen that the IMEP gross matches relatively well between the experimental and simulation results, with a maximum deviation of around 7%. The BMEP also shows a very good match, however such quantity is not a true indicator of how well the simulation model compares to the experimental data. This is because deviations in IMEP gross, PMEP and FMEP having different signs might compensate for each other's deficiencies leading to a mistakenly interpreted good match on the BMEP.

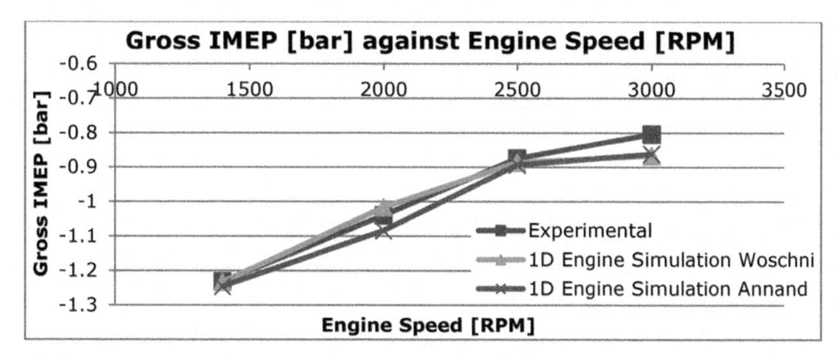

Figure 14. The gross IMEP graph comparing simulation and experimental results.

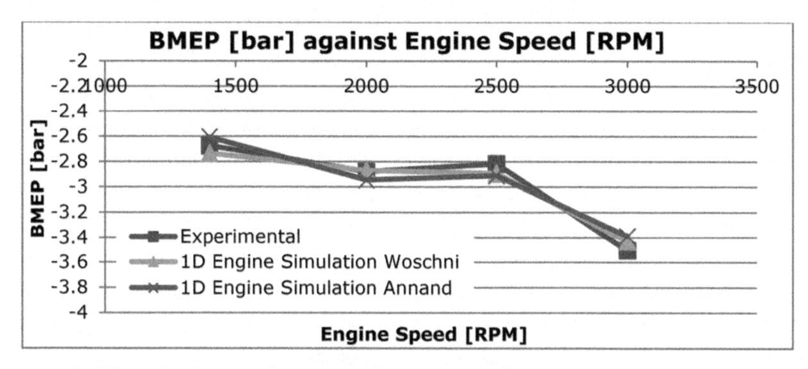

Figure 15. The BMEP graph comparing simulation and experimental results.

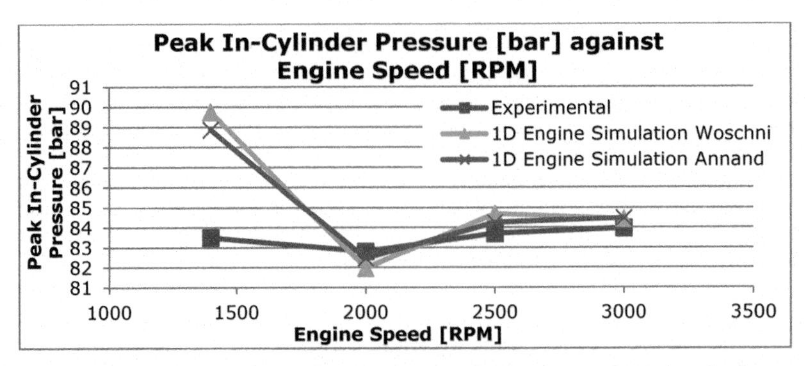

Figure 16. The peak in-cylinder pressure graph comparing simulation and experimental results.

The setpoint which seemed to show the largest deviation from that of the experimental data was the 1400rpm; 84bar. It was noted that this setpoint suffered from an error of 10% on the trapped mass at IVC, which consequently resulted in a higher peak in-cylinder pressure as seen in Figure 16.

Figure 17 shows the PMEP comparison between the simulated and experimentally obtained values. It can be seen that a very good match was obtained on all setpoints, with the 2500rpm and 3000rpm having the largest deviations. Such discrepancy on these setpoints was noted to have originated from a relatively different pressure wave on the exhaust stroke between the simulated and experimental data.

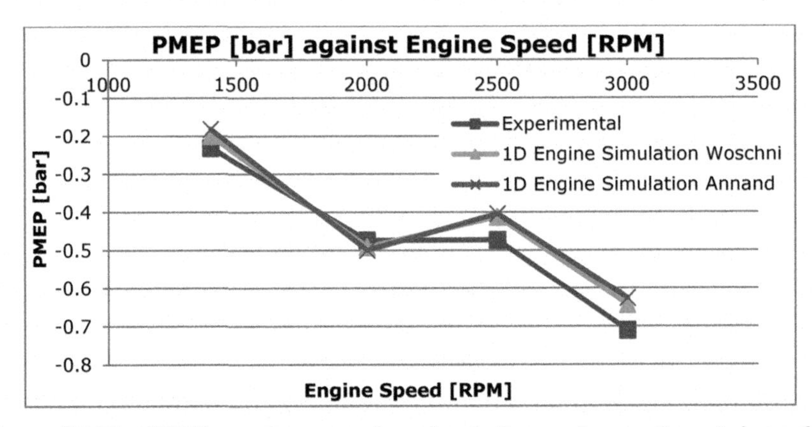

Figure 17. The PMEP graph comparing simulation and experimental results.

3.1 Friction mean effective pressure

As mentioned earlier, the FMEP model used in this one-dimensional study was that by Chen and Flynn (8), given in equation 6. The Chen-Flynn coefficients were found through a regression analysis and documented in Table 2. The FMEP comparison between simulation and experimental data, given in Figure 18 shows an acceptable match with a maximum deviation of 10% and a total average deviation of around

4.5%. The coefficients in Table 2 were inputted in WAVE and used throughout this simulation study.

$$FMEP = A_{CF} + \frac{1}{n_{cyl}} \sum_{i=1}^{n_{cyl}} \left[B_{CF} P_{max_i + C_{CF}} \left(\frac{RPM\,x\,stroke}{2} \right) + Q_{CF} \left(\frac{RPM\,x\,stroke}{2} \right)^2 \right] \tag{6}$$

Table 2. Chen-Flynn Correlation Coefficients.

ACF *[bar]*	**0.5965**
BCF	0.00421
CCF *[Pa.min/m]*	0.0000
QCF *[Pa.min^2/m^2]*	5.4572

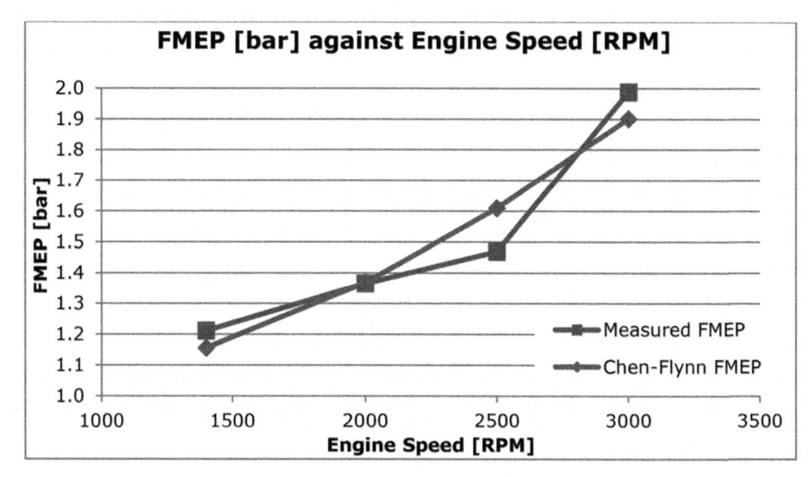

Figure 18. The FMEP graph comparing Chen-Flynn with experimental results.

3.2 Blow-by

As previously shown in Figure 4, a blow-by system was implemented in the one-dimensional model represented by a third orifice in each of the cylinders, as suggested by Ricardo WAVE knowledge centre (8). The orifice diameter was obtained by calibrating the model against the blow-by flow rate measured experimentally and published in (1). To obtain good agreement in blow-by flow rate on all setpoints between the experimental and simulation data, an orifice diameter of approximately 0.6mm was assigned. It was also noted that the peak in-cylinder pressure obtained through the simulation model had a large effect on the relatively small blow-by flow rate, meaning that a small deviation of the peak in-cylinder pressure from the setpoint of 84bar had a significant difference on the blow-by average value.

The blow-by flow rate was also computed on the experimental in-cylinder pressure data using equations 7 and 8 from (9). Figure 19 shows the comparison between the

WAVE generated blow-by flow rate and that obtained from the experimental data, using equations 7 and 8.

For $\frac{p_{crankcase}}{p} > \left[\frac{p_{crankcase}}{p}\right]_{CR} \approx 0.53$

$$\dot{m} = A_{orifice}\sqrt{\frac{2\gamma}{\gamma-1}.m.\frac{p}{V}.\left[\left(\frac{p_{crankcase}}{p}\right)^{\frac{2}{\gamma}} - \left(\frac{p_{crankcase}}{p}\right)^{\frac{\gamma+1}{\gamma}}\right]} \tag{7}$$

For $\frac{p_{crankcase}}{p} < \left[\frac{p_{crankcase}}{p}\right]_{CR} \approx 0.53$

$$\dot{m} = A_{orifice}\sqrt{\gamma.m.\frac{p}{V}.\left(\frac{2}{\gamma+1}\right)^{\frac{\gamma+1}{\gamma-1}}} \tag{8}$$

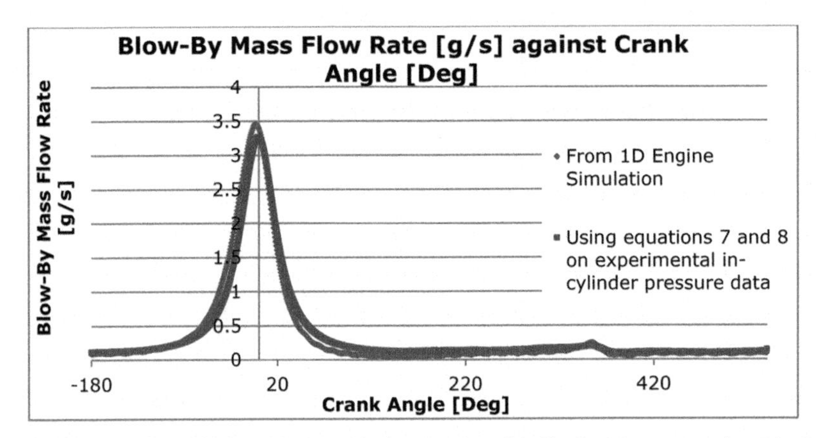

Figure 19. The graph of blow-by flow rate of cylinder 1 comparing that obtained from simulation and that obtained using eqns. 7 and 8 on experimental data - 3000rpm; 84bar.

Figure 20 shows the transient blow-by flow rate for cylinder 1, the total blow-by flow rate as computed by a summation on the four cylinders, and the average total both for the summated and sensor-obtained total flow rate. It can be seen that the blow-by flow rate of each cylinder reflects the in-cylinder pressure trace, meaning that the blow-by flow-rate increases with the increasing in-cylinder pressure on compression stroke. At around 358° crank angle, a small increase in blow-by flow rate is evident, showing an increase in blow-by on the exhaust displacement phase. This is due to the recompression effect. The average values from the summated total and that obtained from the sensor were equal as shown in Figure 20.

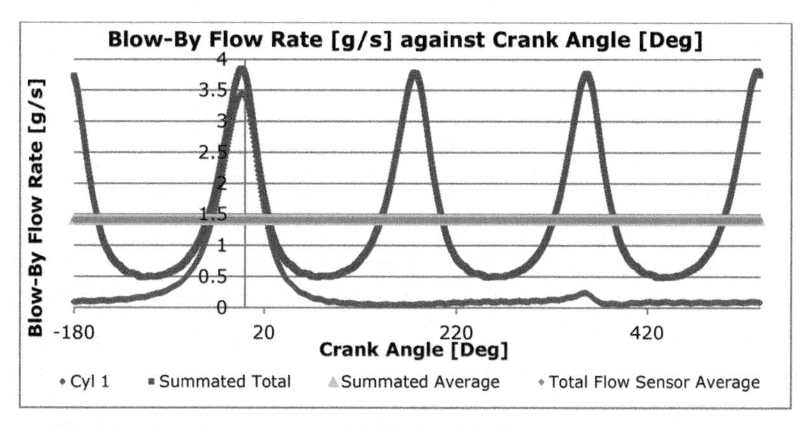

Figure 20. The crank-angle resolved blow-by flow rate graph obtained from simulation for cylinder 1, total, and total average on the setpoint of 3000rpm; 84bar.

In order to investigate the effect of blow-by on the simulation results, the model was re-run but with the blow-by system removed. It was noted that on the 3000rpm, 84bar the peak in-cylinder pressure increased from 84.5bar to 85.5bar. The IMEP gross magnitude decreased from 0.862bar to 0.777bar. Such variations are consistent with theory, since blow-by acts to reduce the peak in-cylinder pressure through leakage, while increases the magnitude of the overall losses. With no blow-by, the location of peak pressure is also expected to be closer to TDC. During this study it was noted that Ricardo WAVE obtains the location of peak pressure by finding the point of $\frac{dP}{d\theta} = 0$ on the polynomial fitted to the generated in-cylinder pressure curve. The crank angle resolution of the generated $p - \theta$ curve in the one-dimensional software was noted to vary with engine speed and was never better than 0.6DegCA. This therefore implies that the location of peak pressure as obtained from the one-dimensional software might be in disagreement compared to that obtained experimentally (which had a resolution of 0.1DegCA). At the 3000rpm; 84bar with the blow-by model activated, the one-dimensional model predicted a loss angle of -0.43DegCA, whereas with the blow-by model de-activated, the loss angle was estimated as -0.37DegCA. The experimental results gave a loss angle of -0.3DegCA on the same setpoint.

3.3 Heat transfer

In this one-dimensional simulation, the heat transfer sub-models used were that by Annand (6) and Woschni (7). The multiplier on closed valve which was found to give a sensible compromise on the peak in-cylinder pressure was around 0.42 for Annand and 1.2 for Woschni on all the setpoints considered in this work. It should be noted that Annand's model was initially developed for a fired engine, and hence using this model for a pressurized motoring setup might be driving the model into unfavorable conditions. As said earlier, to have obtained an in-cylinder pressure trace which compares well to that obtained from experiments, a compromise had to be found between the heat transfer model, the cylinder head, piston and liner temperatures, compression ratio and trapped mass. It should be said that the compression ratio stated by the OEM for the engine tested is 18:1. This value was confirmed through a static compression ratio measurement using paraffin. It is however known that CR values do vary from that given in the engine manual (9). In fact, even if just the difference in thermal expansion is taken into consideration, for an engine with an aluminum cylinder block and steel connecting rod, when the connecting rod is at a temperature of $120°C$ and the block is at $90°C$, the static CR changes from 18:1 to 17.7:1. The engine used in this study does not have an aluminum cylinder block (cast iron block), but nonetheless the mechanical loading effect (due to in-cylinder pressure) on CR would still be present. The value of CR which was found to give the best compromise in this simulation study was that of 17.3:1.

Having used these variables, i.e. heat transfer, temperatures, CR and trapped mass to obtain reasonable pressure-theta curves, it was noted that the temperatures that had to be assigned to the cylinder head, piston and liner might have been different than the real temperature values, had they been measured. This means that the deficiency in the heat transfer models in predicting accurately the heat transfer in the pressurized motored engine might have had to be compensated for by fictitious temperatures in the cylinder head, piston and liner. To support such hypothesis, Figure 21 shows the in-cylinder temperature given by the simulation software and the heat transfer on the same cylinder as computed through Annand's and Woschni's model. According to experimental and analytical works by authors such as Lawton (13), Wendland (14) and Pinfold (15), the peak heat transfer in the cylinder should occur significantly earlier than the peak in-cylinder temperature because of pressure work in the boundary layer. Figure 21 however does not capture such effect, which therefore might consolidate the theory suggested earlier that the temperatures assigned to the cylinder head, liner and piston might have been compensating for deficiencies in the heat transfer models.

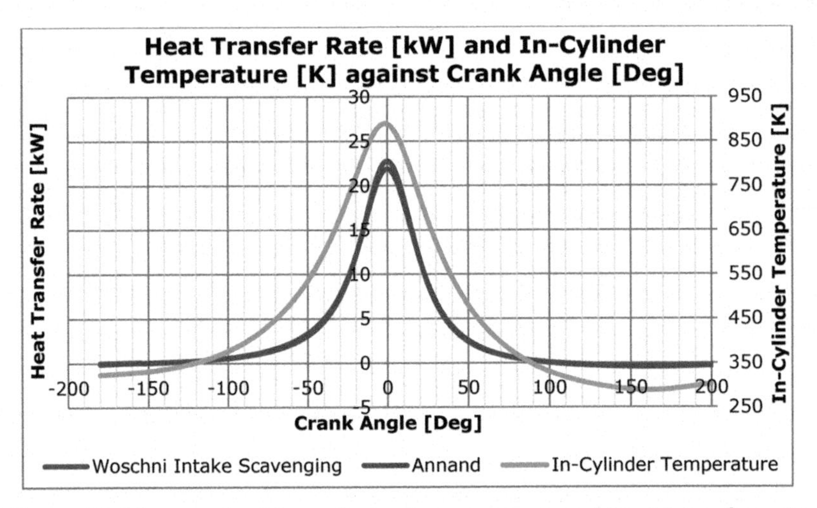

Figure 21. The graph of In-cylinder temperature and heat transfer rate obtained through Annand's and Woschni's correlation.

To further investigate such result, the Thermodynamics first law was considered on the system for the time in which valves are closed, i.e. compression and expansion strokes, as suggested by Heywood (16). Since no combustion is present in the setup being considered, the system resolves into equation 9, which further simplifies to equation 10, where mass flow due to blow-by is neglected. It should be said that since this equation does not account for blow-by, but the in-cylinder pressure on which such equation is used is reduced by the effect of blow-by, the heat transfer result obtained from equation 10 will be artificially greater to account for the effect of blow-by as heat lost. Equation 10 was used on the ensemble average experimental in-cylinder pressure data and plotted in Figure 22 for the 3000rpm; 84bar. It can be noted that due to small serrations in the in-cylinder pressure data, the term $\frac{dP}{dt}$ in equation 10 suffers from large magnitudes of interference. Due to this, a moving average filtering scheme was implemented both on the pressure trace and the heat transfer curve. It was also made sure that with this filtering scheme the shift in the pressure trace was not significant. This can be seen in Figure 23.

To obtain a better heat transfer result, equation 11 presented by Pipitone (9) was used which assigns an additional term to equation 10 to include the effect of blow-by, hence eliminating the error in the result of the heat transfer rate. The result from this equation is also shown in Figure 22.

$$-\frac{dQ}{dt} = p\frac{dV}{dt} + \frac{dU}{dt} \tag{9}$$

$$-\frac{dQ}{dt} = \frac{\gamma}{\gamma - 1}.p.\frac{dV}{dt} + \frac{1}{\gamma - 1}.V.\frac{dP}{dt} \tag{10}$$

$$-\frac{dQ}{dt} = \frac{\gamma}{\gamma - 1}p.\frac{dV}{dt} + \frac{1}{\gamma - 1}.V.\frac{dP}{dt} - \frac{\gamma}{\gamma - 1}.\frac{pV}{m}.\frac{dm}{dt} \tag{11}$$

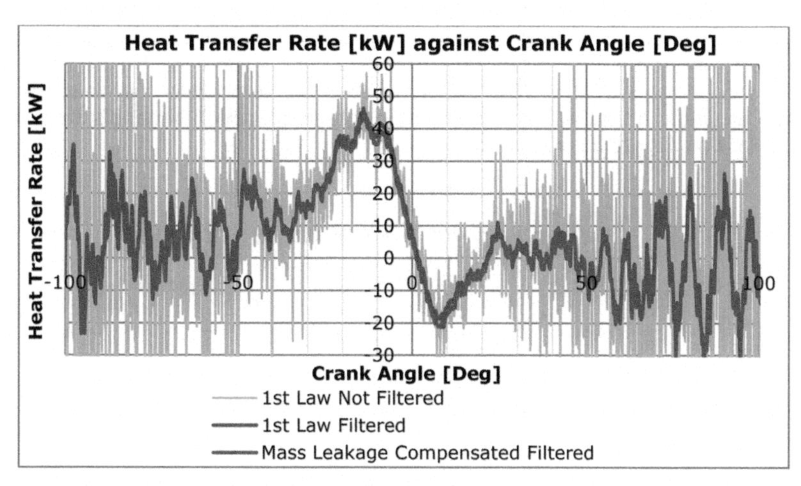

Figure 22. The graph of heat transfer rate against crank angle – 3000rpm; 84bar.

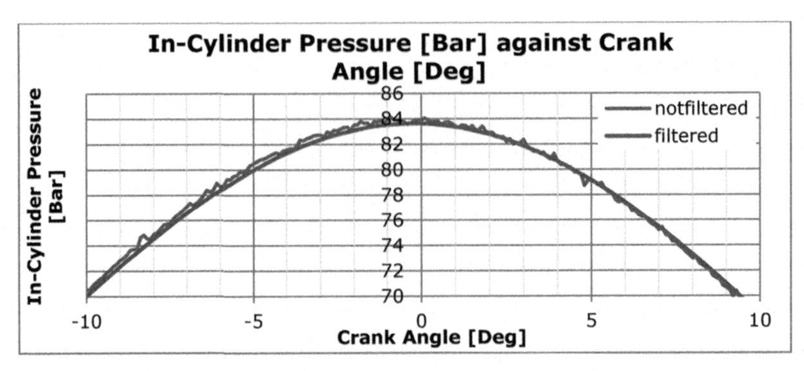

Figure 23. Comparison graph between filtered and unfiltered experimental in-cylinder pressure data – 3000rpm; 84bar.

Equations 10 and 11 shows that the ratio of specific heats is required in the calculation of heat transfer. It is known that the ratio of specific heats vary significantly with temperature, as shown in Figure 24 for the same setpoint of 3000rpm; 84bar. Equations 10 and 11 were used with a varying gamma according to Figure 24. The heat transfer obtained in Figure 22 shows a good match to work published by the previously mentioned authors; Lawton (13), Wendland (14) and Pinfold (15), where the heat transfer from the cylinder peaks at an angle of around 11DegCA before peak in-cylinder temperature and drops down to below zero shortly after TDC, when the temperature is still very close to its maximum value. On the other hand, such result is detached from that of Figure 21 which was obtained through simulation by using Annand's and Woschni's correlations, where the peak heat transfer occurs very close to TDC and no heat flow reversal was shown just after TDC. To show better the comparison between the heat transfer calculated from the first law and that obtained through Annand's and Woschni's correlations, Figure 25 was plotted.

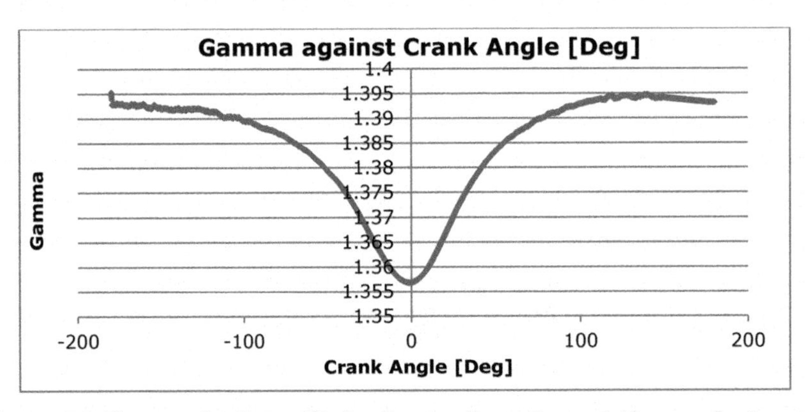

Figure 24. The graph of specific heat capacity ratio variation against crank angle.

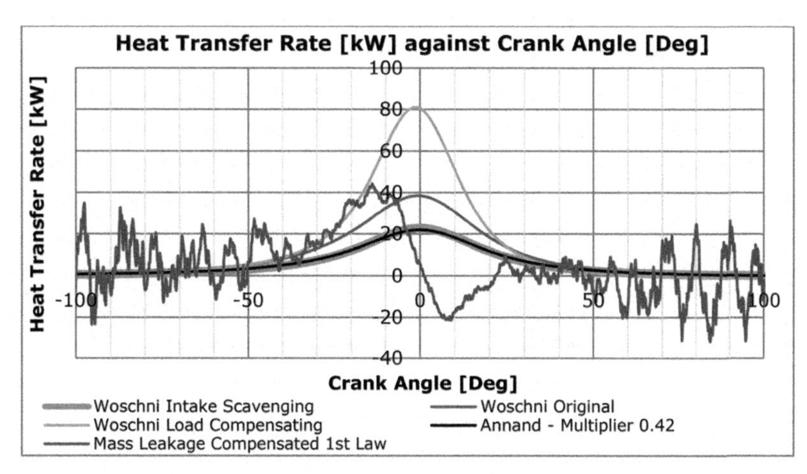

Figure 25. The graph of heat transfer rate against crank angle, comparing different models to the calculated heat transfer – 3000rpm; 84bar.

The heat transfer computation from equations 10 and 11, requires the in-cylinder volume and its rate of change. In turn, to obtain the volume from the crank-slider equation, the compression ratio is required. It had been said previously that usually the true CR for an engine varies from that quoted by the OEM due to tolerances of manufacturing, thermal expansions and cyclic loadings in the engine structure. The sensitivity of the heat transfer determined from equations 10 and 11 due to CR used in the crank-slider equation was investigated. Figure 26 was plotted for the 3000rpm; 84 bar with a CR of 17.3:1. Variations of ±5% and ±10% from 17.3:1 were intentionally imposed on the same experimental in-cylinder pressure data to obtain the resulting heat transfer rate, as given also in Figure 26. It can be seen that for 5% variation on a CR of 17.3:1, an error of around 48% in the peak heat transfer rate resulted. Also, biasing further the CR would result in the loss of the early peak heat transfer and negative heat transfer characteristic just after TDC.

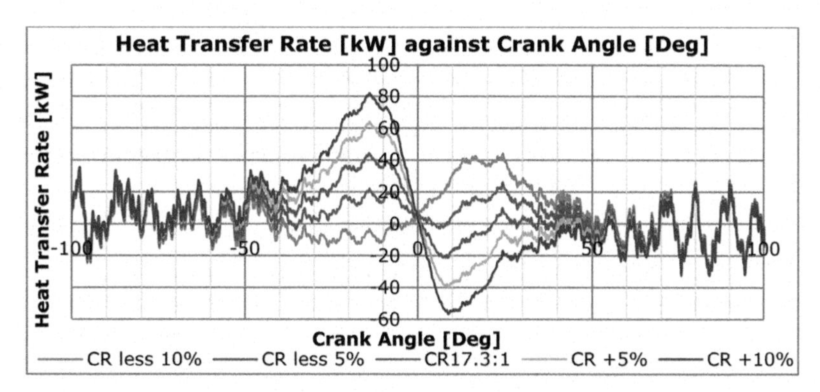

Figure 26. The graph of heat transfer rate against crank angle showing difference in heat transfer due to CR uncertainty – 3000rpm; 84bar.

3.4 Trapped mass

On the experimental pressurized motoring setup, no measurement of gas mass flow was conducted. After calibrating the one-dimensional simulation to the experimental in-cylinder pressure data, the trapped mass was obtained as shown in Figure 27 for the case of 3000rpm; 84bar, where the cycle starts with the compression stroke, and compared to that calculated from experimental values. It should be said that in the calculation on experimental values, since the gas temperature can only be predicted during closed valves period, the gas temperature during the intake and exhaust strokes was taken to be equal to the measured shunt pipe temperature. During closed valves period, the trapped mass was calculated using the initial trapped mass estimated at IVC, and decreasing the mass flowing out as blow-by per crank angle. On the other hand, during gas exchange periods, the entrapped mass had to be calculated using the ideal gas law equation.

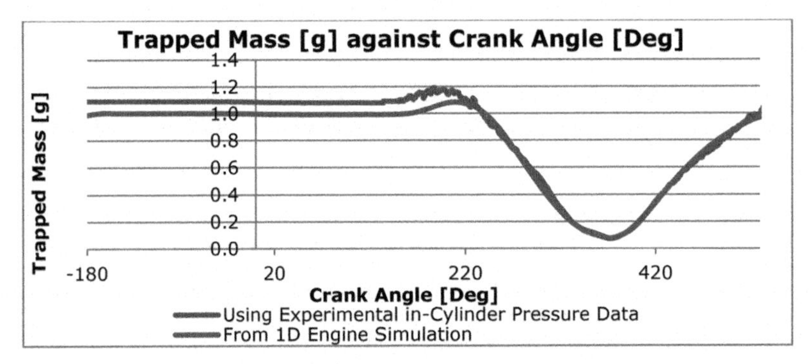

Figure 27. The graph of trapped mass against crank angle for the 3000rpm; 84bar.

4 DISCUSSIONS AND CONCLUSIONS

As stated earlier in this publication, the main aim of this work was to use the experimental data obtained in the previous work (3) to develop and calibrate a one-dimensional model, with the hope of identifying any shortcomings in the experimental data, or simulation strategies. The results obtained confirm that the one-dimensional simulation software is able to model the pressurized motoring setup successfully when operating with air as the pressurization gas. The quantity and quality of the experimental data obtained in (1) (2) (3) seem to provide adequate information for the one-dimensional analyst to calibrate a one-dimensional model.

As discussed in the previous section of heat transfer and namely Figure 25, the heat transfer models still used in one-dimensional simulation were found to deviate from the experimentally obtained heat transfer and are not able to capture the early peak in heat transfer out of the cylinder and the negative heat transfer just after TDC. It was shown that the characteristic where the heat transfer from the cylinder peaks at an angle of around 11DegCA before peak in-cylinder temperature and drops down to below zero shortly after TDC, (when the temperature is still very close to its maximum value) is in agreement with work published previously (Lawton (13), Wendland (14) and Pinfold (15)). However this characteristic is not in agreement with that obtained through simulation by using Annand's and Woschni's correlations (where the peak heat transfer occurs very close to TDC and no heat flow reversal was shown just after TDC).

The FMEP model given by Chen and Flynn (8) showed acceptable agreement with experimental data.

The relatively good agreement between one-dimensional results and those obtained experimentally also puts more confidence in the experimental data obtained from the pressurized motoring setup.

The blow-by simulation proved to be relatively straight forward and gave successful results, which agreed with that obtained experimentally. Such model resulted in a slight increase in computational time, which for large DOEs might be significant. From the simulation it was found that blow-by has a small contribution to the loss angle and peak-pressure magnitude when compared to heat loss.

The simulation generated in-cylinder pressure can be used as an intelligent filter to the experimental in-cylinder pressure, if well calibrated. It should however be noted that due to the deficiency of the heat transfer models which are imposed on the in-cylinder pressure generated from simulation, some deviations will surely be present. The simulation in-cylinder pressure generated was also found to be a relatively good pegging reference for the experimental in-cylinder pressure traces.

5 SUGGESTIONS FOR FURTHER WORK

Having calibrated a one-dimensional model with the experimental data obtained in previous work gave the authors the chance to experience the requirements from experimental data from the point of view of the one-dimensional analyst. This showed that the experimental data obtained was adequate; however one quantity which would have been useful had it been measured was the trapped mass, or the mass air flow into the engine. Such quantity was required when calibrating the intake stroke, but since such experimental data was not available, a simple calculation using the ideal gas law had to be done instead. Such result shows that if the mass flow into the cylinders would be measured in future experiments; a better correlated model can be achieved.

As was discussed, none of the available heat transfer models were able to capture adequately the heat transfer from the engine. Therefore a user-defined heat transfer model, which is easily implemented, could have resulted in much better results from the simulation software. If such difficulty is overcome, future work can include the development of a user-defined model based on equation 11 which would be able to give the crank-resolved heat transfer.

To obtain better confidence in the one-dimensional model, the cylinder head, piston and liner temperatures can be compared to that achieved experimentally had such values been available from experimental sessions. The pressurized motoring setup at University of Malta is being modified at the moment to include two eroding type, transient surface thermocouples which is hoped to address the heat transfer analysis experimentally.

ACKNOWLEDGEMENTS

Mr. Andrew Briffa is thanked for his assistance in setting up the flow bench used in this research.

The research work disclosed in this publication is partially funded by the Endeavour Scholarship Scheme (Malta). Scholarships are part-financed by the European Union-European Social Fund (ESF)-Operational Programme II–Cohesion Policy 2014-2020 "Investing in human capital to create more opportunities and promote the well-being of society".

LIST OF SYMBOLS

$\dot{m}$	Mass flow rate
A_{valve}	Valve Area
N_v	Number of Valves
ρ	Fluid Density
ΔP	Difference in pressure
P_{up}	Upstream Pressure
P_{atm}	Atmospheric Pressure
γ	Ratio of Specific Heats
R	Specific Gas Constant
T_{up}	Upstream Temperature
C_f	Flow Coefficient
C_d	Discharge Coefficient
D	Diameter
L	Lift
V	Volumetric Flow Rate
P	Downstream Pressure
P_{max}	Downstream Pressure
Q	Heat Energy
$\dot{V}$	Volume
U	Internal Energy
m	Mass
$p_{crankcase}$	Crankcase Pressure
$A_{orifice}$	Orifice Area

REFERENCES

[1] Caruana C, Farrugia M, Sammut G. The Determination of Motored Engine Friction by Use of Pressurized 'Shunt' Pipe between Exhaust and Intake Manifolds. SAE Technical Paper 2018-01-0121. 2018.

[2] Caruana C, Farrugia M, Sammut G, Pipitone E. Further Experimental Investigation of Motored Engine Friction Using Shunt Pipe Method. SAE Technical Paper 2019-01-0930. 2019.

[3] Caruana C, Pipitone E, Farrugia M, Sammut G. Experimental investigation on the use of Argon to improve FMEP determination through motoring method. SAE Technical Paper 2019-24-0141, SAE Naples ICE2019 14th International Conference on Engines and Vehicles, Capri, Napoli. 15th - 19th September, 2019.

[4] Camilleri S. Investigation of Common Rail Diesel Engine. Msida: Undergraduate Dissertation, University of Malta; 2010.

[5] Caruana C, Azzopardi JP, Farrugia M, Farrugia M. Common rail diesel engine, fuel pressure control scheme and the use of speed — Density control. In IEEE 25th Mediterranean Conference on Control and Automation (MED); 2017; Valletta, Malta. ISBN: 978-1-5090-4533-4. p. 201–216.

[6] Annand WJA. Heat Transfer in the Cylinders of Reciprocating Internal Combustion Engines. Proc. Inst. Mech. Engrs. 1963; 177(1).

[7] Woschni G. A Universally Applicable Equation for the Instantaneous Heat Transfer Coefficient in the Internal Combustion Engine. SAE Technical Paper 670931. 1967.

[8] Ricardo. WAVE User Manual. Ricardo; 2017.

[9] Pipitone E, Beccari A. Determination of TDC in internal combustion engines by a newly developed thermodynamic approach. Applied Thermal Engineering. 2010;: p. 1914–1926.
[10] Heywood JB. Appendix C: Equations for Fluid Flow through a Restriction. In Heywood JB. Internal Combustion Engine Fundamentals.: McGraw-Hill; 1988. p. 906–909.
[11] Farrugia M. FSAE: Engine Simulation with WAVE. Michigan: Oakland University; 2004, https://software.ricardo.com/resources/fsae-engine-simulation-with-wave.
[12] Randolph A. Methods of Processing Cylinder-Pressure Transducer Signals to Maximise Data Accuracy. SAE Technical Paper, 900170. 1990.
[13] Lawton B. Effect of Compression and Expansion on Instantaneous Heat Transfer in Reciprocating Internal Combustion Engines. Proc. Instn. Mech. Enginrs., Part A, Journal of Power and Energy. 1987; 201: p. 175-186.
[14] Wendland DW. The Effect of Periodic Pressure and Temperature Fluctuations on Unsteady Heat Transfer in a Closed System. Wisconsin:; 1968.
[15] Annand WJD, Pinfold D. Heat Transfer in the Cylinder of a Motored Reciprocating Engine. SAE Technical Paper 800457. 1980.
[16] Heywood JB. Combustion in Compression Ignition Engines. In Heywood JB. Internal Combustion Engine Fundamentals.: McGraw-Hill; 1988. p. 508–5510.

SESSION 8: DESIGN AND DEVELOPMENT OF INTERNAL COMBUSTION ENGINES

Internal Combustion Engines and Powertrain Systems for Future Transport 2019 – Institute of Mechanical Engineers, ISBN 978-0-367-90356-5

Combustion system development of the second generation 'New Engine' family for the Chinese market

B. Waters, M. Joyce, Y. Liu & X. Zhang

Changan UK R & D Centre Limited, UK & Changan Auto China

ABSTRACT

This paper presents an overview of Changan's modular 3 and 4 cylinder gasoline New Engine for the 2020s. The core design exhibits compact dimensions, low friction and a high efficiency combustion system which is described with its specification and the resultant attribute performance of the launch variant. The modular nature of the engine permits low cost and rapid development of additional variants to meet market requirements.

Following a successful handover of the first 1.4 litre base engine to Changan HQ in Chongqing, the UK R&D centre has been focused on future derivatives of the engine. The drivers for engine evolution are improved fuel consumption and emissions as well as the ability to meet the needs of electrified powertrains.

The second half of the paper focuses on combustion system development of the second-generation New Engine and the areas addressed include combustion chamber and intake port design. Use of early and late inlet valve closing as a means to increase compression ratio has been investigated along with a range of boosting technologies. The engine design and hardware specified are chosen to meet the requirements of the challenging Chinese vehicle market. Finally, the performance of this second-generation engine are assessed for fuel economy and performance at a range of conditions which include full load, part load and catalyst heating conditions.

1 NEW MODULAR ENGINE

1.1 Introduction

Changan have responded to the needs of the market and legislation to design and develop a new modular range of gasoline engines for the 2020s. This engine will replace the majority of Changan's existing engines across 5 existing engine families covering the performance range 70 to 140kW. This will dramatically reduce the complexity of engine types and vehicle installations whilst increasing production volumes of core components, both of which reduce production and development costs. The rationalisation of the engine line up enabled by the New Engine (NE) can be seen in Figure 1.

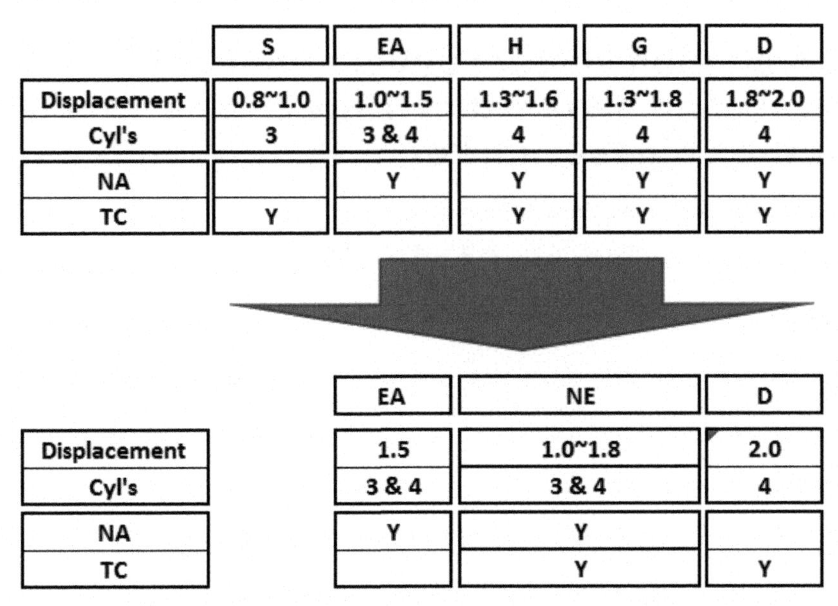

Figure 1. Changan current & future engine families.

Key areas of focus for the engine family were fuel economy, tail pipe PN emissions, efficient catalyst heating, competitive performance, good NVH characteristics as well as compact and lightweight design. To realise these a "clean sheet of paper" engine has been conceived to deliver a wide range of engine variants from a common architecture using alternate technology modules.

All variants use a set of core technologies which support the desired engine attributes; long stroke/small bore, 8mm crank offset, 4 valves per cylinder, integrated exhaust manifold, aluminium block and head, roller finger follower (RFF) valvetrain, dual VVT, chain driven double overhead camshafts etc.

Based on these potential engine variants, component re-use strategies were developed using a complexity matrix to ensure high use of common components. Commonality levels as high as 93% are achieved between variants of the same technology with an alternate displacement but are still in excess of 60% between 3 and 4 cylinder variants. In addition, by standardising interfaces both within the engine and for connections to the vehicle, the effort to design alternate modules is dramatically reduced. Figure 2 illustrates the complexity matrix for the engine family with 10 conventional combustion variants; NA & TGDI with 5 displacements, two 3 cylinder and three 4 cylinder derivates. It can be seen there is high common use of sub-systems and components across the range of variants.

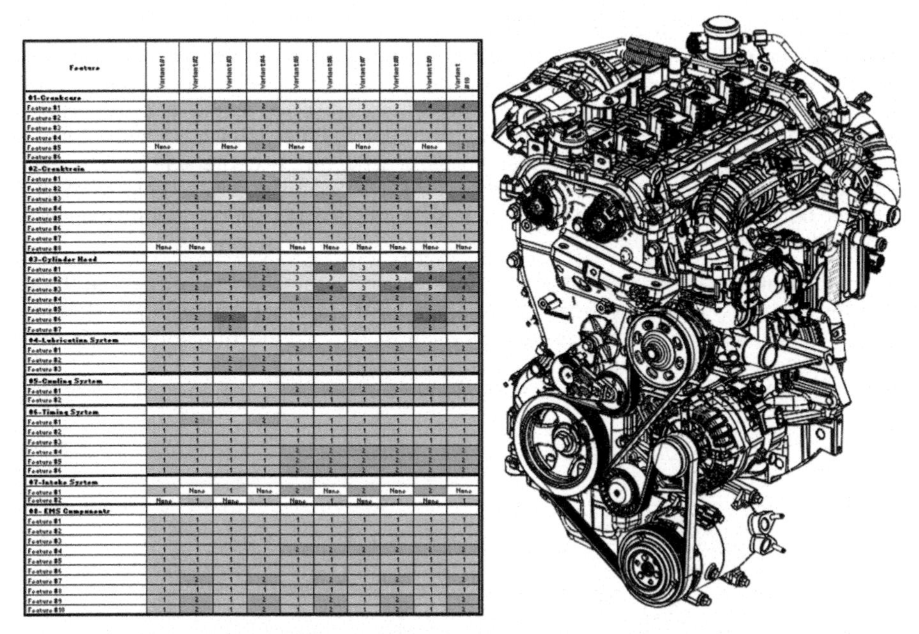

Feature	Variant#1	Variant#2	Variant#3	Variant#4	Variant#5	Variant#6	Variant#7	Variant#8	Variant#9	Variant #10
#1-Crankcase										
Feature #1	1	1	2	2	3	3	3	3	4	4
Feature #2	1	1	1	1	1	1	1	1	1	1
Feature #3	1	1	1	1	1	1	1	1	1	1
Feature #4	1	1	1	1	1	1	1	1	1	1
Feature #5	None	1	None	2	None	1	None	1	None	2
Feature #6	1	1	1	1	1	1	1	1	1	1
#2-Cranktrain										
Feature #1	1	1	2	2	3	3	4	4	4	4
Feature #2	1	1	2	2	3	3	2	2	2	2
Feature #3	1	2	3	4	1	2	1	2	3	4
Feature #4	1	1	1	1	1	1	1	1	1	1
Feature #5	1	1	1	1	1	1	1	1	1	1
Feature #6	1	1	1	1	1	1	1	1	1	1
Feature #7	1	1	1	1	1	1	1	1	1	1
Feature #8	None	None	1	1	None	None	None	None	None	None
#3-Cylinder Head										
Feature #1	1	2	1	2	3	4	3	4	5	4
Feature #2	1	1	2	2	3	3	3	3	4	4
Feature #3	1	2	1	2	3	4	3	4	5	4
Feature #4	1	1	1	1	2	2	2	2	2	2
Feature #5	1	1	1	1	1	1	1	1	2	1
Feature #6	1	2	3	2	1	2	1	2	3	2
Feature #7	1	1	2	1	1	1	1	1	2	1
#4-Lubrication System										
Feature #1	1	1	1	1	2	2	2	2	2	2
Feature #2	1	1	2	2	1	1	1	1	1	1
Feature #3	1	1	2	2	1	1	1	1	1	1
#5-Cooling System										
Feature #1	1	1	1	1	2	2	2	2	2	2
Feature #2	1	1	1	1	1	1	1	1	1	1
#6-Timing System										
Feature #1	1	2	1	2	1	1	1	1	1	1
Feature #2	1	1	1	1	1	1	1	1	1	1
Feature #3	1	1	1	1	1	1	1	1	1	1
Feature #4	1	1	1	1	2	2	2	2	2	2
Feature #5	1	1	1	1	2	2	2	2	2	2
Feature #6	1	1	1	1	2	2	2	2	2	2
#7-Intake System										
Feature #1	1	None	1	None	2	None	2	None	2	None
Feature #2	None	1	None	1	None	1	None	1	None	1
#8- EMS Components										
Feature #1	1	1	1	1	1	1	1	1	1	1
Feature #2	1	1	1	1	1	1	1	1	1	1
Feature #3	1	1	1	1	1	1	1	1	1	1
Feature #4	1	1	1	1	2	2	2	2	2	2
Feature #5	1	1	1	1	1	1	1	1	1	1
Feature #6	1	1	1	1	1	1	1	1	1	1
Feature #7	1	2	1	2	1	2	1	2	1	2
Feature #8	1	1	1	1	1	1	1	1	1	1
Feature #9	1	2	1	2	1	2	1	2	1	2
Feature #10	1	2	1	2	1	2	1	2	1	2

Figure 2. Engine family complexity matrix – conventional combustion applications.

A wide range of optional technology modules were considered during the engine concept design to allow for all needed future variants in the range 1.0l 3cyl to 1.6l 4cyl as illustrated in Figure 3. Some of these modules are for Miller combustion applications and hybrid vehicle applications.

System	Module/feature	Options			
Cylinder Block & Cranktrain	Bore	73.5mm	76mm		
Cylinder Block & Cranktrain	Stroke	78.5mm	82mm	88mm	
Cylinder Head	Inlet Port & Chamber	Naturally Aspirated	Turbocharged	Miller	
Cylinder Head	Inlet Valve	27.3mm	29mm		
Cylinder Head	Camshaft profiles & springs	Naturally Aspirated	Turbocharged	Miller	
Cylinder Head	Variable valve lift	None	Inlet cam profile switch	Cylinder de-activation	
Lubrication System	Oil Pump	2-step flow	Variable Flow	Variable flow with balanceshaft	
Lubrication System	Piston Cooling Jets	None	Pressure switched	ECU controlled	
Lubrication System	Oil Cooler	None	Oil/coolant type		
Cooling system	Coolant Pump	Mechanical - low flow	Mechanical - high flow	Electric	
Cooling system	Flow control	Twin thermostats	Ported Rotary valves		
Intake System	Inlet manifold	Compact plastic	Resonant tuned plastic		
Exhaust System	Turbocharger	3 cyl mono scroll	4 cyl mono scroll	4 cyl twin scroll	Variable Geometry
Fuel System	Injector, rail, pump	PFI	350 bar GDI		
EGR system	Valve, cooler, pipes	None	Low Pressure		
FEAD	Belt, tensioner, accessories	None (hybrid)	Single belt	Twin belt (B-ISG)	

Figure 3. Optional technology modules.

1.2 Engine structure

The main structure of cylinder block and cylinder head is all aluminium; high pressure die cast in the case of the block and low pressure die cast for the cylinder head.

The cylinder block is a deep skirt design with the water pump mounted to the inlet side and oil pump mounted under the crankshaft. Oil pump drive is by a wet belt for variants without a balance-shaft and by gear for 3 cylinder variants with an integrated oil pump/balance assembly. Camshaft drive is by a single chain common to all variants. The windage tray on the 4 cylinder is also structural, stiff-ening the bearing caps. The cylinder bores are honed with the use of a stress plate helping to realise low bore distortion. The crankshaft is offset 8mm from the cylinder bores, this amount being chosen as a balance between friction reduction and block height.

The crankshaft is cast iron on all applications with 4 counter-weight design on the 4 cylinder variants. The 3 cylinder crankshafts have different balancing strategies dependent on whether a balance-shaft is used or not. Connecting rods are forged steel and pistons cast alloy. The piston rings are relatively thin at 1.2/1.0/2.0mm and with a low tangential load for low friction and enabled by the good shape of the bores.

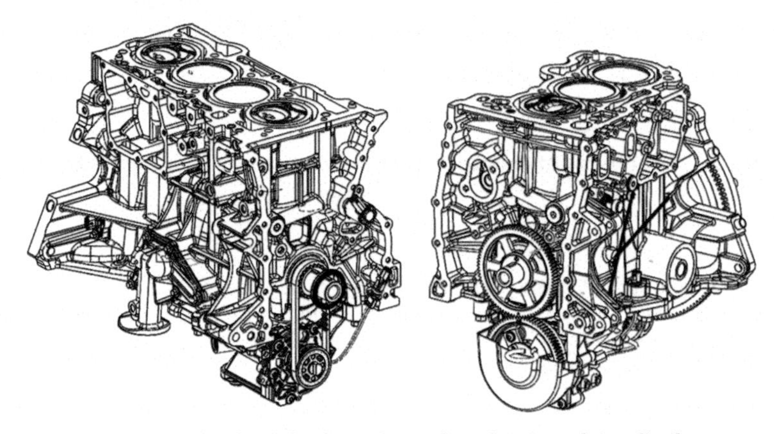

Figure 4. Cylinder Block and cranktrain; 3 and 4 cylinder.

The cylinder head has common major dimensions across all variants and the major-ity of the valvetrain is common to all variants. The cylinder head layout is a classic 4 valve per cylinder, roller finger follower layout with the inlet cam rockers out-board and the exhaust cam rockers in-board for packaging reasons. The included valve angle is relatively narrow at 33° supporting a compact combustion chamber and high compression ratios. All cylinder heads feature an integrated exhaust manifold (IEM), in the case of the 3 cylinder with a 3-1 layout and in the case of the 4 cylinder with a 4-2 layout. Each has been chosen for optimum gas dynamic matching with the turbocharger and in the case of the 4 cylinder supports the use of a twin-scroll turbo-charger use on some variants.

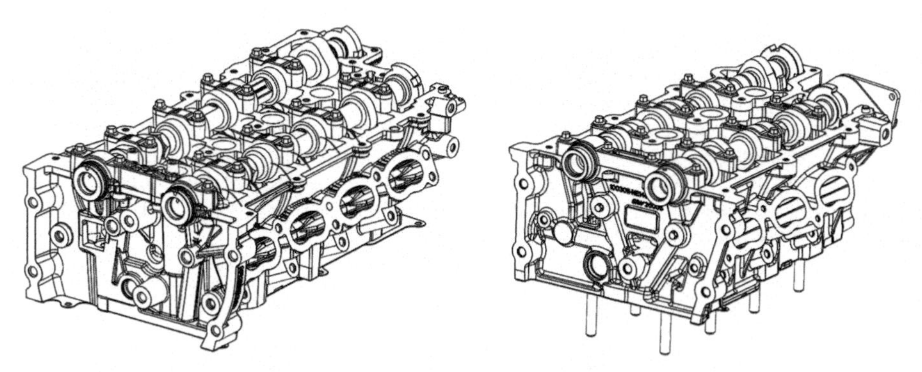

Figure 5. Cylinder head assembly; 3 and 4 cylinder.

1.3 Cooling system

The cooling system is a hybrid longitudinal/cross-flow layout, the layouts are conceptually the same on 3 and 4 cylinder although the detail is varied due to the different IEM structures. The system is split between cylinder block and cylinder head. Two flow control modules (FCM) were developed for the family; a passive arrangement with two thermostats (block & head) with different temperature characteristics and an active arrangement with a rotary control valve powered by a DC motor under the control of the engine control module (ECM). The flow control modules fit to the rear of the cylinder head and are common between 3 & 4 cylinder engines. The water pump mounts to the cylinder block. Two modules have been developed, a mechanical belt driven pump with 2 alternate flow rates and an electric pump. The use of the FCM's is beneficial for fast engine warm up, both for improved fuel economy and the reduction of PN emissions. The selection of which FCM and water pump is based on the engine and vehicle application. The combination of an IEM, low engine mass and FCM results in improvement of 35% in engine warm up from 25 Deg C to 90 Deg C on the NEDC cycle compared to Changan's current engines.

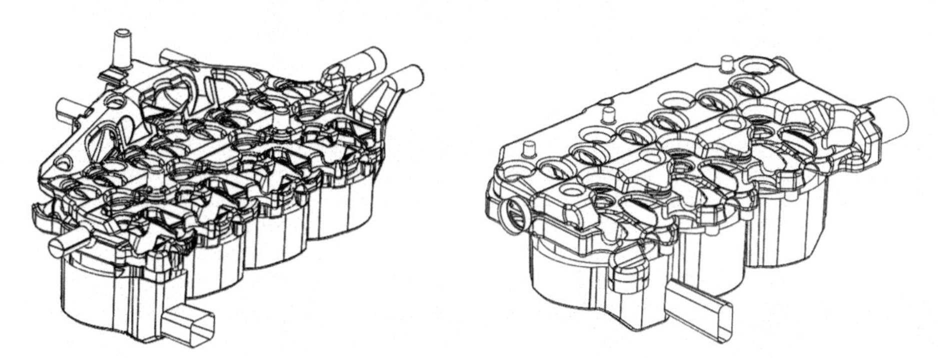

Figure 6. Cooling system layout.

1.4 Lubrication system

The lubrication system is fully integrated with the cylinder block and cylinder head. A fully variable oil pump (FVOP) (vane type) draws oil from the sump and is delivered to the multiple oil consumers of the engine via drilled galleries. An oil cooler can be mounted to the inlet side of the block for applications where this is required. The turbocharged variants of the engine feature a 2nd gallery on the exhaust side of the block for the piston cooling jets (PCJ) which are switched on/off by a solenoid valve. The use of FVOP, optimised for 0W20 oil and switched PCJs also delivers a useful improvement in both fuel economy and PN emissions due to the reduced work by the oil pump and the higher piston temperatures when the PCJs are turned off [1]. Fuel economy is improved by 2~3% on the NEDC cycle compared to current Changan engines with conventional oil pump, pressure switched PCJs and 5W30 oil.

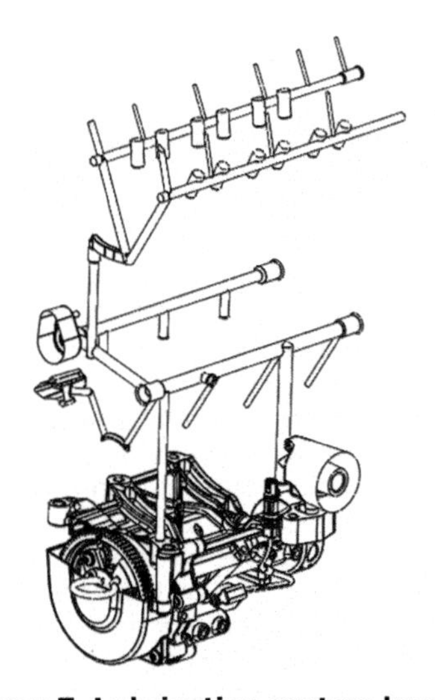

Figure 7. Lubrication system layout.

1.5 Crankcase ventilation

The crankcase ventilation system is highly integrated into the structure of the engine. There are two separators, a part load one mounted to the cylinder block with the flow regulated by a PCV integrated in the separator and a full load one mounted on the exhaust side of the cam cover. At part load, fresh air flows through the engine from the full load separator to reduce the propensity for sludge generation. During boosted operation the flow reverses in the full load breather, which then vents to the air intake upstream of the compressor and the flow in the part load breather reduces to close to zero. The gas passages from part load separator to intake system are fully integrated into the block, cylinder head and inlet manifold with no external pipework. This helps to minimise the risk of icing or sludging.

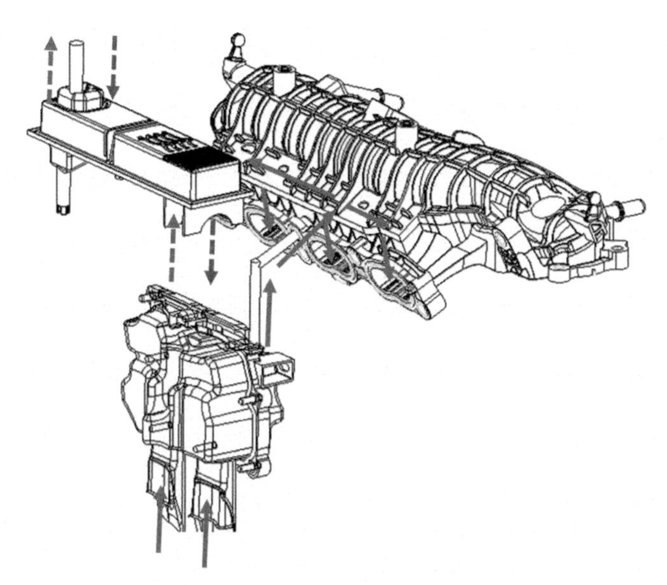

Figure 8. Crankcase ventilation system layout.

1.6 Air path

The air path of the engine consists of the inlet manifold, intercooler, transfer pipe and turbocharger. The intercooler is an air-coolant-air type serviced with a dedicated electric coolant pump and low temperature radiator. The transfer pipe includes a resonator for noise reduction. The resulting air path is relatively compact thanks to the intercooler being engine mounted which enhances engine transient response. Various technologies of turbocharger are used on different variants; mono scroll, twin scroll and variable geometry turbine (VGT). All feature an electric wastegate actuator for fuel economy and transient response enhancements.

1.7 Specification and attribute performance – 1.4l T-GDI variant

The major specifications of this launch engine variant, the 1.4l T-GDI engine, are illustrated in Figures 9 and 10. This will shortly be followed by a 1.5l Turbo GDI engine and a 1.5l Miller Turbo GDI engine which is described later in this paper. The 1.4l engine delivers an excellent balance of attributes.

Bore x Stroke	mm	73.5 x 82
Stroke/Bore		1.12
Bore spacing	mm	83
Displacement	cc	1392
Cylinder Block		High Pressure Die Cast, open deck, Deep skirt
Crank offset	mm	8
Conn rod length	mm	145.6
Cylinder Head		Low Pressure Die Cast with Integrated Exhaust Manifold
Valve Angles		15° Inlet, 18° Exhaust
Cam drive		Silent chain, hydraulic tensioner
Valvetrain		Roller Finger Follow, Dual VCT, Hydraulic Lash Adjustment
Fuel Injection		350 Bar, 6 hole intake side mounted injectors
Compression Ratio		10.5:1
Power	kW @ RPM	116 @ 5000
Torque	Nm @ RPM	260 @ 1400 ~ 4000

Figure 9. 1.4l Turbo GDI engine specification.

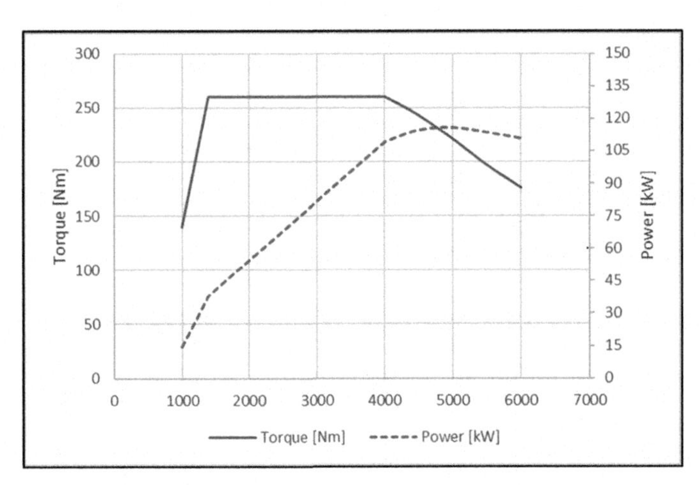

Figure 10. 1.4l Turbo GDI engine power and torque curves.

When compared to industry benchmarks, the engine has competitive power at 83 kW/ litre, equal best in class thermal efficiency at both 2000 RPM, 2 bar BMEP and at the best efficiency point, equal best in class friction and lower radiated noise at maximum power when compared to two leading engines from high volume OEMs.

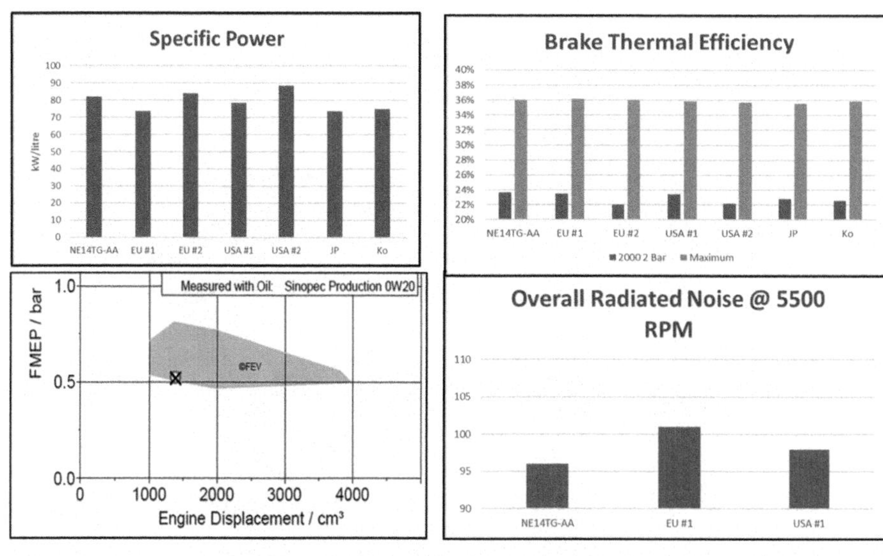

Figure 11. 1.4l Turbo GDI engine attribute performance compared to industry benchmarks.

2 COMBUSTION SYSTEM DEVELOPMENT

2.1 1.4l Turbo GDI combustion system

Central to any internal combustion engine is its combustion system. The NE placed this at the heart of its development. Some of the key requirements of the combustion system for the NE were:

- Allow all performance targets to be met
- Strong catalyst heating performance
- Low feed-gas particulate emissions to minimise the requirement for particulate filtration.

In order to achieve these requirements, the development of the combustion system focussed on air-charge motion coupled with optimised direct fuel injection using a 350 bar pressure system.

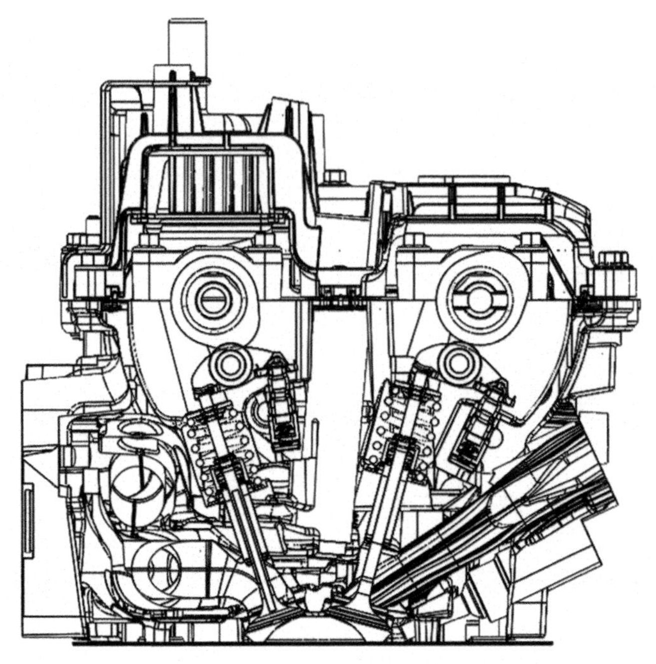

Figure 12. 1.4l Turbo GDI engine intake and exhaust port details.

The intake port designed for the NE was focussed on high tumble generation in order to improve charge motion, fuel/air mixing and combustion [2-6].

The fuel injector is side mounted (intake side) and uses a 6-plume spray pattern. One key attribute of the combustion system is to allow efficient catalyst heating operation which requires good combustion under heavily retarded conditions. The fuel injector spray pattern was developed to allow excellent catalyst heating performance and incorporates two spray plumes in the central spark plug axis of the combustion chamber. These spray plumes allow good localised charge stratification in the vicinity of the spark plug tip using either an air-guided or wall-guided injection strategy to improve combustion stability during catalyst heating operation.

The chosen injector spray pattern was also optimised to balance other key attributes such as particulate emissions and oil dilution with gasoline. Low temperature running conditions have shown that the NE is class leading in terms of oil dilution.

The piston crown design for the NE was developed in conjunction with the intake port and fuel injector spray pattern to allow both air-guided and wall-guided combustion. In order to facilitate the latter, a small bowl on the piston crown is utilised. This bowl was designed to minimise surface area on the piston and to allow good air-motion in the cylinder. A 10.5:1 compression ratio is utilised along with Otto type cam timings.

The combination of the 350bar fuel system, optimised 6-plume injector spray pattern and high charge motion result in exceedingly low engine out particulate emissions.

2.2 Miller-cycle combustion system

The second-generation NE has been developed to improve the fuel consumption of the first series of engines. The engine platform used for the second-generation engine is a 1.5 litre unit. A Miller-cycle has been employed as the key-enabler for fuel consumption improvements.

Numerous previous studies have noted benefits of either early inlet valve closing (EIVC) or late inlet valve closing (LIVC) strategies in conjunction with high geometric compression ratios to improve fuel efficiency [6-11].

The key mechanisms that enable the improved fuel efficiency are:

- Increase in geometric compression which results in a greater expansion ratio.
- Reduction in the end of the compression temperature at high load/knock limited sites through additional external boosting and increased inter-cooling in conjunction with EIVC/LIVC cam timing strategies. This mechanism is the key enabler of the high geometric compression ratio.
- Reduction in throttling losses at low load sites through EIVC/LIVC cam timing strategies.

With a high compression ratio Miller cycle strategy being employed on the second-generation engine, the specific performance of the engine compared to the 1.4l Otto variant described previously was reduced; maximum BMEP target of around 18 bar BMEP, down from the 23.4 bar of the Otto variant. To partially offset this specific performance drop, the engine capacity of the Miller cycle engine was increased to 1.5l, which was achieved by lengthening the stroke of the 1.4l variant. The bore and stroke for the new Miller engine are 73.5mm and 88mm respectively.

2.3 LIVC or EIVC cam timing strategy

The choice of LIVC or EIVC cam timing strategies was a key item to be defined for the second-generation Changan NE product. Key items to consider when defining the cam timing strategy to be utilised included:

- Engine is to work in both hybrid and conventional powertrains and as such still must meet the specific output requirements (~18 bar BMEP) and good low speed torque performance as required by the Chinese market.
- Engine should perform well in key areas of emissions related testing which includes catalyst heating and low speed high load conditions.

In order to understand the trade-offs between the two cam timing strategies, testing was completed on a mule Miller cycle engine. Figure 13 illustrates the full load performance for one EIVC camshaft and two LIVC variants. Note, inlet cam timings referenced are the duration between 1mm valve lift.

It was observed that in a turbocharged engine, the EIVC and LIVC cam timing strategies behave somewhat differently at full load conditions. The first of these is at higher engine speed conditions (above 3000rpm) LIVC cam timing strategies result in volumetric efficiencies closer to Otto cam timing levels and as such the 'Miller effect' is reduced and the engine becomes more knock limited with a subsequent increase in brake specific fuel consumption (BSFC). The two LIVC cam timings (260 and 240 CA degrees camshafts) tested show that to maintain the 'Miller effect' at higher engine speeds increasing inlet cam durations are required. However, increasing the duration of the inlet camshaft was found to be detrimental to low speed torque due to a decreased trapping ratio as a result of charge push-back into the intake port. In order to try to increase the trapping ratio with LIVC cams the inlet cam timing can be advanced, however this can only be done in modest levels as this increases valve overlap and scavenging levels. Scavenging levels of the second-generation engine are required to be relatively modest in order to allow lambda =1 operation in the exhaust gas stream feeding the catalyst.

Another result of the charge push-back with the LIVC cams at very low speed sites where scavenging occurs is an increase in hydrocarbon emissions. This is a result of the charge pushed back into the inlet manifold also containing fuel due to the timing of the in-cylinder fuel injection, which then during the valve overlap scavenging period of the cycle is passed straight through the cylinder and into the exhaust stream.

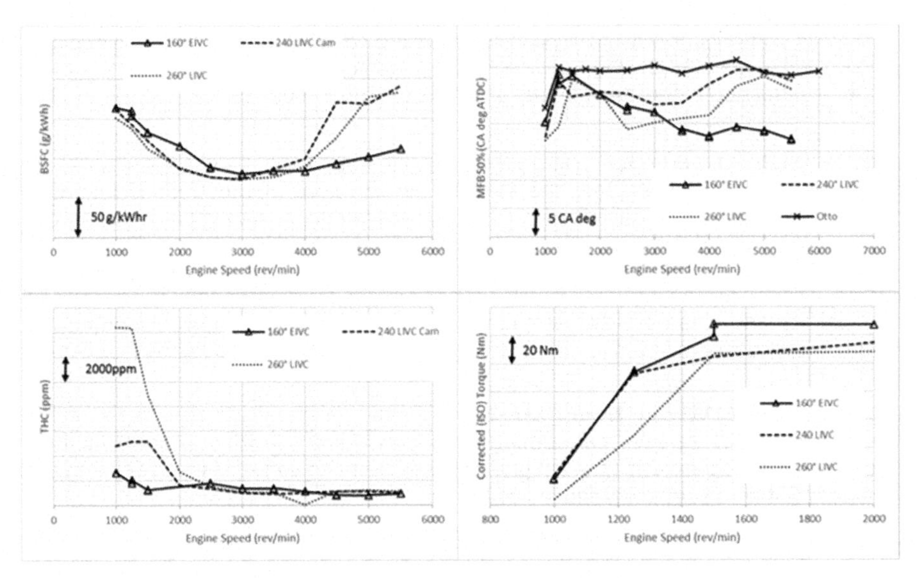

Figure 13. Full load performance of Miller cycle mule engine when employing both EIVC and LIVC cam timing strategies.

LIVC and EIVC cams were also compared at catalyst heating operation. The LIVC camshafts were found to have poorer catalyst heating potential. Figure 14 illustrates that a higher intake manifold pressure is required to send a given heat power down the exhaust during catalyst heating operation. This is once again due to the lower trapping ratio of the LIVC camshaft without having high valve overlap levels due to having to advance the inlet cam timing.

Based on the full load and catalyst heating performance, an EIVC timing strategy was deemed most suitable for Changan's second-generation engine.

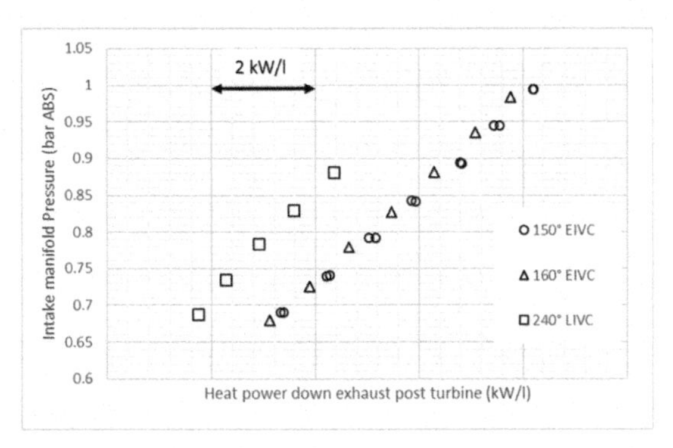

Figure 14. Intake manifold pressure required to produce a given energy down the exhaust during catalyst heating operation.

2.4 Combustion system requirements for EIVC operation

A key change with an EIVC strategy is the camshaft design. In order to satisfy allowable valvetrain accelerations/loads a shorter duration camshaft will result in a lower intake valve lift. There are two key air-charge characteristics associated with the EIVC cam profiles:

- Lower valve lift results in relatively lower tumble flow generation as illustrated in Figure 15 (see un-masked results for the baseline intake port).
- Early intake valve closing results in lower transient tumble/air motion and hence total kinetic energy (TKE) at spark timing as illustrated in 16.

As a result of the reduction in charge motion, in order to maintain satisfactory combustion, modifications were required to the combustion chamber. The change adopted was to use valve-masking around the inlet valves which significantly improves tumble motion at low valve lifts as illustrated in Figure 15.

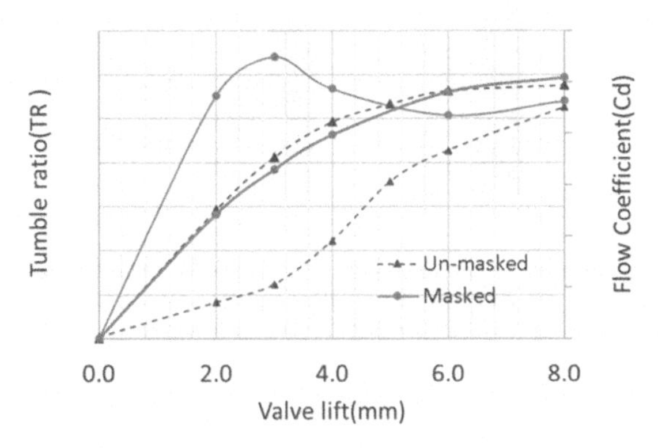

Figure 15. Flow coefficient and tumble ratio for the standard intake port and the Miller engine developed masked inlet port which uses inlet valve masking in the combustion chamber.

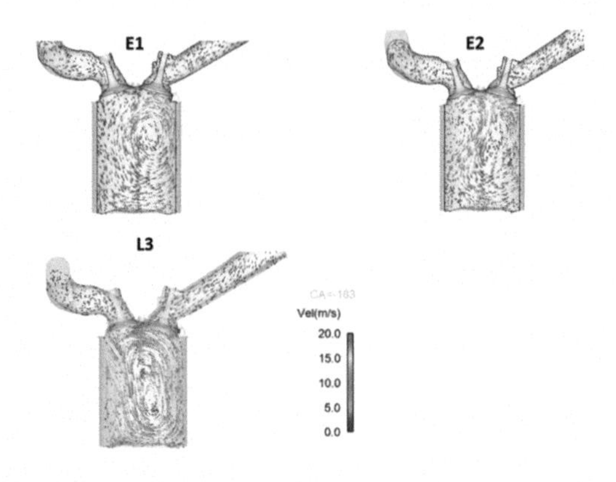

E1= 155 CA degrees 7.2mm lift
E2 = 145 CA degrees 6.3mm lift
L3 = 230 CA degrees 8mm lift

Figure 16. Charge motion comparison for three different intake camshaft profiles (two EIVC, one LIVC).

Figure 17 illustrates the benefit of inlet valve masking on actual engine running conditions. It was observed that combustion stability (defined by CoV of net IMEP) was improved on the cylinder head that utilised valve masking. The improved combustion stability was a result of faster combustion due to the higher in-cylinder charge motion. The greater margin to combustion stability limits with valve masking can be exploited to achieve improvements in fuel consumption by allowing more 'Miller' cam timing strategies (earlier IVC point).

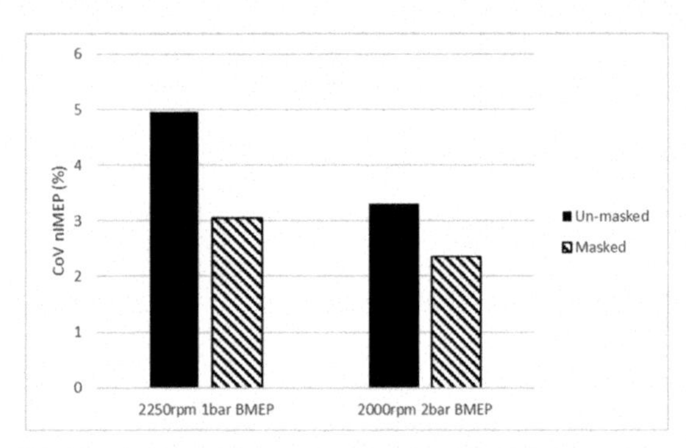

Figure 17. Influence of valve masking in the combustion chamber on combustion stability at two part load conditions. Note all other engine settings (cam timing etc.) are common.

Due to the good performance of the combustion system of the base NE Otto engine when operating in an air-guided catalyst heating mode it was decided that the same philosophy would be utilised on the second-generation engine for the rest of the combustion system, namely: The use of the same 6-hole fuel injector which uses two spray plumes in the central spark plug axis of the combustion chamber. The high air-charge motion of the combustion system allows effective air-guided combustion system performance under extreme operating conditions.

This combustion system methodology allows the use of a relatively flat-top piston which is especially beneficial when trying to achieve high compression ratios required on the Miller cycle engine. The flat top piston also benefits thermal efficiency (lower surface area for heat transfer) and emissions of particulates (thinner fuel film accumulation) [1,12].

2.5 Turbocharging requirement of the second-generation engine

When selecting turbochargers for the second-generation engine, one of the key performance indicators was low speed torque which is deemed very desirable in the Chinese market. Cost versus performance benefit is another key factor for defining the technology utilised.

Assessments were made of various turbocharger technology for the second-generation engine. Technology analysed focussed on the turbine wheel and housing design and included mono-scroll, twin-scroll and variable geometry turbines. Some key observations from the testing are (see Figure 18):

- Variable geometry turbo allowed excellent low speed torque and relatively low pre-turbine pressures at high engine speeds.
- The twin-scroll turbo had improved low speed torque over a mono-scroll turbo but had higher pre-turbine pressure at peak power conditions (4500rpm and above).

The pre-turbine pressure at peak power conditions was found to be a key factor determining combustion phasing and hence combustion stability through changes in trapped cylinder residual gas fraction levels.

Whilst considering all factors the variable-geometry turbocharger had the best performance but was not selected for the second-generation engine due to its higher cost. As such, for the engine project a mono-scroll turbocharger was selected as the most suitable component rather than a twin-scroll part due to the performance at peak power conditions.

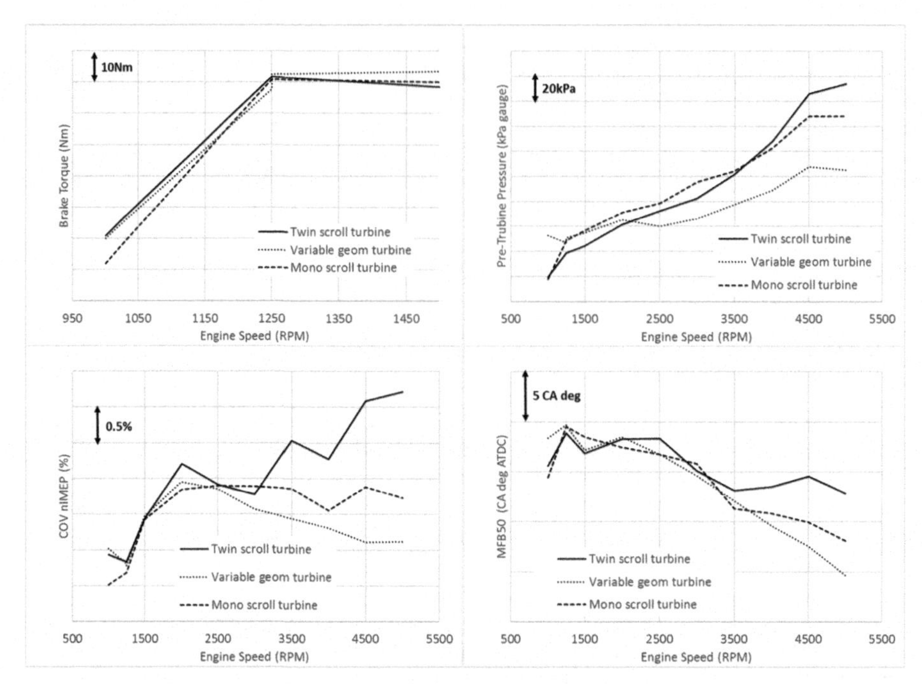

Figure 18. Turbocharger testing investigating different turbine technology on a 12.5:1CR EIVC engine.

2.6 Fuel octane rating, compression ratio and cam profile

The primary market for the second-generation engine will be in China. One of the key differences to the European market is the quality of fuel in the China market. The calibration fuel used by Changan UK has a 92RON octane rating and the engine design needs to be centred around this assumption. This impacts the choice of key items such as the engine compression ratio and cam profile. As an illustration of the influence of fuel octane rating on engine performance, Figure 19 demonstrates that the higher octane fuel allows improved combustion phasing before the knock limit is reached; approximately 4 CA degrees advancement in the 50% mass fraction burnt location at full load conditions. This advanced combustion phasing results in improved engine fuel consumption and combustion stability.

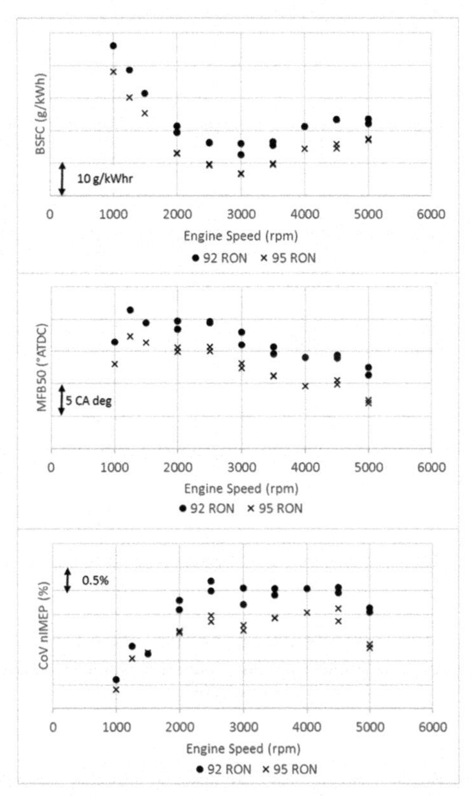

Figure 19. Influence of fuel octane rating on full load performance of the second-generation Changan Miller cycle engine.

It is well established that higher compression ratios will increase the theoretical efficiency of an engine, however, the selection of a compression ratio must consider other key factors such as engine performance at full load and higher load fuel consumption which can become influenced by the knock propensity of the engine.

Figure 20 illustrates the full load performance and BSFC influence at part load conditions of different compression ratios on the second-generation Changan engine when fitted with a 155 CA degree EIVC inlet camshaft. With the highest compression ratio tested (13:1CR) the full load BSFC is poorer because of more retarded combustion phasing due to greater knock propensity. The more retarded combustion phasing also results in poorer combustion stability at peak power conditions. The highest compression ratio engine also has to limit low speed torque; for example, peak torque of 210Nm cannot be reached at 1250rpm due to the occurrence of low speed pre-ignition.

The fourth plot in Figure 20 illustrates the influence of compression ratio on the fuel consumption at 6 part load sites compared to a 12.5:1 engine baseline. At the loads below 7bar BMEP, the elevated compression ratio reduces the BSFC. However, as the load is increased above 7bar BMEP the engine starts to become knock limited and the higher compression ratio has an adverse impact on the BSFC over the 12.5:1 CR baseline. In a similar manner a lower 12:1 compression ratio has a negative impact on fuel consumption at the loads of 10bar BMEP and below. At the most knock limited site (2250rpm 14bar BMEP) the 12:1 compression ratio has the best BSFC as a result of lower knock propensity.

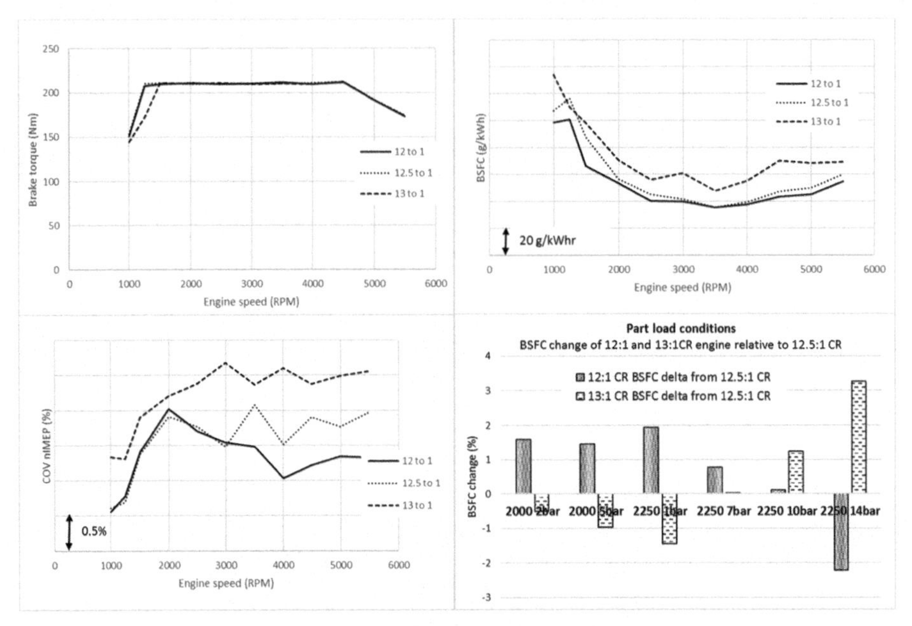

Figure 20. Influence of geometric compression ratio on full load performance and part load fuel consumption on the second-generation Changan Miller cycle engine.

The decision for optimum cam profile for the second-generation engine is linked with the engine compression ratio and octane rating of the fuel to be used. Figure 21 illustrates the influence of the extent of EIVC cam timing on full load performance.

The shorter duration 140 CA degree camshaft results in the inlet valve closing earlier Before Bottom Dead Centre (BBDC) which has an adverse impact on low speed torque. Closing the valve earlier BBDC reduces the trapping ratio of the engine which consequently results in lower mass flow through the turbine of the turbocharger and hence lowers the achievable boost pressure from the compressor. These two factors result in a drop in brake torque at the lowest engine speed range up to 1250rpm.

At higher engine speeds (e.g. 4500rpm) the shorter duration inlet camshaft requires higher inlet manifold pressures to achieve a given load. At 4500rpm where there is the biggest difference in the manifold pressure to achieve a given torque, there is also an increase in CoV nIMEP with the 140 CA deg inlet camshaft which shows combustion stability deteriorates.

Figure 22 illustrates the influence of the extent of EIVC cam timing on the brake thermal efficiency at three part load sites. The shorter duration 140 CA degree camshaft can result in improved brake thermal efficiency at low load sites such as 2000rpm 2bar BMEP through a reduction in pumping losses. At the two higher load sites the mechanism for improved thermal efficiency is slightly different with the shorter duration camshaft. At these sites, the inlet valve can either close earlier in order to have a larger Miller effect for knock suppression, or for a given IVC point the shorter camshaft has less valve overlap which results in lower levels of scavenging and hence improved thermal efficiency.

Whilst the shorter 140 CA degree inlet camshaft offers fuel consumption benefits at part load conditions, on the balance of all attributes including full load performance a 155 CA deg inlet camshaft was selected for the second-generation Changan engine. From referring to Figures 22 and 11 it can be illustrated that engine specification proposed for the second-generation has allowed considerable brake thermal efficiency improvements over the 1.4l launch model. For example, the peak break thermal efficiency has increased by 2% to be over 38% on 92RON fuel. These thermal efficiency improvements have been achieved without using additional high-cost technology on the second-generation engine.

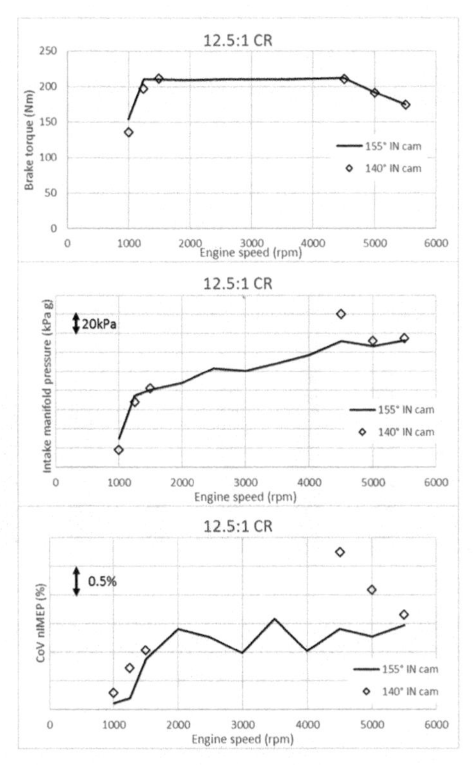

Figure 21. Key full load performance indicators of a 12.5:1CR Changan 1.5 litre Miller cycle engine with two different duration EIVC camshafts.

	Miller cycle engine: 155CA deg IN cam	Miller cycle engine: 140CA deg IN cam
Peak torque at 1000rpm	152 Nm	136 Nm
Peak torque at 1250rpm	210 Nm	197 Nm
Brake thermal efficiency 2000rpm 2bar BMEP	25.8 %	26.0 %
Brake thermal efficiency 2250rpm 10bar BMEP	36.9 %	37.5 %
Brake thermal efficiency 3250rpm 12bar BMEP	38.0 %	38.3 %

Figure 22. Key performance figures from a 12.5:1CR Changan 1.5 litre Miller cycle engine with two different duration EIVC camshafts.

2.7 Further assessments of the combustion system of the second-generation engine

In order to assess the combustion system of the new 1.5l Miller cycle engine, the simulated catalyst heating performance of the engine was compared to other Changan GDI engines which utilise wall or air-guided combustion systems. In order to assess the catalyst heating potential, catalyst heating efficiency has been used which is a measure of the heat power that goes into the catalyst as a function of the fuel flow into the engine. As illustrated in Figure 23, the 1.5l Miller cycle engine shows excellent catalyst heating efficiency. This is impressive as the Miller cycle engine is utilising an air-guided combustion system with its correspondingly low particulate emissions during catalyst heating operation. The high catalyst heating efficiency of the second-generation NE will be beneficial to achieve fuel consumption benefits in vehicle on drive-cycles and in real-world conditions.

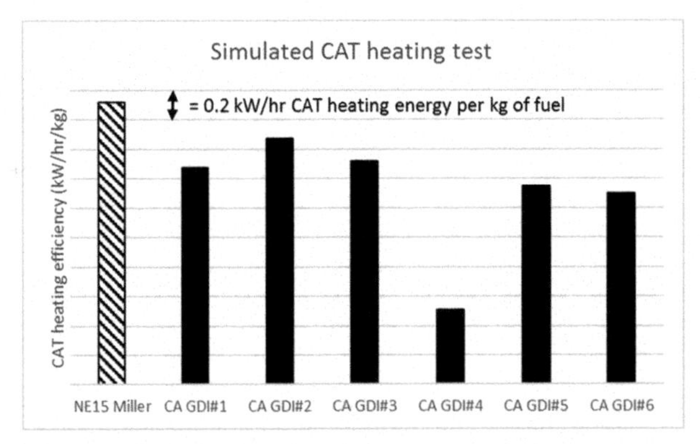

Figure 23. Catalyst heating efficiency for a range of Changan GDI engines. Engine running condition: 1200rpm 3bar nIMEP, 35 degrees C coolant out temperature.

The thermal efficiency benefits of the second-generation engine have already been established, but in order to show the combustion system performs well in terms of low feed-gas particulate emissions, the engine out part load emissions of PN are illustrated in Figure 24. The results from the new Changan engine are also compared to the class leading European 1.5l TGDI Miller cycle engine which also utilises a 350 bar fuel injection system.

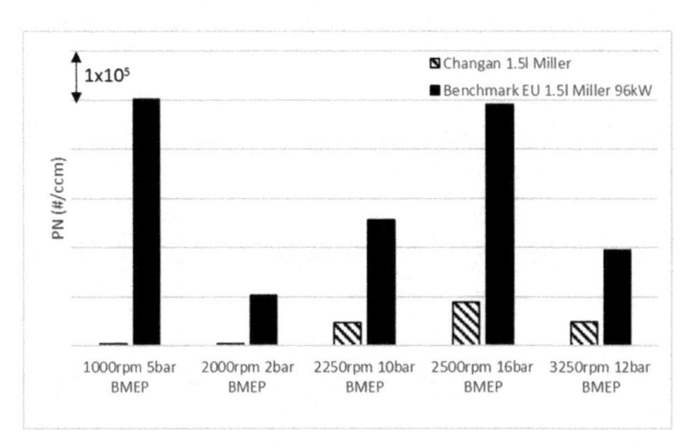

Figure 24. Engine out particulate emissions for the second-generation engine at an array of part load operating conditions.

3 CONCLUSIONS

An overview of Changan's modular 3 and 4 cylinder NE for the 2020s has been provided.

Following this, the combustion system of the base launch model has been detailed before focussing on the combustion system development of the second-generation NE.

The second-generation engine uses early inlet valve closing combined with increased compression ratio in order to increase the thermal efficiency of the engine. Combustion system developments for an early intake valve closing Miller cycle strategy have been presented which includes illustrating the benefits of intake valve masking to improve air-charge motion at low valve lifts.

Turbocharger technology has been explored for the second-generation engine which has shown the best performing turbine design has been a variable geometry part. However, given the cost sensitive nature of the Chinese market, a mono-scroll turbocharger has been selected for the engine.

The trade-offs between compression ratio, octane rating of the fuel used, and valve timing strategy have been presented. Whilst high compression ratios and earlier intake valve timing strategies have been shown to improve part load fuel consumption, the sometimes-adverse effects at full load such as low speed torque and combustion stability have been demonstrated.

With the specification of engine presented, the second-generation has been shown to be able to make all full load performance targets (max BMEP of ~18 bar) as well as achieve good brake thermal efficiency improvements over the launch 1.4l NE variant. This has been achieved on the second-generation engine without using additional high cost technology and as such makes the new engine very competitive in the cost-sensitive Chinese market. The combustion system of the second-generation engine has also been shown to be robust as it has been shown to perform well in some of the most severe operating conditions, such as catalyst heating.

DEFINITIONS/ABBREVIATIONS

NE	New Engine
PN	Particulate number
NVH	Noise vibration and harshness
RFF	Roller finger follower
VVT	Variable valve timing
IEM	Integrated exhaust manifold
FCM	Flow control module
ECM	Engine control module
FVOP	Fully variable oil pump
PCJ	Piston cooling jet
VGT	Variable geometry turbocharger
GDI	Gasoline direct injection
NA	Naturally aspirated
LIVC	Late inlet valve closing
EIVC	Early inlet valve closing
IVC	Inlet valve closing
BSFC	Brake specific fuel consumption
TKE	Total kinetic energy
IMEP	Indicated mean effective pressure
BMEP	Brake mean effective pressure
COV	Coefficient of variation
RON	Research octane number
CA	Crank angle
CR	Compression ratio

REFERENCES

[1] Whitaker, P., Kapus, P., Ogris, M., and Hollerer, P., "Measures to Reduce Particulate Emissions from Gasoline DI engines," SAE Technical Paper 2011-01-1219, 2011, doi: 10.4271/2011-01-1219.

[2] Wada, Y., Nakano, K., Mochizuki, K., and Hata, R., "Development of a New 1.5L I4 Turbocharged Gasoline Direct Injection Engine," SAE Technical Paper 2016-01-1020, 2016, doi: 10.4271/2016-01-1020.

[3] Yoshihara, Y., Nakata, K., Takahashi, D., Omura, T. et al., "Development of High Tumble Intake-Port for High Thermal Efeficiency Engines," SAE Technical Paper 2016-01-0692, 2016, doi: 10.4271/2016-01-0692.

[4] Han, S., Qin, J., Lin, M., Li, Y. et al., "Simulation Study of Injection Strategy and Tumble Effect on the Mixture Formation and Spray Impingement in a Gasoline Direct Injection Engine," SAE Technical Paper 2014-01-1129, 2014, doi: 10.4271/2014-01-1129.

[5] Arcoumanis, C., Hu, Z., Vafidis, C., and Whitelaw, J., "Tumbling Motion: A Mechanism for Turbulence Enhancement in Spark-Ignition Engines," SAE Technical Paper 900060, 1990, doi: 10.4271/900060.
[6] Lee, B., Oh, H., Han, S., Woo, S. et al., "Development of High Efficiency Gasoline Engine with Thermal Efficiency over 42%," SAE Technical Paper 2017-01-2229, 2017, doi: 10.4271/2017-01-2229.
[7] Hakariya, M., Toda, T., and Sakai, M., "The New Toyota Inline 4-Cylinder 2.5L Gasoline Engine," SAE Technical Paper 2017-01-1021, 2017, doi: 10.4271/2017-01-1021.
[8] Cao, L., Teng, H., Miao, R., Luo, X. et al., "A Comparative Study on Influence of EIVC and LIVC on Fuel Economy of A TGDI Engine Part III: Experiments on Engine Fuel Consumption, Combustion, and EGR Tolerance," SAE Technical Paper 2017-01–2232, 2017, doi: 10.4271/2017-01-2232.
[9] Luisi, S., Doria, V., Stroppiana, A., Millo, F. et al., "Experimental Investigation on Early and Late Intake Valve Closures for Knock Mitigation through Miller Cycle in a Downsized Turbocharged Engine," SAE Technical Paper 2015-01-0760, 2015, doi: 10.4271/2015-01-0760.
[10] Eichler, F., Demmelbauer-Ebner, W., Theobald, J. et al., "The New EA211 TSI evo from Volkswagen," 37th International Vienna Motor Symposium 2016, Vienna, 2016
[11] De Marino, C., Giovanni Maiorana, G., Pallotti, P. et al., "The Global Small Engine 3 and 4 Cylinder Turbo: The New FCA's Family of Small High-Tech Gasoline Engines," 39th International Vienna Motor Symposium 2018, Vienna, 2018
[12] Kim, J., Cho, Y., Park, S. et al., "The Next Generation 1.6L Naturally Aspirated Gasoline Engine from Hyundai-Kia," 26th Aachen Colloquium Automobile and Engine Technology 2017, Aachen, 2017

Author Index